A HALF
AND HALF
AFFAIR

CHRONICLES OF A
HYBRID DON

D. R. DENMAN

CHURCHILL PRESS

First published in 1993 by
Churchill Press Ltd
Number One, Wardrobe Place, London EC4V 5AH

© Donald Denman 1993

British Library Cataloguing in Publication Data
Denman, D.R.
Half and Half Affair: Chronicles of a Hybrid Don
I. Title
941.082092

ISBN 0 902782 28 2

Designed by Sebastian Carter
Typeset in Sabon by Goodfellow & Egan
Printed by St Edmundsbury Press

Printed and made in Great Britain

Donald Mecann

Jr. 10, '53

A HALF AND HALF AFFAIR

DRD with Asantehene, Prempeh II

CONTENTS

FOREWORD   7

A CONFESSION   9

1   A HALF AND HALF AFFAIR   11

2   THE LONG WEEKEND   33

3   BLITZKRIEG AND BEAUTY   56

4   EXPECTANT YEARS   80

5   THE VISION AND THE FALLACY   105

6   TIDE AND PREJUDICE   129

7   POLITICS IN BLACK AND WHITE   158

8   CLIMACTERIC   189

9   UNDER THE PEACOCK THRONE   208

10   TITLE WIVES AND OTHER HONOURS   231

11   DEVOTION AND PROMOTION   253

12   NEITHER EAST NOR WEST   278

13   TAM SOLI QUAM AQUAE   303

14   ST CHAD'S EVE   326

INDEX   351

FOREWORD

by Sir Oliver Chesterton MC, FRICS
Past President of The Royal Institution of Chartered Surveyors

Amongst all his many friends and admirers, there must be some who, like me, have marvelled at the skill with which Donald Denman has contrived, against all the odds, to play the two incompatible roles of *éminence grise* and *enfant terrible*. It is yet another example of what must, in deference to the title of this book, be called 'Half and Halfery'. I draw attention to it because, in a sense, it encapsulates the character of the man.

As a chartered surveyor I am conscious of the debt which my profession owes to Donald. He had the intellect, the insight and, some would say, the impertinence to reveal an academic basis which has revolutionised the former conception of the profession, widened its horizons and generally turned it inside out.

In the course of the Land Economy story, Donald sometimes appears like St Paul, sometimes John Bunyan and sometimes Don Quixote. I know that the calculated incivility of some of the entrenched Cambridge establishment was deeply hurtful to him. My own feeling is that Harold Samuel's wonderfully generous endowment was not only a source of great strength to the infant Department, it was also the innocent cause of much jealousy in certain University quarters. Quite a number of people were hoping to get their sticky fingers on the cash.

But those were stirring times, and I am most glad that Donald has set down the whole story – or, at any rate, as much as may decently be recorded.

He emerges from his life story as a witty, wise, and loveable person, inspiring equal measures of respect and affection; in short, as an *éminence grise* and an *enfant terrible*.

Oliver Chesterton

A CONFESSION

Oh, 'tis a glorious thing, I ween,
To be a regular Royal Queen!
No half-an-half affair, I mean,
But a right-down regular Royal Queen!

The Gondoliers, W.S. Gilbert

Somewhere in the recesses of my memory lay the lyric of an unwritten song. Composed of the events of life, of triumphs and disasters, of woes and wonders, kindness and hurt, its keynote, nevertheless, was an overtone of joy. To coddle my curiosity, if for no other reason, I wanted to write the words out, could I but find them. Over a retrospect of three score years and twenty, the beginnings proved less difficult to uncover than they were to believe. The measurement of a life lies not in time but in change and by that calibration I had come unbelievably far. What is more, the accompaniment has a descant on moieties of various kinds running through it; half saint, half sinner; the economist crossed with the legalist; the professional turned academic; these variations are best understood through the interchanging half-orbs that make up the fullness. A pre-natal sense of moiety had, also, seized me from my mother's womb. That uterine condominium I had shared with my twin and on that score in particular life has been a half and half affair.

This testimony to memory, then, is first a self-indulgence, the gratifying of a desire to bring into lasting focus the woolliness of passing memories. Secondly, it has been written to meet the requests of others. At table or clutching the driving wheel of a car or *tête-à-tête* by the fireside or urgently over a leather-topped desk, wherever conversation is wont to flow – and even in the pleas of letters – friends and colleagues, academic and professional, and less intimate acquaintances have asked me to put on record some of the stanzas of the song. This brief chronicle is written primarily to cater to their personal amusement but also to meet more serious ends: histories, anecdotes, stories of battles fought and won, of acorns sown to grow into oaks, of kings and witch doctors whom to know makes life more colourful. Thus, I have written, as best I knew how.

Self-satisfaction alone came too close to presumption for it to move me. The requests were crucial to the justification of my pen. They came from the patriarchal and from the rank and file of the chartered surveyors' profession, notably here in Britain but with no less sincere importunity from overseas. They came from politicians, Ministers of the Crown and other varieties, Tory and Whig in equal mix. The academics were pressing for reasons of their own, some for the pith of the story, others that 'evil be not interred with the bones of dead men'. But most persuasive of all has been the refrain, 'so that was how it started!' on the lips of the current generation who know too little of yesterday's truth. Lest there be some apprehension among you, O my readers, be of good cheer – I give no secrets away!

Although the words are my own, the imprint and arrangement of them, the editing out and removal of errors and infelicities, indexing, and the final arrival of the result between a jacket and two hard covers are jobs no one could do with a do-it-yourself-kit: I am in deep debt to many kind friends and willing aides who have helped me. In especial, my thanks are due to Sir Oliver Chesterton for contributing the Foreword; to His Honour Judge Kolbert for a penetrating critique of the original manuscript; to Dr Leedham Green who in the early days helped me find my way through the labyrinthine maze of the University Archives; to John Raybould for his invaluable expert publishing assistance; to Audrey Clark for undertaking the painstaking task of compiling an Index; to Iris Elkington for preparing original type-scripts; to Sebastian Carter for his inspired touch which shaped the book; and to George Davies, Paul Gray, Diane Williams and Michael Carter of the RICS for practical wisdom and help in marketing and distribution.

D.R. Denman

1

A HALF AND HALF AFFAIR

In form and feature, face and limb,
I grew so like my brother
That folks got taking me for him
And each for one another.

The Twins, Henry Leigh

Sidney, my twin, and I were born on a hillside with the west in our eyes. The sun went down over Dollis Brook, a 'Jordan Water' dividing the fields and farmlands of London's northernmost fringe at Finchley from rural Hendon. Our house was new, along with the new King and the trams now running up Ballards Lane. Expectancy was in the air. My family were proud of their doings, the virgin streets of neat, tightly-packed houses taking the dwellings of Finchley down valley to the brookside. Westward, over the brook, lay a land of unexplored promise. Southward, the Great Northern Railway Company's branch line from the countryside station of Woodside Park terminated in an ultimate dark mystery called Kings Cross. That Stygian terminus lay beyond the march of our childhood wonderlands; they reached only to the penultimate station at Finsbury Park where we would get off the train – just in time. Northward, in contrast, the branch line meant heavenly hours in the bluebell woods of Hadley.

Eastward, unknown and unexpected, doom threatened; out beyond Tottenham, beyond Wanstead Flats and the Mappin Sands, away in far-off Germany, Kaiser Bill was working his dark intentions out, plots and policies which in three years time were to shatter all that promised fair.

We were pre-war children – just; and, thus, Empire-born. Despite the fuss and fury over the Parliament Bill, which in that year of 1911 was in the lap of the legislators and was greatly disturbing our elders, the stability of Pax Britannica wrought a grace in our nascent souls and infused in them an abiding sense of a 'right order of things' – something later generations could never know. Domestic peace was also our lot. Our parents were contrasts in love. The chronicles of

Christ's Hospital, before it left its City home for rural Horsham, carry records of my father as Head Grecian and of a momentary glory when, as a lad of fifteen, he had won a Medal of the Royal Humane Society for jumping off Southend pier to rescue a drowning girl. My infant memory of him has no clear images but from the hearsay evidence of my mother, he became, for me, personified idealism, a manhood to which I would pattern my super-ego. Actually, his 'likenesses', as my grandmother called photographs, show a truly handsome fellow; and the love tokens he sent Letty Barnes, later to be my mother, show nothing lacking of passion and human kindness. In 1901, he carried her 'likeness' with him to Glasgow along with an aspiration to become a qualified marine engineer, an aspiration he duly satisfied with academic honours.

Education was a vogue in the days of King Edward, despite the newly-found levity and touch of hedonism. All parties wanted progress. The race was won by Arthur Balfour with his Education Act of 1902; a measure which was to touch, eventually, my own chequered schooling. In the run-up to the Balfour reform, teacher training was *à la mode*. My father, Robert, had two sisters, Ruth and Bessie. Both fell into line and went to a seminary called the Home and Colonial College for training young ladies how to teach. There, among the alumni, was Letty Barnes. So it came about that Robert met Letty. The early datings were clandestine and furtive; Letty suffered the *sobriquet* 'London Bridge', a destination to which Robert would, from time to time, repair!

The Home and Colonial College had, in Letty Barnes, disciplined a bright, natural wit to become an informed, shapely intellect without denying her open-eyed countenance its innocent impertinance, its blend of pixie mischief and probity. Characteristic of her, and to the confusion of her friends, she called her first home *Emohym* – an anagram of 'my home'. Emohym was, in fact, a small place, the last of a row newly built by her brother-in-law, Ellison, and looking westward down the valley overlooking Dollis Brook. When birth pangs were signalled, the couple moved to Woodside Park Gardens and there the traumas began. Letty's spirit was willing but the flesh was unyielding. Had not my brother and I graciously agreed to share a single zygote and, hence, a close identity, my mother would probably not have survived the birth. When, in the war-red year of 1914, she was in labour with Roland, our young brother, Woodside Park Gardens was thickly carpeted with straw, lest the clatter of carts over cobbles perturb the confinement.

In contrast to his unwavering love for Letty, Robert's devotion to marine engineering was unsteady. His eyes turned elsewhere, for the tide of fortune was flowing landwards. Well before his marriage in 1906, he followed his brother Ellison into the Finchley housing market. They built feverishly and well, free, at first, of planning controls and hindrances of that ilk; for Parliament was yet to try an apprenticed hand at making planning laws. Its first successful attempt to do so was in 1909 at the height of the Denman housing drive – maybe because of it! The effect, however, was benign, pastoral even and helped to hold off commerce and other ugliness from the brookside.

Finchley had always been toffy-nosed and reactionary, taking its stance from the social legacy of the few grand houses which had occupied the Common and woodlands after inclosure in 1816. At the turn of the Century the local housing market was slow and selective, restrained by the influence of subsidised rail-fares which were thrusting forward the darker side of north London at Tottenham and Harringay. Finchley preferred it so. After all, Finchley was Finchley, chestnut-wooded and prim! The residents of that new suburbia disdained streets and roads, to live in granges, crescents, gardens and parks. My father's efforts up-staged somewhat his brother's in housing eloquence. Both succeeded, however, and advanced in pace with the mortgage market.

Our contribution was 'Clarence Gardens' in the rural vicinity of the parish church of St Mary at Church End. According to the Ordnance Survey for 1912, its graduated dwellings which were kerbed in from an 'unadopted' trackway by the pride of Aberdeen granite were by then fully in place alongside the new railway line and running westwards to end abruptly a field's width off the brim of Dollis Brook. There, in Clarence Gardens, grandmother Elizabeth Jane, was housed by her two builder sons, Ellison and Robert, in an ultimate residence of para-mount superiority, as became her pretensions. And there my infant memory shares its first awakenings along with the straw on the road at Woodside Park Gardens, two miles distant at Finchley North End.

≈§ §≈

Time and sequence have little place in the prologue of my memory. Events are a pageant of unconnected snapshots. My father is there in most of them, a hazy figure, yet by his presence giving them historical authenticity as being not later than mid-summer 1915. Unbiquitous throughout these early impressions was the more solid memory of my twin brother. Form, voice and temper were matched with my own in confusing identity. The confusion was exacerbated by my mother's inveterate whim to dress us, top, middle and bottom, button for button, pocket for pocket, as exact replicas of each other. My very earliest memory is of an act of protest. I cracked my head open crawling under a massive chair. The scar never fully healed and remained to distinguish one twin from the other all our lives.

The identity rig-outs never irked us. They were accepted as part of a natural oneness which at times was near absolute. Not only play but, often, wickednesses were acts of unpremeditated unison. One bedtime, in order to get us 'down', we were promised tiny, glittering jeu-jeus made of rainbow colours, if 'like good boys' we ate up our porridge oats. Tucked abed there was no escape. Our reactions were in concert and spontaneous: as soon as the door shut, and without a word, we tipped the loathsome gruel under the pillows. The jue-jues were won and our parents went purring with contentment to their supper. Dawn brought its retribution. It was probably very light. Our hands were too tender and our bottoms too small for the whackings such joint sins justified in later years.

Among the disassociated memories of my prologue years, one stands out more vividly than the rest. My brother was beside me but the unity then sensed was not the unity of twin-brotherhood. It was a oneness of all being. The experience was transcendent, out beyond my infant self, transported by the enchantment of a woodland glade of verdant greens, slanting sunlight and the noiseless frolic of baby bunnies in the grass – an ecstacy caught in a dew-drop of suspended time. The place is uncertain: maybe Ireland, maybe Devon or in the Norfolk woodlands that hem Runton to the sea. Only once again did a like experience overcome me. It was much later when six years old and the nuns of St Mary's Convent were farming the meadows of Finchley. While walking in velvet sunlight through an expanse of shoulder-high, gold-green cocksfoot and timothy grass stretching endless to the sky, joy held up for a moment the passage of time. Field, sky and the heat-haze between were one in ethereal harmony with me and the opening wonder of life.

Tuition in 'letters' came early. The unconventional seat of learning was a push-cart. We twins were expected to recite together an alphabet which ran from 'W' to 'S' sitting side by side, our eyes just level with the road name-plate – 'Woodside Park Gardens'. Starting at 'W', the orthography laboured on to the final 'S'. My mother was proud of this achievement, apparently oblivious of any limitations it might impose on later alphabetical comprehension. The process, like the rest of our schooling to come was, to say the least, singular.

It was so with other disciplines of daily life. They were few but regular, strict and odd, whether practised by the maternal hand or by

Left A basketful of beginnings *Right* Ten years later, with Roland

the housemaid. One oddity resulted in the game of 'battleships' invented by my brother, Sidney. We were spoken to, as occasion required, as if involuntary bowel movements were ours to command. Failure to obey meant the insertion of well-proportioned, homemade suppositories called 'nit-nees' and a twin enthronment on chamber pots. At this stage we became battleships in combat, twisting and corvetting at increasing speed over the nursery floor. After all, everybody was at war, so why not we?

These games belonged to the wild joys and other excitements of toddlerhood. There were plenty such with an indulgent father and eccentric uncle at hand. It was probably our third birthday, the April of 1914, when over the threshold of the French-windows of the new drawing room in Clarence Gardens, we passed into a garden unlike any other known before or since. Uncle Ellison had filled the place with battalions of hyacinths. No other flora in God's creation had a showing. Hyacinths filled every wide bed, border, nook and cranny: save for a space on the lawn where my father had built a toy railway of exquisite carpentry propelled by gravitation. We were intoxicated with perfume, colour and a touch of foreboding. But we shared these wonders with ourselves and the grown-ups. It was all part of the awakening process of a natural, infant world, far removed from the

15 *A Half and Half Affair*

moment of ecstasy in the sunlit woodland glade. Hyacinths and train fitted the concrete reality of everyday.

Just after Christmas in 1915, my father's discarded marine engineer's credentials came into their own. Ships of the mercantile marine were taking to port in thousands to be fitted with armour and guns against the German menace at sea; and he, impatient to play a part, left for Belfast. The family followed soon after. Our going is lost to memory, but not the returning. That is indelible: the deck of the Larne packet-boat circled in light stands out from the indigo of the dockland night and, in the glare of shaded arc-lamps, a messenger is calling my mother's name. On shore her husband, recently admitted to hospital, is fighting for his life against the deadly Yellow Fever imported by Chinese coolies to the Belfast docks. Desperation and despair call her to his bedside. We twins embarked some nights later, now in care of Aunt Ruth. Again, the indigo of dockland night and the spot-lights, holding as on a floodlit stage the boards and hatchways of the moored deck, welcome us. Later, out at sea, in an eerie darkness, the whisper runs of German U-boats in pursuit. Excitement banishes any childhood fears we might have as they chase the packet into Stranraer.

~§ §~

The fabric of our infant family was torn apart by my father's death. Roland, two years old, moved sideways into the maternal branches of the family at Finsbury Park where Grandfather Barnes had a small establishment. We, the twins, settled down for a while with Grandma Denman and Auntie Ruth who in love and adoration ranked with us second only to our mother. These three women shaped our lives. Our father gone, we lacked all male influence within the family circle for the rest of our childhood. Uncle Ellison, odd, remote and rich, never talked to us but, at best, only at us.

The credentials of the care and kindness of my Grandmother's home were reflected in our happiness. With the coming of spring in 1916, however, an ominous shadow arose to darken the sunlit days. The shade of William Ewart Gladstone and the provisions of his Education Act of some thirty six years back were prescribing school. Graduates of the Home and Colonial College had a nice conceit of themselves. They were the new professionals and struck a hauteur over lay governesses and dames of small, private establishments. With three professionals – two aunts and a mother – *in situ*, we were, without question, set for the elementary school system, to which they belonged. St Mary's at

Finchley had been running for nearly a century, from the earliest days of the old national schools. By the time our fifth birthday had arrived, the school was under the hand of the local authority and integrated in the elementary school system. Its portals were ours to enter but we found them threatening and its interior worn, shabby and scary. Amid tears of distress, I recall trying to draw pictures with my fingers in trays of silver sand. We stuck it for only a term, an ephemeral experience of misery which did nothing to endear us to William Gladstone and his education policy.

By late summer, life again had found its gladness. We had a new home – not Ireland, not Finchley, but Salcombe in Devon. From Belfast our mother had returned to Devon, to where the rias of the coast draw inland waters to the sea past labial bays of golden sand. From that time onwards, for us children, there was no place on earth to match Salcombe and the mighty bluff of Bolt Head.

Salcombe had a smell of its own; an aroma of seaweed, shellfish and tar mingled with the effluent of surface drains whose vents poked eyes in the harbour wall. In our young nostrils, the perfume was a powerful drug, an intoxicant which hooked us back to the place year after year. Salcombe's orthodoxy was unorthodox. Even the sweetshop, Cranch's, seemed an import from fairyland. The high street, called Fore Street, twisted and wound through a narrow defile of old harbourside buildings to hook under an archway, leading to a paddlesteamer beyond. Into this thoroughfare, round unexpected corners, would enter hillsides criss-crossed by terraced lanes and zig-zagging walks. On one such hillside we lived, guests in 'the best room' which Mrs Murch, the wife of the postman, had to offer. Her eldest son, Fred, brother of Jim and two sisters, was a sailor, one of the many thousands braving the war-wracked seas to help the nation feed itself. Jim was our contemporary, ally and accomplice. Even in this haven of delight, however, we were pursued by the shadow of Mr Gladstone – we all had to go to school, the native and the twins.

Jim Murch was with his 'ain folk', but for us strangers the Infants' School, as local usage called it, was more terrifying than ever St Mary's had been. We were dragooned by the teacher, a formidable Miss Gunn, under whose shafts of wrath we trembled in misery. She seemed blind or indifferent to our childhood fears. One notorious morning, Sidney and I synchronistically were caught in dire conflict between nature's necessities and our mother's strict admonition not 'to go near the boys lavatories'. Sidney's resistance broke first; a relentless stream meandered from his sodden pants within the Gunn sights of the teacher's desk. He was thundered out of the room amid threats of dreadful mutilation. I, shamefaced, followed walking quickly and to some purpose. We never returned. Maybe there was some psychological link-up between Miss Gunn, our enemy, and the guns we heard of every day which aroused such anguish and anxiety among the grown-ups around us (it was the year of Verdun and the Somme). On Mrs Murch's kitchen wall there hung, to remind the unmindful and the forgetful, a dismal but treasured photograph of the armed merchant-man, Kaisar-I-Hind. 'That be Fred's boat; he's carpenter; fighting the Germans he be'. We would nod our heads. Fred became our hero overseas. Had he been home, we were certain he would be silencing Miss Gunn.

❧ ☙

Anxiety, sailors in peril of the enemy, warships in the estuary, food shortages and a half-baked blackout were the score in wartime Salcombe. When we returned there, among the outward signs was a huge hospital ship, the Red Cross visible against the white of the battered hull swaying in the tides off the rocks of Bolt Head where, in dodging the ferocious submarine onslaught, it had foundered. Between the visits to Salcombe, home was not the dower house at Church End but one of similar dignity at Finchley's northern pole. Part of my father's heavily mortgaged estate, it stood empty in a state of wartime neglect. Its air of abandonment cast a spell of unrestrained freedom over us. The back garden was a thicket, a pampas prairie, adult high where we were quickly lost to sight. Fortunately for all concerned, its lure kept us kids from getting under the feet of the removal men. Alas, the place ill-fitted the family penury and before long its delights had to be exchanged for more limited *Lebensraum*. Small by comparison, the new abode had its own fascinations, not least among them a regal chestnut tree by the front gate (the tree-stump still stands after eighty years). For a time, life would follow a spasmodic triangular migration between Teigngrace, the dower house at Church End; Strathleven, our new home by the conker tree; and Church Street, Salcombe. With the war still raging and although pinched for money and fatherless, we were by comparison among the luckier ones.

Understandably, the Finchley orbit was the busiest. Its linkway was Nether Street, a lane of fields and copses with sprawling houses at either end. Seven years old, we were tramping it unaccompanied; and often when food was short were humping home weighty bundles from Church End, the better supplied extremity.

On one occasion, Sidney was apprehended by a possessive female whose raucous voice threatened to call the police. He had plucked an overhanging rhododendron from her front garden. Seized by panic, I fled out of sight with two food bundles; she might think we had pinched them also! White and trembling, the child criminal was released with a wigging. We never ventured that way again.

In our one-parent family under the conker tree, my mother's warm, self-giving love bonded us together in mutual dependence. She was protective but not possessive; we had more liberties for our age than, today, would be thought prudent. Her sanguine aim was to seed in us the generous virtues of courage, loyalty, truth and kindness; and she tended to make the memory of my father a touchstone of all virtue. Whatever he may have been, we were certainly no paragons. 'A good talking to' was the standard reproof for naughtiness. Chastisement, however, was not eschewed. But there were limitations. By the time our mother, who was slight and unmuscular, had spanked one culprit suitably derobed over her knee, she would run out of ire and energy and be too exhausted to deal out even-handed justice to the other. 'It hurts me more than it does you' was, I believe, a true enough epigram, particularly on one occasion. At table I was holding a knife and fork after the manner of the giant in Jack and the Beanstalk. Intending to administer a disapproving rap with the flat of the carving knife, she failed to turn the blade. The scar ran deep and remains to jolt my manners at high-table and elsewhere. Mother's tears were copious and far more genuine than my own.

So itinerant a childhood would have been less possible if our schooling had not been so footloose. One more try was made with an elementary school; this time a grimy place of local board school origin in Albert Street. Disaster followed – we came home with ticks in the hair. Now, at that time towards the north end of familiar Nether Street, the nuns of St Mary's Convent offered to catechize and tutor the very young. So to them were entrusted. Their benign care bent over us. In contrast to previous experiences, here was a haven of tranquility and light, surrounded by the meadows where for me time stood still, in a transport of ecstasy. The kindly nuns doubtless taught us the beginnings of the ABC and other fundamentals. Nothing of them remains in my memory. What is recollected are frames of bootlaces on which we were taught to tie the bows of shoes. A tearful maiden of some six summers was my partner. Her plight demanded chivalry. I took her laces, tied two granny knots, pulled out the ends to look like butterfly wings and assured her we had tied a bow. The tears stopped, to evoke my first heart throb. Mother, however, was in despair; a hagiology of saints days, strictly adhered to, meant numerous half-day holidays. So the Convent was left for a spell in Salcombe.

❧ ☙

Sidney and I had fecund memories and a restless, importunate curiosity, a combination which could evoke both endearment and exasperation from the grown-ups. A saucy 'Why shouldn't I ask Why?' brought its deserved reprimand. But, often enough, the plea was genuine. We wanted to know; we got to know, sometimes without knowing what it was we knew. Although far from being precocious, these lively, natural endowments, plus want of regular and sustained schooling, generated in us a paradox of wisdom in ignorance and the proclivity to become childhood autodidacts. It meant also that, throughout the earlier years, we were more 'home-boys' than schoolboys. Each domicile had a domestic tone and style of its own. Differences were peculiar not comparative. The conker tree home was the mother base, a place of unpretentious cosy comfort; from thence we would go and return on shorter and longer migrations. Here through the central years of childhood, we became somewhat protective towards mother. Much beloved, we depended upon her absolutely. Our dependence, however, was counterbalanced, more and more, by a growing realisation that we were the other half of the family unit on which she relied. We pulled together and learnt a lot in doing so: how to mop, how to scrub, how to knit, how to tend the poultry which had turned the garden into a quagmire; and, sitting either side of her, how to draw the landscape, for she was no mean amateur artist. Tone-deafness, a twinned defect, nipped in the bud any flowering of the musical arts in us.

As naturalists we were inclined to be zoo-keepers rather than bird-watchers; sticklebacks, newts and other water-life fauna were captured to populate a galvanized bath which served as a stagnant pool amid the chickens and the mud. At Strathleven, then, life was home-centred. Young Roland had joined us and traipsed around in a sailor suit. We loved him, and counted him a second claimant to the twin-powered protection we provided our mother. He was, to start with, a kind of refugee from Finsbury Park but soon became naturalized under the conker tree. Sundays often meant a trip to Finsbury Park, a humdrum sabbath day's journey relieved only by the night train ride at the end of the jaunt. The Barnes colony also had roots in the Garden of England. An Aunt Jane, who by our reckoning had reached an immense old age, lived at Staplehurst in a magic garden where white, red and black currants grew in fragrant and tempting clusters level with our eyes and noses. Beyond lay meadows with liberal supplies of half-stacked hay to bounce in.

Salcombe was more carefree, less self-conscious. The house had a primordial feel about it, especially the loo by candle-light. While there in the summer of 1919, the family graduated from being lodgers to itinerant residential status. There was little option. The railwaymen had refused to take us home and went on strike because Eric Geddes of the famous 'Axe' was preparing to hand back the railways to private companies. Although the enforced sojourn meant yet another school, a grey seminary called 'the Boys' School', we were more or less *en fête* for the winter and spring. Of schools, the Methodist Sunday School, 'down Island', left a benign mark upon us far more enduring than any lay school imprint. 'Stories of Jesus' became stories of Salcombe and the basis of a childhood mythology which haunted the place. The topography of Salcombe estuary was renamed with the place-names of the Scriptures. Across the water, the highest hill was, in our book, Mount Ararat and one reached it by boat from Bethel Steps. Hand in hand with Jim Murch, we were allowed great liberties; enough rope to hang ourselves which on one occasion nearly happened. On Mount Ararat, I had fallen out of the topmost branches of a sizeable tree. The other two sped home to report me asleep beyond arousal. Night had fallen before the rescue party had lifted the concussed body from the bracken. Salcombe itself, not its schools, taught us the lore and practice of the out-of-doors: how to handle boats, scale rocks and climb trees, fish in pools and the open sea, to search out and sometimes eat caterpillars and larvae, scrump apple orchards and get home in the dark. These things brought out the Tarzan.

We would sport naked in the bracken on Mount Ararat, but never on a Sunday when the reins were short, taut and irksome. Unconscious of the process, we were in fact learning, in that wonderful Devonian beauty spot, the rudiments of faith and fearlessness.

Teigngrace, the dower house, meant heavy, quilted bedspreads, deep arm-chairs and the opulence of over-hanging eves. We would step over the threshold with bated excitement to a promise of out-of-the-ordinary happenings, provided our shoes were properly wiped on the doormat. There was an unwritten code of right behaviour associated with ambrosial scents of floor polish, potpourri, a selective cuisine and the whiff of mothballs in the deeper cupboards. Loyalty to the code was repaid in the coin of generous sanctions. Ours would be the run of the house from the billiards room under the roof rafters to the wine and coal cellars underground. Beyond the garden fence it ran to a kind of war-time *rus in urbe* farmyard where free-ranging hens and badly penned pigs threatened to turn the entire neighbourhood into Flanders' mud. My earliest memories of Christmas are lodged under this ample roof with its wide chimney-stacks, obviously built for the occasion. On Christmas Eve, a colonnade of coloured candles would ring the chimney-breast of our bedroom. Sleep under heavy eyelids challenged determination to see the last candle snuff out – sleep invariably won. In the morning, the Glorious Morning, there would be gigantic Christmas crackers slung across the ceiling. Pulled apart after dinner they would empty a cornucopia of presents on to the carpet. To us, there were no toys like these toys in all man's creation. Also, at Teigngrace, as nowhere else, came the excitement of night air-raids. Carried from our beds and wrapped in eider-downs, we would join the grown-ups for sandwiches and milk in the cellars. On one occasion *en route*, flames in the sky lit a bedroom window as a zeppelin plunged blazing to the ground over Enfield.

The trusting naïveté which allowed freedom within the behaviour code was wont to turn a blind eye on what 'the boys' were up to. One summer's evening, supposedly in bed, we had climbed through an open window, dropped down on the glass roof of an expansive conservatory and were daring each other to tread the glass to the other side. Fortunately, the brittle roof held under our four stone weight long enough for a return traverse to the window ledge. Adult lenity, also, enabled our self-educationary curiosity to outmatch the establishment. Under their very noses, one day, in high summer jam jars filled with mysterious contents were smuggled into the lower reaches of the best china cupboard in the drawing room. The jars represented an early essay in applied science. Each jar contained the entrails of dismembered frogs, pickled in pond water. We had seen such, so our ignorance supposed, in the Natural History Museum. So why not have our own? After some weeks the stench was intolerable. The scientists emulating Brer Rabbit laid low. Eventually, after ripping up the floorboards and scouring the drains, the scientific experiment was disclosed. Morals in this case are confused. Whether we were pernicious or precocious was an open question. At all events, the drawing room became barred territory.

Life at Teigngrace served to intensify home-love in us. Nevertheless, it was there that we first became aware of greater constellations of power and authority beyond the home-circle and of the whirl-pool of good and evil into which all simple lives were eventually swept. Above us all, second to none in supremacy was, The King, superior, indeed, to a divinity who, in the dialect of our Grandmother, was reverently spoken of as 'Gord'. Both were on the side of the good, to which we all belonged without question! So with literature: clear in the lead among the 'good' were *Gord's Word* and *Pilgrim's Progress*. In contrast, there was an evil text which seemed to infuriate Uncle Ellison, called the *Rent Restriction Act*. And at seven and a half years old, in the days of the 'Coupon General Election' that followed closely upon Armistice Day in the autumn of 1918, there was pinned on our juvenile breasts enormous red rosettes. These gifts from Aunts Ruth and Bessie stood us publically on the side of goodness.

Finchley, at the time, was caught up in the acute political fever that attacks all virgin Parliamentary Constituencies. Opposed to our 'goodness' were the demon legions. They wore blue rosettes and were led by a Mr. Robertson who made marmalade in a place called Dundee. He labelled himself a Liberal. This political game was played almost every year for the next five years. Only once, for a brief spell of twelve months, did the marmalade manuafacturer represent Finchley. Normally, the incumbent was Sir Edward Cadogan (C). By the time wearing a red rosette meant our having a vote, Conservative dominance at Finchley was ingrained and had become part of an accepted natural order of things.

꿿 ꙮ

Three more schools lay ahead after the fleeting passage through the Boys School at Salcombe. Whatever William Gladstone may have intended and barring the final years, enforced schooling did, in our case, no more than provide an unstable grasp of the three Rs. Of the worthy Victorians who shaped society at the turn of the Century, it was Robert Baden-Powell whose vision and its application helped to banish shyness, engender self-reliance and fashion in us the extrovert and a proclivity to make natural lasting friendships. Socially, compulsory schooling had failed, probably not from some inherent incompetance but because our sampling of it had been too spasmodic, too sporadic to allow Gladstone's insistence a fair trial. Running through later childhood and into adolescence, the contrast and sometimes conflict between the emptiness of school life and the warm wholesomeness of scouting did a great deal outside the natural tutelage of a loving home to chisel, round and fit our psyches for the years to come.

As if contemptous of the abolition of fees for elementary state education in 1881, two private fee-demanders opened their doors the following year to serve the middle classes of Finchley. One of these, a self-preening establishment, went so far as to call itself the Finchley High School for Boys; the other, with greater immodesty, claimed to be Christ's College – also for boys. Unlike good fortune, we were to attend both. Samuel Vernon and his more competant wife made up respectively the higher and lower divisions of the Finchley High School. To our lack of formal learning they added nothing. It was, therefore, with heads empty of the essential requirements that we came to sit the entrance examinations to Christ's College. Vernon's High School had added naught to naught, a useless sum in all circumstances! According to family legend, however, the Head Master of Christ's College averred never to have interviewed a couple of youngsters so woefully backward in elementary schooling yet in command of an amazingly liberal range of general knowledge. We were unripe, too raw to benefit from any schooling he could give us. He had other matters to puzzle him than the enigma of the Denman twins. We were rejected, banished from his sight and put back into the elementary schooling system to attend the recently-established Squire's Lane School. A year was enough. Admission to Christ's College followed soon afterwards but more as an act of grace than reward of merit. If this meant school on a sound and steady base, we had wandered for six crucial years down a crooked lane to reach it.

Fortune forsook the Finchley High School with the death of Samuel Vernon's wife. Our childhood wits were not slow to mark and match the change. The tone of the place lost the resonance of her presence. Unknown to us, the old man with little intention of going it alone, was ardently wooing our mother to fill both gaps. Unavailing, his passion merely deepened the pathos.

School mornings, in the days of decline, were spent surreptitiously devouring twopenny thrillers. Other, more necessitous deceits took place in the lunch-hour.

Behind the rose bushes we were wont to bury chunks of bleeding, underdone, inedible mutton which had been 'eaten' at lunch by slipping them unnoticed into a jacket pocket. Afternoons were short, mainly taken up with getting home, either as 'Indians' or 'Cowboys' which, in reality, meant something little short of hooligans. Although day-boys, there were occasional weekends when we were inmates with the boarders. The resident pupils gave freely a not unprofitable back-stairs education; nothing beyond a lightsome lewdness at break of day, aimed at initiating us and the Devon housemaid, Jessie Applin, who came to open the curtains, into what little and big boys are made of!

It was the year of the police strikes in Liverpool, when we first felt the terror of the police. The Cowboys and Indians had invaded the quiet residential precincts of a nearby close called Thyra Grove. At its approach a wrought iron lamppost stood sentinel, supporting a gas-lit glazed lantern. Under the guise of a Red Indian, I had lobbed a 'fire brand', a sizeable stone, at the cowboys gesticulating the other side of the lamppost. The missile went wide and shattered the street lantern and my make-believe with it. A wretched little hooligan, I made for home as fast as a pair of terrified legs could carry me. The police never came. Nevertheless, I suffered agonies of trembling anxiety cowering in the scullery whenever the front door bell rang.

The damaged lantern was naked sin and deserving of just retribution. Yet, about this time and in a matter truly for the police, my brother and I became victims of the power of casuistry to turn wrongdoing to illusory righteousness. A fellow called Green, a rogue twice our age in his middle teens, had seen in us a pair of gullible, identical twins – as handy a shop-lifting tool as he could want. From the upper ranks of Finchley High School, he was the evil potter, we the pliable clay. Mesmerised by the force of his persuasion, we were dutiful to his assurance that nothing was amiss as long as we kept to the letter of his instructions. So alike were we, that one of us could enter a shop for a commodity which plausibly could be in store but not on display. As the assistant disappeared to search the back of the shop, the enquirer would nick a perfume or some other loot from the counter and carry it in triumph to the waiting crook outside. Simultaneously, the other twin would counter-march his brother and be waiting in the shop when the assistant returned to announce his disappointment at not being able to satisfy the patient lad. Many years later, remembering our misdoings, I went to confess and redress the crime, only to find the erstwhile chemist had become an estate agent! The end of these deplorable episodes came near to mortal tragedy. Green had coaxed us to lift a full can of petrol from the running board of a stationary car. With this lethal cannister, he committed arson and set going a huge conflagration. The fire nearly finished the lot of us, as it scorched the maple trees down Lovers' Lane off Nether Street. Back home, singed jerseys as well as our manifest fright gave the game away. We were duly admonished. 'Never, never, never, to see that wicked boy again'. And so it was. The 'Green-movement' of those days, stopped there.

❦

By our ninth birthday, the two romantics, Robert Baden-Powell and Rudyard Kipling, had created a world of fantasy and inspired practical make-believe for the youth of the new century. Outpost trekkers, the soldier-scouts of the South African veld under his command in the Second Boer War, had aroused in Baden-Powell a vision of peace-time boy scouts, trained to manhood, to self-survival and service. Baden-Powell was marching under Mafeking skies, in the wake of the 'elephant's child' in Kipling's *Just So Stories*. Probably his kit-bag carried also a dust-grimed and well-thumbed copy of its companion, the recently-published *Jungle Book*, out of whose pages came Mowgli, the man-child among the wolves and his 'brother' cubs. Mowgli hunted with the pack under Akala, the prototypes of the wolf-cubs, the junior recruits in Baden-Powell's vision of his Scout movement.

On 9 October 1920, the number of the nation's wolf cubs was increased by the intake of two apprehensive Denman twins. Years soon fell away. By the next summer, less than a year in the pack, we were roughing it in army bell-tents, far away over tossing seas in the island of Jersey. Route marches, iron rations and the relentless dawn reveille were accepted rigours which shaped and stiffened the weak-kneed and put the man-stuff into us. Real life, however, let it now be known, was met after dark, unauthorised, clandestine and ruthless. Raw cubs were spared no pity. To flinch was to fall. In the darkest hour well before dawn, whispered commands to obey woke selected 'initiates'. Sleeping bags and pyjamas were stripped off and rough but kindly hands smeared the shivering, naked bodies with boot blacking, carefully annealing the crevices. Struggling but silent, the victims were dumped into a trek-cart. The muffled wheels, guided by the serious torches of the conspirators, pounded the rattling tumbrel over ploughed stubble to a distant pond. The cart was there upended and the apprehensive, naked trios pitched into the silent waters, heavy with duckweed and slime. Masters of ceremonies would then haul the 'baptised' to the bank and scrub and cleanse them fit for return to the tents. Suppressed 'Bravo' and the smell of hot cocoa filched from the cook-house would cheer the rites. In the morning, only the glint of pride in the eyes of the 'twice-born' revealed what had happened in the night hours. The silence held.

The Tenth Finchley Scout Troop, to which as cubs we now belonged, had a fully-orbed social existence far removed from the grubby knees and billy-can image of normal scouting. The boy scouting of Baden-Powell's fancy, aping the soldier-scout, was, in the Tenth, metamorphosed into something approaching adult theatre. Although on occasion the troop would camp in 'the green fields of Tyrone', its 'heart was in the Highlands'. With an *amour propre* worthy of a royal regiment, the Tenth were the *Finchley Scottish* and played their destiny out to the full. Under the patronage and grace of the Duke of Richmond and Gordon, this signally exceptional group of cubs, scouts and rovers were cadets of clan Gordon, by tartan, kilt and plaid. The Duke gave the nod. The architect of the achievement, however, was Captain William Barclay, a public relations genius over six foot tall, who had mastered the art of tempering successful determination with discretion. There was nothing ostentatious about the Tenth, only a splendid ornate eliteness that would make the votaries of todays' left-wing false social values squirm.

A two-rank pipe band, complete with the whirling mace of the drum-major, would sound the 'pibroch' through Finchley's genteel suburbia on church parade of a Sabbath morning. Unintentionally, but, most certainly, the pipers evoked a spirit of theatre both in semblance and fact. Indoors, band and highland dancers were footlight attractions. They not only supported and drummed up 'in house' show business at Annual Concerts and for the Gordon Dramatic Club but boasted of a public fawning of fans at fairs and fetes. In the Jersey camping years, the organisers of St Helier's Summer Fete would synchronise their pro-grammes with the coming of the Finchley pipers and dancers. By the end of the decade, our notoriety had spread from the Pyrenees to Liechten-stein. The Pyrenees venture struck a note of bathos. Exuberant with two weeks under a Spanish summer sun and the ice cold of the high Pyrennes, the Tenth had stopped off in Paris on the journey home. It was 1926 and the pound sterling was an over-valued currency against the French franc as Winston Churchill, then Chancellor of the Exchequer, had put Britain back on the Gold Standard the previous year. With pockets bulging with centimes, we toured the City scattering the worthless coins out of open taxis to lure street arabs to run in pursuit – there was room for such manoeuvres in the Paris streets of those days. Called to order next day, the Troop had thought to honour its French hosts and France's fallen heroes by placing a wreath of remembrance by the Eternal Flame within the Arc de Triomphe. Slowly, to the pipe strains of *Flowers of the Forest*, the Union Jack and Cross of St Andrew were carried up to; into; and through the arched sanctum; thereby, though quite unwittingly, dishon-ouring every Frenchman dead and alive. The flags should have been dipped and left draped, never borne over sacrosant ground. Next morning *les journals* were black with indignation.

The Troop inured two World Wars and in 1989 celebrated its 75th birthday. So remarkable a history is testimony to the strength and worthiness of the foundations laid, in brick and brotherhood.

To training, camping, drama, prowess in the sporting arena and high quality, home-brewed journalism, the Tenth added property development. Unknown to us twins, Pop Barclay and Uncle Ellison clubbed together in 1921 to buy some three acres of land, sunken, wet and badly drained, between the Great Northern Railway and Nether Street. Here, guided by 'hands' experienced in crafts and trades, the cubs, scouts and rovers, on a do-it-yourself principle, erected an impressive wooden *Gordon Hall* standing on brick pillars. It symbolised the sheer energy and cooperative spirit generated by Pop Barclay and a dynasty of head families who ardently supported the scouts. Our family played its part. Besides the land purchase, my mother donated the Robert Martyn Denman swimming cup in 1924 – for us youngsters a most embarassing gesture. Second to none, however, among the Troop followers was the Hughes family. Len Hughes, a contemporary, was, in those days, the Troop 'sparks'. He lit electric bulbs and blew electric fuses with the assurance of a virtuoso. His mother ran whist drives and other popular events. And, sixty years on, sister Catherine still reigns, a kind of local 'Queen Mum' among them all.

Scouting and show-business suited me well. I had nursed for some time a secret aspiration to go on the stage. My friend, Ron Grose, a skilled piper whose father ran a vast sports outfitters in Ludgate Circus, made up with me a quasi-professional duo. Now and again, we were joined by Peter Dawson, a bass-baritone singer who was associated with the Troop and whose name was high on the popular HMV recordings of the day. We found ourselves performing in dining clubs, on concert platforms and at other events far removed from the world of scouting.

Drama and theatre in the Tenth and my part in it, kept my loyalty steady. Sidney, however, at sixteen was showing signs of boredom. Alas, at one solemn meeting of the Court of Honour he was demoted in rank for refusing to take orderly duty. Although schooling and scouting ran in parallel lines and never touched, my heart went out of scouting soon after the school gates shut behind me in 1928. These contrasting paths belonged together in time but never in temperament. Scouting, undoubtedly, had for me, provided a happier human arena in those formative years than schooling ever did.

⋘ ⋙

John Tindal Philipson, Headmaster of Christ's College, Finchley, was a scholar of no mean stature. He acted by faith rather than by reasoned conviction when, in the late autumn of 1922, he yielded to our mother's petitioning and opened the gates of Christ's College to her ill-tutored twins. We were eleven years old and a bit, an age which generations later Rab Butler recognised as a criterion of educational destiny. Christ's College, in those days, owed its status to Arthur Balfour's Education Act, 1902. The place was a proud, early example of the successful independent grammar school which the Balfour Act had assigned to the financial care of county councils. Balfour's educational beneficence, however, was not a form of socialist wanton largess – only schools which demonstrated determined promise were counted worthy of it. As far back as 1895, John Philipson had come to the rescue of Christ's College, then a struggling, privately-owned boarding establishment open only to those who could afford its fees. By 1903, he had disciplined and polished the school to shine with a peculiar luminosity that flashed in the eyes of the newly-endowed Middlesex County Council. Whether dazzled momentarily by Philipson's lustre or from genuine lasting satisfaction, the Council took the school under its wing.

We knew nothing of all this, nor anything much about proper schools and orderly schooling. Once, when as a new boy timid in his paces, and on being asked what House I was in, I gave the answer 'Strathleven'. Coming from Vernon's domestic one-man seminary, I accepted the fact that schools could be in houses. To find a 'house' within a school ran against the grain of the known world and to discover the truth through suffering the pain of the ridicule my *faux pas* had aroused was a sore kick on the shins of my new school legs. Another confusion, also, quickly overtook me. There were no rule books giving guidance on expected salutations when addressing masters. With few exceptions, the beaks at Christ's College were recently demobilised army officers. Some carried their service titles for a while, some dropped them and others held on to them illegally. So was 'Sir' a Mister, a Major, a Colonel or what?. Finding the answers provided a profitable lesson in tricky diplomacy. Personal rapport was rare, the distance between master and pupils was palpable.

We were taught in classrooms of terraced desks mounting from the front to near the back ceiling, a veritable auditorium after the fashion of a Greek theatre. The masters were performers, looked down upon, especially from the back row of the classroom. Some were gentlemen, others brutes, now and then a comic would take over the form and rarely and by good fortune a scholar would appear. There were hybrids among them, notably a comic brute, a fiery Welshman, Jenkins, in charge of physics. With the wrath of Thor and the energies of Wodan, he would flog a junior across his knees with the rubber hose of a Bunsen burner, threatening all the while to tear the pants off him. Jenkin's comic antics were only funny at a distance.

Pre-eminent among the rare scholars was the much respected and feared John Philipson. He had two sons of his own at the school who took after him; one left with a scholarship to St. Paul's and destined for holy orders and the other, the younger, eventually occupied the Chair of Veterinary Medicine at Cambridge University. As Head Master, John Philipson kept a calculated distance. Fear of his tread stalked the classrooms. His was the source from which corporeal corrective ran throughout the school. On one occasion while we were there he directed a public flogging and formal expulsion. The event had all the trappings of a Speech Day, repleat with Governors, teaching staff and servants marshalled on stage to witness the execution.

Christ's College is a local landmark; anyone can spot the building. Its architect, a man of local repute, chose the highest eminence above Brent Water to build there an aspiring, circular tower topped out by a elongated cone and standing in the vanguard of a train of high-pitched gable-ended classrooms. The *tout ensemble* resembles a long-funnelled locomotive of George Stephenson vintage on the Stockton and Darlington railroad suddenly called to a halt by dousing the smoke-stack with a gigantic candle extinguisher. This monument to secondary education still towers over the Finchley skyline. By 1922, numbers attending the school had far outgrown its accommodation. Expansion, however, was to wait another three years as conditions became more cramped. When the time came, a few old favourites, first among them the college chapel, were swept away by a con-servation-blind Board of Governors with a modernising programme on their hands. Like other institutions where authority insists on squeezing today's quart into yesterday's pint, the school's interior walls were badly chipped, dirty and unkempt. Nothing, however, matched in horror and nastiness the repellent slime-grimed floor of the cold water swimming bath after a winter's hibernation of neglect. Within the school's austerity, the pupils of Christ's College were fed a wholesome varied diet of crammed knowledge but never taught how to masticate what was given.

Neither Sidney nor I made an outstanding contribution to culture, cricket or cadets. Being twins, we were almost morbidly individualistic and misfits in team games and collective events. Our individuality, however, paid well on the racing track and in cross-country mar-athons. A pair of legs beneath me could run faster and for longer than others in the place. So with *usque proficians*, the school motto, singing in my ears, I set out to dominate the penultimate Annual Sports Day and take the Victor Ludorum.

To the very end and not excluding the six last and most fruitful years, my school days were a progression of no clear purpose. Even the eventual success of matriculation to London University was no more than the end delivery of a production line process. To accuse me of indifference to the schooling process would be a misjudgment. Schooling was a daily event suffered for the moment, devoid of any sequential long-term aim. Commendation, however, certainly meant much to me; as when old Davis, the form master, having received a package from London University announced in his sarcastic way: 'I have the examination results. There are a few happy, most unexpected results. I'm, however, mentioning no names at this juncture – Denman!' I gave him a grin, a wink and a warm heart to share. Realisation that perhaps there was something lacking came with retrospect when what should have been there was found wanting. At no time throughout my schooling had there been a sense of commitment, conscious mind-slogging and the pressure of mental discipline. No one ever made me learn by severe self-exertion. Schooling was taken in its stride, like cross-country running, save that no records were broken; my name came in with the mainstream lot.

Nevertheless, it was while at school and in the English class to boot, Davis's lot, that I hit upon a trick of intellect which has served well ever since. Faced with a problem or a specific task: by-pass the obvious, the sane, the conventional and the ready-to-hand. Throw what has to be done straight at the upper layers of the subconscious mind. Tell it to get on with the job. Wait; don't grope and go searching introspectively for the answer, the outcome of the impact. Here's an example: old Davis, in his role as English master, had set an essay – to write up an imaginary conversation between Mary Queen of Scots and Good Queen Bess who had met each other on the Royal Yacht in waters just off the Isle of Man. Convention expected talk of Crowns and Kings, of Dudleys and Dauphins, of Popes and priests. My subconscious, so it transpired, jettisoned the lot. Suddenly, as in a vision, there were the two women being most violently sea-sick. Their speech was of their dire straits, of their common dyspepsia, of leeches and physicians. So it was written. Top marks were awarded along with an invitation to display my originality by reading the essay out before a grinning class of second-raters!

Brother Sidney also learnt lessons at school in unorthodox ways. He was wont to oppose authority for its own sake. Years before, the College had acquired a stretch of ground for new playing fields sloping down to Dollis Brook. To protect both the public and the fair name of the school, the Governors at the time had gravely announced a safeguard: 'the pathway which bisects the fields will be bounded on both sides by unclimbable iron railings'. Those railings still stood. As became his character, Sidney Ellison Denman made bold to challenge their declared purpose and essayed to climb them. His foot slipped and he fell spread-eagled on the wicked spikes. The severity of his wounding and the pain inflicted were outmatched by the great good fortune of a narrowly missed fatality. Future evidence suggests the lesson was of no lasting benefit, as my strange self-illumination had been. However that may be, whether by orthodox or other means, we had in our hands the Certificates of Matriculation to London University. In June 1928, the school gates closed behind us for the last time.

2

THE LONG WEEKEND

Something hidden. Go and find it. Go and look behind the Ranges –
Something lost behind the Ranges. Lost and waiting for you. Go!

The Explorer, Rudyard Kipling

The workmen's train from Liverpool Street drew into Ingatestone
station in the early dawn of a late September morning. For all the
excitement, expectancy and curiosity fluttering my nerves as I jumped
the moving carriage, the destination might well have been Outer
Mongolia. Professional life had just begun. In total ignorance of what
was going on, I met and accompanied an unsmiling, solemn-faced
Stanley Woodcock in rustic jacket and gaiters to what he called a 'farm
valuation'. There, an opposition party pitched into verbal confrontation conducted in a vernacular which sounded like pure Mongolian for
anything my noviciate ignorance could make of it.

My mother, now wedded to our Uncle Ellison, had been on a
house-hunting jaunt into Essex with him the previous summer. Woodcock & Son, a family firm of estate agents from Ipswich and in those
days closely woven by tradition and long service into the social history
of south east England, had helped in the house search. As an aside
before the day's hunting was over, a contract of pupillage for her two
boys had been proposed and agreed. So came about my rendezvous
with the solemn senior partner, Stanley Woodcock, on that Michaelmas morning at Ingatestone.

Houses for our family meant bread and butter and now and again a
spot of jam. Uncle Ellison, solely in charge of all family prospects since
our father's death, was vehemently opposed to middlemen, their tricks,
trade and fees. Howbeit, he couldn't beat them; so he had proposed
that Sidney and I, the rising generation, join them and become
professionally qualified practitioners to manage the family housing
estates. Button-holing Stanley Woodcock on the outskirts of Sible
Hedingham was a fortunate but by no means a fortuitous event. Here
was opportunity to be seized in pursuit of premeditated design.

Stanley Woodcock was a man of exceptional parts. His primary concern was the Celestial City, allegiance to 'an inheritance that fadeth not away'. Farm valuations, estate brokerage and land management were, like St Paul's tent-making, merely a means of livelihood on his pilgrimage to Bunyan's delectable mountains. Stanley was a devout member of the Exclusive Brethren, yet for all his religious fervour, he was a likable, full-blooded fellow, a cross between a parson and a butcher. Fully qualified by examination to be a Fellow of the then Surveyors' Institution, the privilege of fellowship was foregone lest he should be found 'unequally yoked with unbelievers'. In contrast, his brother Frank, the pin-striped and urbane Head of Woodcock's London connections in Conduit Street, was far less inhibited. Sidney was his pupil. My tutelage was in farm dung, hay and harvests, while Sidney passed the days counting aniseed balls in jam jars to facilitate take-over transactions of general stores in humble urban alleys. Sadly, my days among the dungheaps were few and far. Paul, the youngest of the three Woodcock brethren, chained me to a desk in his small office in the West End. From thence trips to the leafy lanes of rural Hertfordshire were sparing and seldom. Life for any professional man, especially an estate agent and surveyor, must have been precarious in those days of mounting depression. House building, unlike its fate seventy years on, was an exception to the shrinking economic activity. Even so, the property market was slipping into decline. Britain, her allies and her erstwhile enemies stood on the brink of severe economic recession and financial turmoil. Stanley Baldwin and Ramsay MacDonald were playing musical chairs at being Prime Minister and the deadly Stock Exchange crash of October 1929 was just round the corner.

In the ignorance of youth, Sidney and I had answered a half-cast vocation, a mongrel mix of businessmen and professionals. Auctioneers, vociferously sold chattels and lands under the hammer; high stewards would manage their principals' landed estates; land surveyors measured and mapped acres; and valuers put a price on them. In the confusion of this harlequinade each performer in one or more of the chosen modes was cousin-germane to the estate agent. The mix divided further in a cultural cleavage of aesthetes and Yahoos. Now and again, a lone aesthete, graduate from a University or a graft from a learned profession, would inadvertently bring a semblance of scholarly accomplishment to estate agency. Higher learning, however, was not universally a trait. So it fell to lawyers and civil engineers to induce serious academic discipline into the schooling of any who professed to manage, measure and market land. In the early years of the new Century, Parry, Adkin and Parry, a teaching partnership of civil engineers and lawyers, became the solar plexus of a system of professional education for the uncoordinated pragmatists who hand-led land. Woodcocks were no exception. Like other leading firms, business nous ran the show. No deep professional erudition, no mysteries, no arcana informed the young. Pupillage was a farce. A bright youth could in a few days master the intellectual content of the daily round; and with a pair of sharp eyes and a pair of open ears pick up the tricks of the trade. To rise among the aesthetes of our chosen, yet still very immature profession, other schooling was necessary. And the way we took was as odd and exceptional in the realms of higher learning as our childhood schooling had been at lower levels.

On the benches of Middle Temple and in his office in Victoria Street, in the year when Robert and Letty Denman, newly wed, were laying carpets in their first home, a young barrister, Beniah W. Adkin, wrote a text-book on the law of tenures. As the years passed, he became an oracle. Through war and peace, over twenty years of unrelenting writing and unabated energy, in company with two civil engineers called the Parry brothers, he inspired, staffed and equipped the prodigious private seminary of Parry, Adkin and Parry. In 1920 its virility fertilised the embryonic vision of one Sir William Wells whose desire was to conceive, deliver and nurture a College exclusive to teaching the skills and knowledge pertinent to the professions of the land. After four years, the older Parry, who had been the first Principal of the new educational enterprise, was succeeded in office by Beniah Adkin. Beniah's genius brought to full-term the new venture. From a leather-topped bureau overlooking the greens of Lincoln's Inn Fields, Adkin came to exercise an Olympian command over the education of surveyors and their ilk. Anyone who aspired by learning to enter elite circles among surveyors and auctioneers had best seek tutelage of this guru and his staff at the College of Estate Management. By the time the Denman twins met him in the late 1920s, Beniah Adkin was in the full spate of his years. To us and many other untutored youths he came to rank among the unforgettables. Heavy jowls appeared to carry the full weight of his scholarship in a pendulous solemnity and, like a judge's full-bottomed wig, to contradict his age beyond its years; an illusion at once dispelled by the lively twinkle of his Argus eyes. There were, he told us, two ways to go: the high way and the low way. Both brought one to the promised land of professional qualification but the high way was the more arduous and led, also, to a London University degree. Sidney opted for the cap and gown and signed me up in the bargain. Beniah nodded and with the calculated courtesy of impatience under pressure made for the door. We left, amid nervous bows, to ponder the outcome of this latest blind commitment to the unknown.

True to their word and contract Beniah Adkin and his colleagues sent their thoughts to us month by month through the post. To the initiate, the papers provided useful introductions to basic knowledge. Annexed bibliographies, however, proved more helpful than the texts themselves. The annexes were clues to accepted works of substance and comprehension. These were acquired. We shut ourselves up with them for long evenings and longer weekends. The deeper the study the further the remove from the shallows of the daily practitioner's world. Student stature grows through the stimulus of argument. Office rote was too thin a diet to feed it; office time too precious and office intellects too torpid. So my brother became my rubbing post, and I his. We pursued dialectical parleys pacing up and down a small suburban garden whose extremity touched the railway line that ran as a backbone behind 'Dollis Park', the new name for Clarence Gardens. There the home that had replaced the childhood nest under the conker tree was housed, along with a growing complex of teenage tensions, in a four-bedroomed semi-detached villa.

In the year when Britain's financial woes drew all political parties into Ramsay MacDonald's 'lifeboat', the 1931 National Government, the swelling ranks of the unemployed had two more idle 'hands' added to their millions. The beehive of the economy was busy enough but it was deplorably shrunken. Healthy worker-bees once outside could not get back into its comfy industry. With Woodcock days over and the farewells said, Sidney and I had stepped across the border that divided active, well-to-do Britain from the realm of enforced idleness and despair. For us, the half-qualified, the only possible way back, if any, was to buy a place among the boss class, provided one knew how to do it and had the money to command.

Cash in small measure was there but not the know-how. A skinflint attitude towards paying for professional advice would time and again land Uncle Ellison in trouble and put him at the mercy of plausible charlatans. Now he was to be once again taken in. To our acute embarrassment, his own undoing and exemplifying his self-generated imprudence, he bought out a back street estate agency in Eastbourne. If ever there was a Yahoo in the business, the man behind Leslie Abott & Co. was surely he. Named 'Wyborn' he traded as 'Abott' to gain first place in the local Kellys Directory. Wyborn perched vulture-like at the end of a narrow passage euphemistically alluded to as 'the office'. The crimson and ochre of high blood pressure and an ill-used liver discoloured and pinched his rodent features. For all that, Wyborn's *sang-froid* camouflaged the trickster within him. The family made a sucker's payment for proffered 'goodwill' which no accountant had been invited to investigate, only to find we had bought nothing but accumulated losses and the criminal embezzlements of clients' monies. Wyborn, alias Abott, escaped into oblivion and left the mess behind.

The fiduciary obligations of that Augean stable took priority over the Herculean task of cleansing, staffing and enlarging the place. Beniah Adkin and his merry men with their excellent tutorial scripts had never taught us how to transform the bad faith of sustained roguery into genuine professional respect. Those first months in Eastbourne were the hardest school yet. Failure meant an ignominy which would far out-shame mere bankruptcy. Studies, final examinations and professional qualifications had to be sacrificed to the redemptive pursuit of a good name. We both quickly matured in a self-reliance which was to brook no man as master and well before our twenty-first birthday.

Wyborn, ever bluffing the public and eager to present an 'Under new Management' image, had changed the name of the infamous firm to 'Owen & Co.' – as near a bizarre pun on the true state of business as the wit of any slick comedian could wish for. The name irked and offended our susceptibilities. As soon as prudent, Owen & Co. was buried. Under our own name, offices were taken on the further side of Terminus Road, the main thoroughfare – the nearer to Beachy Head, the better the social standing. Youthful optimism also set up two satellites, one at Heathfield and the other in High Holborn behind the famous Tudor façade.

Among the unaccountable banal traits of Yahoo estate agents is the propensity to apply excessive hyberbole to describe the virtues of properties portrayed in their shop windows and to keep silent about defects. The alternative is to point out the defects to would-be purchasers and leave the virtues to speak for themselves. This principle became cardinal in our dealing. It offended the Wyborns and others of like demeanour but those who sought our guidance did business with us in confidence and respect. This *glasnost* paid off, especially in a buyers' market. Another harmless ploy was to exploit the latest musical hits. My secretary just then was Paula Green. As lovely as she was useless, Paula eventually left to do *Paula's Half Hour* with Tommy Handley's ITMA show on the BBC. Sitting on my desk, she would croon, 'Have you ever seen a dream walking' to inspire adverts under the caption, 'Have you ever seen a dream dwelling?'. The current dull markets responded.

Experience with Woodcocks had, at least, taught the benefits of wider regional markets. Eastbourne, Heathfield and Holborn were regarded as home bases. Our *de facto* domain stretched from Devon to Derbyshire. One early morning a drive from Eastbourne with breakfast *en route* had made as far as the lavender fields of Hunstanton. There a cry of anger and dismay shattered all tentative hope of further business. 'Why didn't yer holla?': a Stan Holloway husband was admonishing his Eastender wife who had brought up her breakfast all over the back seat of our Armstrong Siddeley. The couple were looking for a smallholding in East Anglia. To such lengths our sales techniques drove us – go anywhere to sell anything. The car sickness journey, however, proved to be an unrecognised portent of a future yet to be for it had taken in Cambridge, a city of medieval splendour and the lure of learning, never seen before and never to be forgotten.

Roland, the junior in our brotherly trio, had stuck to the family lathe. As an apprentice, he was not home-taught. Practical instruction in the trades was found in other men's yards by which he became a proficient master builder with the gait of a professional. In the doldrums of the economic slump, the three of us joined up in Eastbourne.

On the eastern flank of the Eastbourne Road as it leaves the market town of Hailsham, there lay some sixteen acres of oak and hazel coppice designated on the map as Bolney Wood. Today, the woodland is part of the local Green Belt, inviolate against the house-builder, but in those days it was within any man's grasp. Hailsham District Council welcomed our house-building proposals. Uncle Ellison owned the land; Denman & Denman, the renamed agency would handle the marketing; John Roland (Builders) Ltd, constituted under Roland's management, would do the building. Happy fate had taken care of the finances: my father's cousin Hannah had left her not insignificant fortune to be divided between us at her death. She obliged in 1934, the year work was scheduled to start in the woodland. Thus, nearly all the essentials were there for a successful joint family enterprise. Only one requisite was wanted – an architect. A natural, ethereal loveliness ran through the moss-carpeted rides and gladdened the oak and hazel thickets. A professional architect would bruise its beauty. One evening, therefore, while reading poetry in a Willingdon cottage, my pencil wandered to the blank fly-leaf of the anthology and sketched houses to match the faerie of the woodland. High-pitched roofs as generous in expanse as traditional thatch; sand-tiled walls in keeping and broken here and there by brick-chequered exposures of ochre; asymmetrical fireplaces; the individuality of each dwelling caught in heraldic insignia worked into the oak of the entrance doorways compounded as faerie an inspiration as ever the woodland held. Roland translated the sketches into brick and tile. The results either enthralled or appalled. Enthusiasts commissioned replicas in far away Lewes and Wanstead but the style never carried with it the ethos of Bolney woodland. Selling dream dwellings and erecting faerie homes left precious little time for anything else. Conscience disturbed this combination of poetry and commerce. Relentless as the Hound of Heaven its nagging pursued us over the question of the London University degree. Time and years were apportioned. Sidney went first, graduated and left Eastbourne for the rarefied uplands of major firms. My turn came later, under the foreshadowing of world war. Once graduated and qualified, war service recruited me.

Robert Graves and Alan Hodge wrote in 1940 of the twenty years 1919–1939. They called this inter-war interlude 'The Long Weekend,' a respite between the normal bouts of fighting the Germans. By the time we had turned our backs on Christ's College, the 'long weekend' was nearly half spent. As with a real weekend, this metaphorical one lacked both programme and purpose. Days and weeks passed into years of drifting, first of frolic, then of frustration and ultimately of fear. Like all sensible weekenders, the British folk tried to get away from it all. 'Butter before guns' was Baldwin's anodyne for the masses. But as time ran out many carried in their hearts an unease, an apprehension made more acute by dismissing a future none could fathom. There was a vague, haunting sense of pantomine, of a world playing peace with maniacs – Hitler, Stalin and Mussolini with Franco thrown in as a jester. The *mise en scène*, however, lay too far back from the everyday to disturb its doings; it was there, yet no more than a distant rumble of traffic across the daisy fields, where the 'weekenders' were picnicking.

Cars and motorbikes, the new status symbols, came out in their hundreds at weekends making their way across the new London Green Belt to the sea. Uncle Ellison's veteran Bedford would sometimes join in, batting up to 20 mph on the down slopes. A near neighbour had given me, at sixteen years old, clandestine driving lessons. By next birthday, this motor-mentor reckoned his teaching well and truly done. Fortunately, driving tests were still two years in the future. The grand old Bedford, Anglo-Saxon from every line of her bluff-faced radiator, had been replaced by a humble, Semitic-nosed Morris Cowley. At daybreak one Sunday morning, I secretly mounted this vehicle and with thumping pulses and the grinding of gears drove it from Finchley to the West End. Nobody knew – the Morris was back in the garage before breakfast. But I knew! And with high-blown pride some days later had another go. The weekday roads were busier. Drivers in those days would anticipate each other's movements, watching for hand signals. A lorry driver in front of me stuck his arm out to the right while his mate countered by waving to the left. Taking my cue from the driver, I accelerated, as the lorry turned left, and rammed it amidships. Pieces of Morris Cowley littered Finchley High Street. There was no blood. Only mighty wrath.

The battered machine got its own back some months later. Uncle Ellison who fancied himself as a mechanic was, nonetheless, no lateral thinker. Common sense, as he understood it, moved forwards never sideways. Because the Morris Cowley was pinch-nosed, its radiator and cooling system were also pinched. Logic suggested there was need of a supplementary water tank. An auxiliary was fitted above the engine and had a miniature funnel poked through the bonnet as a vent. All went well on the flat. Going up hill, however, the disregarded laws of science demanded respect. On one occasion, tilted on the long, winding steep of Cotswold above Weston-sub-Edge, the water in the auxiliary accumulated at one end to submerge the mouth of the vent pipe and a dangerous steam pocket formed in the opposite corner. Boiling point was reached long before the summit. Mounting steam pressure forced through the vent a piercing whistle followed by a jet of boiling water. This ferocious spout arched high over the windscreen to fall with deadly accuracy on my unprotected head. A reflex of fear and pain shot me clear of the torture, over the back of the open motor to land scalded, shaken and shattered on the receding tarmac. The folly never happened again. The booster-tank was dismantled. The Morris Cowley had the last laugh.

There was a hierarchy among our family motors. The hard-done-by Morris Cowley was reckoned decidedly lower middle class beside my mother's Armstrong Siddeley. It had a built-in hautiness, a look of high disdain resembling a well-bred camel. The latent power of the engine was never fully taxed; driver and passengers would bowl along looking down from an elevated superiority upon the lower orders running beside them, albeit at eye-level with the perched drivers of horse-drawn brewers' drays. The Armstrong was treated with utmost circum-spection, except by the lady who was sick in the back outside Hunstanton.

At the other extreme, more lowly even than the Morris, was a killer motorbike which belonged to me. Lawrence of Arabia had a machine of similar breed which, acknowledging the terror it caused, was called it Boenerges.

The previous owner of my bike when I bought it from him, was a near fatality in hospital having been thrown from its untamed saddle. This infamous bike carried me into the press headlines. Pevensey Road in Eastbourne crossed Terminus Road to form a corner occupied in both directions by Bobby's plate glass display windows. Inside the glass was a generous array of women's lingerie. One morning a lad riding a bicycle and wheeling two others crossed from Pevensey Road in front of the almost uncontrollable fury of my Coventry Eagle Flying Eight. Options were sudden and stark: either the cyclist or Bobby's windows. Its throttle jammed wide open, the motorbike leaped beyond the cyclist straight through the plate glass windows, to end up in a snow storm of flying lingerie. The roaring machine like a whirling wounded lion was spilling petrol over the shop and grinding beneath its spinning wheels an elderly female swept off the pavement by the impact. The poor victim was the wardrobe mistress of the Matheson Lang Theatricals who were performing on the pier. Trembling and shaken, I lifted the Flying Eagle off the frightened woman, dodged under the six-foot dentures of broken glass and trundled home. Fortunately, the injured woman fully recovered after a short spell in hospital. The ensuing court case, however, meant further unwelcome notoriety. The judge lost his temper over my illegible signature. Purple placards told of his fury: 'Bobby's window smash verdict – judge slams defendant's straight line signature'. When I next rode the motorbike, it ran into the back of an applecart in Parliament Square. That was enough! From then on, mine was a pushbike.

❧ ☙

Eastbourne pier divided the lowly seaside lying to its east from the lordly west, bedecked with grand hotels and the coroneted freeholds of the Duke of Devonshire. Opposite the pier, on the meridian between these two hemispheres, stood 'Berrys', a hybrid, half hotel, half boarding house. While struggling with the Wyborn débâcle, Sidney and I lived there. We were never more than birds of passage. The Grand Parade, Eastbourne, that heart of holidayland, never took us to its own heart – it had none. Until the family moved in 1936 to northern Eastbourne, Finchley was home.

The Finchley house, the semi-detached against the railway line, was a modest place with long roots in family history. Built by the family for the family, it remained so committed for another thirty years, a period that covered my mother's sojourn in Eastbourne. The two domiciles reflected the love-hate separation between Uncle Ellison and his wife. He lived in Finchley, she in Eastbourne. The Eastbourne home never survived the war. We all went our several ways, mother back to the old house at Finchley. Ellison, a species of loner was not, paradoxically, in any way a slippers and fireside recluse. He had a strange penchant for the social circle of family life. Following that streak in him, he tried to turn the incommodious semi-detached into a mini Toad Hall and squeezed on to the small garden, between the railway and the back door, a mediocre glass-roofed structure with a polished teak floor which he dubbed 'the Dance Hall'. This appendage was used avidly as a honey-pot to attract female and other friends in harmless mixes of revelry and relaxation, welcome breaks in the strict swotting regime of our later teens.

The last of the family flings were organised from under the Finchley roof. One of these was a party, the other a romantic sea cruise. Both were to celebrate that arbitrary social climacter, our coming of age. The party, as was customary with the truly Finchley-born, was held in King Edward's Hall – a local auditorium whose pretence was more palpable than its qualities. Guests came in contingents from the south coast, from the Midlands and from the North to join the indigenous folk from Finchley's lang syne. Throughout the night, saxophone and drum throbbed out the new hits – *'Just one more chance'*: *'You are my heart's delight'* between Grandma's polkas and highland reels. Entertainers and caterers, amateur and professional, supported perforce by the Pipers and Dancers of 'the Tenth', filled up the programme. The glimmer of a spring evening dusk passed to the strains of *'Goodnight Sweetheart'* and with the dawn *'The sun had got his hat on'*. Wonderful! Here we were on the threshold of manhood. On the bookstall across the street, Aldous Huxley's *Brave New World*, selling by thousands, avouched the future.

People's cruises in luxury liners at popular prices were features of the escapist culture of the Long Weekend. In the summer of 1932, our birthday celebrations meant sailing north to the Norwegian fiords, and the Baltic capitals. The cruise was not all seafaring. Parties disembarked to explore the mountains. On one such expedition, nature demonstrated how intense cold can burn naked flesh. A Welsh lad and I had wandered to a glacier face high above the fiord. From the glacier, an azure torrent studded with huge white floating dentures of ice challenged the path. A Welsh voice dared me to 'take the waters'.

Leaving my trousers, pants and woollies with the Celt, I plunged for the nearest ice-shelf. Naked, the only protection left me, and that by default, was a watch, a birthday gift still on my wrist. It fell into the blue-cold water but was not the sole memorial to folly to be left in those glacial wastes. Seated on a floating iceberg, my naked frozen flesh was 'scorched' by the intense coldness of the hard ice and the skin came away in raw blisters. Limping back the wiser, Taffy and I pondered how sitting on a red-hot stove or an iceberg were the same thing when a naked bottom encounters them.

Two sisters from Edinburgh, daughters of a Presbyterian minister, brought much merriment to those cruising days. Annette was the older of the two. Aided by the gliding silhouette of high mountains moving against a Prussian blue star-lit night, she had, so it appeared to Sidney's susceptible heart, captured the star-light in her eyes. From the outset of the voyage, Annette and my brother were regularly lost after dinner each evening, locked in the shadow of the lifeboats and in each other's arms on the upper deck. When the last evening came the party were housed ashore. Sidney enticed the Welshman and me into a plot designed to test the practical humour of the two 'manse bairns'. A length of string tied round a pillow at the bedhead was dropped behind the bed and the dressing table and poked under the carpet to end an inch beyond the outside of the bedroom door. When at last the prey were abed, a heave on the string instead of evoking shrieks and giggles as anticipated, detonated the most almighty crash followed by unearthly silence. Next morning, tragedy showed on the white faces of two speechless women at breakfast. The string had moved, not the pillow, but the dressing table, brought down a heavy bevelled mirror and smashed all the ornaments. There were no 'Goodbyes'. Weeks after, however, kisses at a reconciliation party in the Finchley 'dance hall' made good the damage. Latent love revived and Sidney returned to Eastbourne with a fiancée in tow.

❦

At this stage in our twin lives, Sidney's inner experiences differed somewhat from my own. Youthful idealism, for him, found an outlet less compounded of religious commitment, more practical, better balanced and less intolerant. Friendship, genuine and lasting, had grown up between him and the son of the Rabbi of Eastbourne and, indeed, with the Rabbi himself, the Rev. Louis Wolfe, a Lithuanian émigré. Hitler was on the rampage in Germany and more and more friends, relations and other refugees would come to find help, guidance and shelter from the paternal Rabbi. Over the years, Sidney, who would often be invited to the Passover Feasts, did much in the cause. The sum of his giving and help will never be known till the annals of heaven are opened.

Divergence in social circles never parted us in the deeper recess of our being. Each knew what the other meant by a nod and a gesture and even by signals of silence. For some years and more acutely so in the stressful, early days at Eastbourne, we shared a common anxiety neurosis. Looking back, it had its funny side. There was a time when the irrational *angst* gripped us both simultaneously. We lay on twin beds, perspiring in every pore and shaking with fear till the casements above the Grand Parade rattled in unison. The spasms never lasted long. Sidney threw them off by cultivating rigorous extroversion. My Christoversion both helped and hindered. I became more fearful than he did of the unknown cause of the anxiety and so aggravated the malaise. Fortunately, a sense of humour would lift me out of these self-fears. When, with the onset of war, real fears stalked the streets, the neurotic ones subsided. But the memory of them left me disturbed and worried to understand an experience I couldn't fathom, for neither Sidney nor I had sought professional counsel.

❧ ☙

Of all the mysteries of creation in which aspiring youth needs instruction the most enigmatic is woman. On thriving to manhood, our knowledge of this enigma was as bare as Mother Hubbard's cupboard and like the cupboard had been left solely to the care of women. Of the four who fashioned our understanding, our mother, Aunts Ruth and Bessie and grandmother Elizabeth Jane, none was 'feminist' in the idiom of today, although the younger women were all career girls. A hard commonsense guarded them from feminists' notions. Women were different from men: that was the first and greatest lesson and the second was like unto it, namely, that men were required to respect women. Women were above the low level of mere male zoology and at a hand's stretch from the angels. They were made of finer stuff, of a different clay, cast to be admired, adored and handled with extreme care as a connoisseur handles the most exquisite Dresden.

My first infatuation was Nancy Kerry, the girl next door but one. Not only was my worship devout – 'she appeared so high above me, she appeared to my abasement' – it was always conducted from afar. To lean out of the window gave me a side view of her garden where she would be tending flowerbeds. The day she put her hair up, sent me berserk. Later, when more bullish, the calf love faded but not the reticence.

Something serious happened when Aunt Ruth welcomed Winifred to her elocution class. This damsel at first also had her hair in a pigtail. Finally it stayed up and to adorn a head and face of exceptional beauty bejewelled by eyes of sapphire blue. These would behold me in mirthful suspense, as if she were looking at the most curious thing she'd ever seen. Under my Aunt's tutelage, Wyn had made an amateur stage début somewhere. It was reported in local gossip news replete with a picture of the teenage star. The picture was nicked from Aunt Ruth's bureau. With a cumbersome camera propped up on the billiards table, I endeavoured to capture it. All the while, the subject of the portrait was quite oblivious of these antics. With her 'likeness' in my jacket, I would shadow the girl home from school to linger a lovesick fool on the pavement opposite till dusk or snow saw me off.

It was New Year's Eve; our social circle, including close friends Ron Grose, the piper and Ken Campbell, were getting up a party for the Albert Hall. Having read the signs aright, my mother formally invited Wyn to help make up the posse and to come as my partner. This three-way diplomacy was *à la mode* in those days. Wyn, through the mediation of her mother, put up a smoke screen about having nothing for the ball – a touch of the abject human in one so divine! She came nonetheless.

For seven years, in the language of Robbie Burns, we 'were acquaint'. Hand in hand but with hearts held in reserve, ours was a ceramic love, as physically inert and undemonstrative as two china dolls. The inertia was undoubtedly my fault. I was crippled by fear of the 'Dresden' factor. Wyn had a father, an admirable fellow of straight flung speech and ready wit and an uncle who ran a chain of popcorn shops, the Caramel Crispy enterprise, from under our offices in High Holborn. The new agency, Denman & Denman managed the Caramel Crispy premises at Eastbourne, an enterprise which eventually closed down and gave place to a milkbar vendor called Charles Forte with a wider vision of the future than popcorn. Sidney left Wyn and me standing when he became engaged to Annette. We two stick-in-the-muds were ashamed into following suit. A ring, alas, proved to be more a symbol of negation than of commitment. It forced reality upon us. We knew that ours was a marriage not to be, yet a friendship never to be forgotten.

Slowly, and as a pupil of far less aptitude in the school of life than my brother, it dawned on me that although women may be heavenly beings, the kingdom of heaven to which they belong is inherited of flesh and blood. Sidney grasped this essential and its implications for the marriage market far more quickly. Influenced, like me, by the Dresden china imagery, he realised that even Dresden china expects to be scrutinised. To do otherwise came near to insulting rare creations and showing rude indifference to beauty and true worth. Decorum was all very well but it could give an impression of aloofness or worse and even imply that the lady was off-putting; a disparaging posture that becomes not gentlemen. These awakenings did not run counter to the earlier precepts but helped to reinterpret what respect for women ought to mean. They came also not from some transcendental revelation but from instruction close at hand. One of the liabilities of so pragmatic a school was that the 'teachers' were apt to become jealous of each other. A touch of this and the four hundred and fifty miles between Eastbourne and Edinburgh served to sever the link between Annette and Sidney. The pith of the truth was that neither he nor I were looking for a marriage partner. We needed simply the companionship of women and the lessons only they could teach.

Among early educational experiences at Eastbourne when boarding at Berrys on the seafront was a demonstration that the female of the species could lead an assault, a fact all the more revealing when wholly unsuspected. A libidinous manageress of the hotel, noting she had two young fellows as guests, thought it behoved her, for the sake of their welfare, to wander the landings at night in diaphanous night apparel. If the bedroom doors were ajar she considered it incumbent upon her to make an immodest approach and enquire if everything was alright and whether there was anything more she could do to add to the comfort of the guests. Should the bedroom key be missing, as on more than one occasion it was, repulsion of the 'comforter' could take a violent turn.

When it became apparent, that in our view, we had not contracted for 'extras', the lady from Berrys tried a vicarious approach by introducing me to an alleged cousin called Connie. Connie, let it be understood, was at least twenty years her junior and hence fitted well my age bracket. Moreover, she was not unbecoming, if voluptuous curvature softened by powder and make-up to give a general Turkish Delight overtone can be counted fair. Rumour had it that Connie was an heiress of no mean promise; but I judged that, like the 'pretty maid' in the nursery rhyme, the bottom line of her fortune was her face. She was perplexed and pretended a vexed indifference at the insistent introduction. On some pretext whose nature now escapes me, a trip to

46

Kensington introduced me to a well-to-do 'aunt' who had a fixation for Harrods, in her vocabulary, 'the General Shop'. Later, a visit to Hornchurch was on the agenda, this time to meet the parents. They lived between one crumbled wall and the next. Daylight and night chills invaded the sitting room from a six inch fissure open to the east winds. Why the rich 'aunt' could not find a few pence for repairs was a depth I never fathomed. Needless to say, the Connie take-over failed: its intent was all too patent, as clear as the daylight through the gaps in the dilapidated wall. Overtures ceased soon afterwards. But as Kipling's mariner would have had it 'I learned about women from 'er'.

The same mariner having sampled and classified the daughters of Eve from Burma to Meerut summed up his taxonomy in the observation:

> 'When you get to a man in the case,
> They're like as a row of pins –
> For the Colonel's Lady an' Judy O'Grady
> Are sisters under their skins.'

More often than not, this sisterly uniformity is well camouflaged and at the same time confirmed by a universal vanity 'to be different'. What flames of fever beset the breasts of two women at dinner inadvertently dressed in matching gowns. The Colonel's Lady, as seen in her own eyes, would have no affinities whatsoever with Judy O'Grady. How absurd! All differences, also, must meet that other criterion of the universality and see to it that no one variation out-reaches the common propensity to attract the male, whether the difference be in countenance, compliance or duplicity.

The art of duplicity has always seemed to me a self-defeating guile. Connie dulled the lustre of her natural attractiveness by parading the rich 'aunt'. So it was with little Marjorie. Her countenance was a bounteous gift of nature, a beauty of distinction no one could take from her. She was the daughter of a widow living in sorely reduced circumstances and running a small printing shop. Something of her circumstances was not unknown to me and conveyed its own commendation. Amazement, disillusion and a sense of let-down damped my ardour when, having cajoled her to come to a party for the first time, I was told to pick her up at 'home'. The address was an impressive, detached residence screened by fragrant rose gardens. She came tripping through the roses, an exquisite petiteness buoyant with what could be taken for wide-awake intelligence. The occasion was artfully contrived. The well-cultivated 'intellect' turned out to be unsown fallow ground. All attempts on my part to cultivate it were met either by a pretended interest or a nonchalant shrug of the shoulders. Had the girl the wit to accompany her physical magnetism with an iota of genuine compliance with my hopes of her, our futures might have converged. No! She preferred assault tactics but in a manner far removed from the crudity practised by the hotel 'comforter'. A trip to Wales included a picnic for two in a beechwood thicket. There my lady feigned acute indigestion and excruciating tummy pains. To relieve the symptoms, her impromptu curative was to roll off her roll-ons there and then. She speculated on the chances of physical allurement; and I on didactical persuasion and a Pygmalion transformation. Each had misjudged the other. Of the two, mine was, perhaps, the worst offence. Marjorie's reluctance to cultivate the 'fallow' in the early stages of our encounter should have been caution enough to assuage my aspirations.

The calm serenity in my mother's face and the sombre certainty in my father's eyes, as portrayed in their early twenties, told of love in tranquil harmony. Family records support the view. At the same time, only those two, alone from among the close relations of their generation, had followed a conventional path to matrimony. No consensus of guidelines, therefore, based on experience was available to point the way. Were the bumps and bruises of our track records to be accepted as a pattern of humanity's common lot and expectations; or was the steady love between our parents a condemnation of our switchback performances?

The poets for the most part inclined to our side. Pope's introduction to the *Rape of the Lock* might have been Sidney's soliloquy on the turbulence he encountered on love's flight with fair Lilian.

> 'What dire offence from amorous causes springs;
> What mighty conflicts rise from trival things!'

My part, in this, was to be ground staff, responsible for ambulance and medical support when the crash came. Lilian was Sidney's serious rebound after parting from Annette. She, *inter alia* had been his private secretary, alert, forthright, demanding and, above all else, possessive. How he got himself into the tangle is a secret not in my keeping. An explosion was ignited when he tried to escape. Fresh vistas lay before him, welcome foot-loose freedoms beckoned, if only his passport were clear of encumbrances. Lilian's possessiveness, as with Marjorie's duplicity, had blunted the arrows of Eros. Days in Eastbourne were closing in. Ties with the past and the present had to be severed. He would do it by letter! A letter of valediction was handed to me to deliver by hand one evening. Darkness fell. Could it be a symbolic final, discreet curtain? Alas, nothing of the kind. The darkness was but a prelude to a crescendo of wrath and tears. At eight o'clock that evening the telephone rang. The falsetto voice of a mother in distress shook the wires: 'Sidney must come at once. My daughter has taken an overdose and has fallen in a coma on the floor of the conservatory.' The summons was an imperative. Later, when evening had turned to past midnight, the telephone rang again. Agitation, urgency, despair: Sidney was demanding a doctor at once. Dr Snowball, a stalwart friend of mine, as cool a character as his name implied, rose to the late call and came post-haste, either to revive life or to fold the eyelids of the broken-hearted for ever. Within ten minutes of entering the house of tragedy his mission was discharged. He stalked out of the 'death' scene in tight-lipped anger – a case of female hysteria and calculated histrionics. Apparently, the high note of the 'wronged' woman's libretto, barely audible, had been a plaintive, 'Do you know if suicides are received into heaven?' Sidney was agnostic; anyway his lack of experience limited certain knowledge. Dr Snowball had assured him that whatever the truth might be, Lilian was in no danger of finding out and would have to wait many years for personal, empirical evidence. The medical man was right. She lived to become the wife of the Headmaster of Dulwich College and never again sought occasion to find an answer to her questioning.

❧ ☙

Along with us, there was among the residents at Berrys Hotel a middle-aged spinster called Alice. She was a kindly soul of blotchy complexion and a wistful sadness with its roots in a foreshortened love affair. The blotches she put down to a deformed epidermis and the other malaise to the First World War. Burning in the heart of this remarkable woman was an odd missionary zeal. It placed her among one of the many fringe cults that grow like barnacles on the ancient hulk of the Anglican Church. Foremost among Biblical promises and prophecies is the Lord's pledge to Abraham to make his seed 'a nation and a company of nations.' The promises were being fulfilled, so Alice affirmed, in the Twentieth Century by the phenomenon of the British Empire. My response to her startling exegesis was not to suppose my friend to be a latter day Alice in Wonderland nor to acknowledge the Union Jack as the Star of David. Alice was clearly not psychotic. She evoked in me a curiosity to pick up the Scriptures and, for the first time in my life, read them seriously. The key texts she cited did not make me into a British Israelite. Nevertheless, from then on I became aware of a spiritual dimension of existence beyond rational comprehension.

Curiosity opened the doors of hitherto unsuspected bookshops. The mass of devotional literature for sale astonished me. Somewhere, when ferreting and browsing, a gamboge back with the title *Once I was Blind* craved my attention and was purchased for my own meagre bookcase. One summer's evening about that time, in the empty billiards room at Teigngrace (the table had gone to Eastbourne) the oak panelling brindled golden in the flame of a westering sun slanting through the open window. The peace and the silence were disturbed only by the rustling of the pages of the gamboge book on my knees. Consciousness shifted in dimension, from what Martin Buber would call an 'I-it' awareness to his 'I-Thou'. I knew I was no longer alone. Another was by me in the glimmer of the shadows. My mind was on the book; my emotions were inert. But in spiritual consciousness there was a quickening more real than flesh and blood. I closed the book, the familiar oak door and descended the long flight of stairs. My heart broadened within me with a conviction, more vivid than life itself, that Christ was alive.

At that time I knew little of theology, of Biblical criticism and exegesis, of church history and doctrine, of the Fathers and the Schoolmen, of the Creeds even; and yet in that moment in the stillness I had encountered in deeper consciouness what my poor vocabulary can only denote as a Thou at the Centre of things. The experience was real in the sense that neither I nor any other could gainsay it. That Centre, over the months to come, was to pull at my will until the latter gave way, broken and surrendered to the will of God. Mine was a sudden, existential theology. Years later in William Temple's *Readings in St John's Gospel*, his simple, profound theological assertion made sense of what had happened to me. 'For there is only one sin' he writes, 'and it is characteristic of the whole world. It is the self-will which prefers "my" way to God's – which puts "me" in the centre where only God is in place'.

The French have two words for our verb 'to know': *savoir*, distinguishes the act of knowing about things and people in an objective way from the personal, inner – 'I know you' (*connaître*). As a schoolboy in Scripture lessons and attending Crusader bible class in Finchley, I knew about (*savoir*) Jesus Christ as a figure in history, as I knew about Julius Caesar. But in the billiards room twilight and, some time later, on the downland above the Saxon township of Ovingdean in a final battle of wills, I came to know Him (*connaître*) in the intimacy of soul. There was nothing of planned intent nor of merit in all this. A chain of events, so it seemed, had swept me into a crisis of faith. Ostensibly, the chain started with Alice and her Bible. Its later couplings were provided by a leading Eastbourne dental surgeon and a bank clerk who one afternoon pushed open the front gate to our new abode. These two strangers had been tipped off by friends in Finchley. The suggestion came from the Crusader Class attached to Christ's College. The callers ran the Eastbourne counterpart. I responded; Sidney didn't. With them and their friends, I found many others 'in the know', whether in church, chapel or meeting house.

The new life tossed me about like a beginner on an unbroken stallion; time and again to falter and fall, to remount and charge forward unthinkingly. A commitment to obey God's will is one thing; to discover what it is day by day, another. Fledgling and foolish, I tried to pattern life on the experiences of others. Instead of waiting upon God, I tried to plan His will for myself. In a book, *God in the Shadows*, Hugh Redwood had told of his dramatic, miraculous adventures in God's service in the slums of Bristol. So, why not I? Never mind the office duties in Eastbourne. A telephone call to the Salvation Army, and I was over in the danks of Bristol with a carload of kids making for the Cheddar Gorge. God's will undoubtedly would have had me keep to the daily beat and to attend my professional obligations.

Cranmer's 'all sorts and conditions of men' describes well the faithful at Eastbourne. Among their company was one of the most complex characters ever to cross my pilgrim path. He was a burly Irishman, massive and square in frame and features, imperious and laconic with strangers but truer to his Irish affability with those he knew better. At worship, he would be found, if anywhere, with the Brethren. His was destined to be among the most infamous names known to Twentieth Century medicine. Crusader meetings were held in his house and it was there that we would meet, although his presence was more often felt than seen. Years later, the newspaper-reading public, radio listeners and TV viewers the world over would hear of him as the notorious Dr John Bodkin Adams accused of murdering rich widows for their money.

My friend, Norman Gray, the beloved dental surgeon who had run the Crusaders, stood surety for Bodkin Adams on bail, never doubting his innocence. Gray was right: Adams was a fool but no murderer. His infamy made many lawyers famous, especially Lawrence, the counsel whose brilliance moved the court to acquit him.

No one will ever know the real truth. Professional jealousy hounded Bodkin Adams and tried to adduce from the evidence of his professional record, the notion that he lured aged women to bequeath handsome legacies in exchange for feigned kindness and consoling medication. His potions, it was averred, were nothing more than calculated lethal doses. My mother-in-law to be was in those days a friend of Bodkin's mother. As the two genteel ladies toured Eastbourne, cruising up and down the Parade in the doctor's limousine, the dowager Mrs Adams would say in the course of an afternoon chat, 'Bodkin has been left another legacy today'. A plausible explanation was that he took no fees because they attracted Income Tax. Legacies were not Income Tax prone.

Although never used, there is in my possession still a testimonial addressed to 'whoever it might concern' assuring the world at large of my impeccable character, solid integrity and latent genius! The revelation was penned in July 1937 at Kent Lodge, Eastbourne and signed 'John Bodkin Adams'. This devalued assurance is evidence enough of the nugatory worth of personal references.

Mother Church like any other good hen has ample wings. The Evangelical wing prides itself on born-again chicks without whom the old bird would turn broody. Their attitude is irksome at times. The evangelical twice-born in self-assured fellowship join hands across denominational boundaries, whether Anglican, non-conformist, Roman Catholic or other. The bond is cemented in the 'Word of God', to be read and acted upon as if it were a Statute of Parliament. When St John in his Gospel affirms 'He that honoureth not the Son honoureth not the Father' there is for these fundamentalists only one way to understand and follow the precept: Christ is paramount Lord over all ways and worship. The theology is implacable, neat and tidy. I ascribed to it with certain latitudinarian reservations.

Now there was at Finchley an old boys' Masonic Lodge attached to the complex of Christ's College's post-school social life. Many a friend of school days practised the Craft locally and particularly in a Masonic Temple off the Strand. Two of these good folk, showing equal zeal with the Crusaders, chased up Sidney and me in Eastbourne with urgent pleas to join. Sidney and Roland had no qualms. I stumbled at the idea. The secrecy which beset the path to Freemasonry irritated me and blocked my curiosity. However, from a promise made to my mother many years earlier and out of respect for my friends but, certainly, with no burning desire to join the Craft, I deferred to their wishes. Disaster followed. 'Ye cannot serve two masters'. In Freemasony, a craftsman climbs by Degrees from being a lowly apprentice to become a Master Mason and higher still and higher. From the start I was inquisitive and ill at ease. There were philosophers of the Craft who had knowledge of its history and mythology. Their writings, however, were little known and even less heeded. This I discovered from reading widely in the esoteric books. By the time my progress had taken me to the threshold of admittance to Master Mason the mythology had confirmed for me that it was at variance with St. John's precept. Like Agag of old, I went delicately until the day when, with one trouser leg rolled up and a self-appointed aide, the door of the Masonic Temple was ceremonially shut in my face. The aide announced 'a candidate in a state of darkness' seeking admission. Six paces later, I was the focus of a be-aproned assembly of worshippers. To be a candidate 'in a state of darkness' stabbed my conscience; the assertion was unexpected and arresting. What concord had 'darkness' with the light of Christ! Standing central to the assembly with the Grand Master calling me to answer standard ritualistic questions, I was alerted to the exigency of the moment. To answer by rote meant declaring before them all that God's promises to His church had been addressed to Freemasonry. This I would not do. So bowing to the Master and to the aprons left and right, I denounced the words they would have put in my mouth, turned heel and walked out of the Temple leaving behind an uproar of dazed confusion.

A stroll up the Strand quietened the poundings in my breast. Later at supper, a cleric in the cast of Friar Tuck advocated with a patronising smile the grace of patience. We come to Christ, he explained 'in the Fourteenth Degree of Freemasonry'. Worse and worse; his assurance only confirmed my fears. Had the rites not been secret but open and known beforehand, much embarrassment would have been avoided. A quiet resignation would have ensued outside the Temple, instead of martyrdom on a carpet flanked by old school friends in blue aprons. Maybe, I should have bowed in the House of Rimmon and left by the back door. But that was not my way.

Dressing up in uniform gives an outward image to our inner desire for difference. All civilised peoples do it, as if to better the sameness of nature. The British have a noted proclivity for uniforms but exercise a nice distinction between the respectable and the ridiculous. Cassocks for the clergy; wigs and silk for the learned law; cap and gown for dons; khaki and scrambled egg for the soldiers; navy blue for the navy and sky blue and wings for the fliers; these and other uniforms like them earn solemn respect. They are understood. What in Britain have never been understood or counted among the respectable are uniforms for politicians and their parties. So when Oswald Mosley dressed himself and his British Union of Fascists in black shirts and leather belts, it did his cause no good at all. They looked 'funny' to John Citizen and, irrespective of politics, courted the brickbats, stones and bad eggs showered upon them by a jeering populace.

That contingents of uniformed, politically motivated thuggery could arise in Britain was symptomatic, along with the National Government of the day, of our insecurity and drift through the Long Weekend. The Black Shirts, however, were something far more sinister than a bad joke. They militarised politics, marching with jack boot and swagger and proclaiming 'vitality', 'manhood' and 'the enforced mobilisation of the energy of youth'. To punch home their views, their boasted weapon was 'the good old English fist'.

Against the setting of sober, genteel, law-abiding Eastbourne, in the high noon of holidays and in its winter hibernation, the regimented Black Shirts appeared exceptionally absurd. But when the laugh was off, they were menacing and very intimidating. Only with greatest difficulty and relentless persuasion had it been possible to prevent local Jewry organised by Sidney's friend Hymie Wolfe and Jewish refugees from Germany from attacking Mosley's meetings in the back streets. On one occasion, the hailed Leader marched his *fascisti* into the dignified Eastbourne Town Hall. From close quarters, I watched it all. Up either side of the Grand Staircase, standing with raised Nazi salutes, two to a stair, the repugnant army formed a black guard from the Entrance Hall to the upper council Chamber. As Mosley spoke, cheers rang out from the faithful and arms shot up like the limbs of robots. Woebetide any protesting citizen who dared to heckle or failed to take his hat off in the presence of the Leader.

Mosley and his Black Shirts probably did much to unite the country behind the war against Hitler. They demonstrated the nature of the evil. Certainly meetings like the one in the Eastbourne Town Hall brought home to me and other ignorant street-bred people what it would be like to suffer a regime where the fascist Black Shirts were masters. Ironically, when Hitler over-reached himself and plunged Britain into war, he helped this country to rid itself of Mosley and his movement. At war and aided by Defence Regulations, the 'good old English bobby' had at hand a law to make short shrift of the Leader and his men. Oswald Mosley went to prison in May 1940. He left his mark on Eastbourne and on me. At first it was hardly noticeable. My political reflexes were superficial, springing from the childhood loyalty behind the red rosette of the Conservative Party. Mosley's haranguing had fathered in me a deep abhorrence of centralised power, of State curtailment of private liberties and of State *dominium* over individual freedom, whether from the left or right. Throughout the war and its immediate after years, my convictions, shaped and sharpened, became opposed to any policy or doctrine which elevates the State above the individual.

Eastbourne those days had a double life, and still has, divided between tourists and residents. The residents can do without the tourists; but the tourists cannot do without the residents. Someone must provide the elemental body of shops and houses, the staple of any well-groomed town without which there would be no Eastbourne to visit; and for love of the tourists, the burghers of the town hang flashy finery and necklaces of fairy lights along the promenade, light up the glittering pendant of the pier and the tiara of ornate ducal hotels. We belonged to the unadorned corpus, the hinterland of avenues, drives, goffs and meads, unconcerned with boats, beaches and sunbathing.

Through an indent in the skyline, our house in Bedfordwell Road looked eastwards towards the Pevensey flats. To my mother it was satisfaction. The size, style and setting fulfilled her ambitions. Like Pevensey itself the house was a gatehouse to destiny. Voices of portent came over the 'wireless' through the walnut grill of the loudspeaker, and came in never-to-be-forgotten speech: King Edward VIII, to tell of love and an abandoned throne; Stanley Baldwin in dilemma; the Archbishop of Canterbury in certainty; the voice of the new King; of Neville Chamberlain with the assurance of 'peace in our time'; and within a year his contradiction – 'To-day, Britain is at war with Germany'. Even so, the most vivid among the memories were the domestic battles of my mother with an exasperating, obstreperous housemaid. Tempers were high one day, as the house was left in the servant's keeping. On returning, the key was facing outward from a locked front door. Inside, snaking out from the kitchen oozed a lava stream of coals, coke and cinders. The contents of grates and cellars had with vicious force been shovelled out and tossed over every carpet and polished border. The housemaid had burnt her boats and left us the ashes.

World events, alas, were to shorten the family tenancy. Within four years, the house had become little more than a pavilion of passage. Each one of us, in his several ways, was to leave it, never to return. There was, however, no pathos. We were high on the spirit of adventure. The world was brave, indeed, but not in Aldous Huxley's sense. After my mother's departure, as if waiting for her to go, the Luftwaffe razed the house to the ground.

3

BLITZKRIEG AND BEAUTY

Four things greater than all things are, --
Women and Horses and Power and War.

The Ballad of the King's Jest, Rudyard Kipling

A warm September sun high in the southern sky gilded the ancient settles of Holy Trinity Church, Eastbourne. The faithful knelt at Matins that Sunday morning as they had done a hundred mornings before. The sun, the worship, the quietness were solid familiar things. Then it happened – the Wail. A metallic howl of alarm rent the intonation and screeched to the rafters. Barely fifteen minutes earlier, the Prime Minister, Neville Chamberlain, had proclaimed to the world and to the nation his declaration of war:

'This country is now at war with Germany. We are ready'.

Outwardly, that first air-raid warning is no more disturbing than the first call of a cuckoo in Spring. Canon Warner, benign, unfaltering, goes to the pulpit. War or no war, the sermon is the same: 'God's little ships, discipleship, stewardship, worship . . .' They are all afloat. Suddenly he stops. A note is passed up and lost for a moment in the white folds of the preacher's surplice as it billows over the pulpit like a sail from one of God's little ships. By the time the missive is retrieved, the congregation are expectant. But none is more electrified by his announcement than I am: 'If Donald Denman is in church, will he go to the Air Ministry in London at once'.

My country's call! If pride be sin, I surely sinned before them all in that moment of drama as I walked with measured steps down the southern aisle through shafts of sunbeams.

Time was precious, the war posters were saying so. My point, therefore, was made straight for the station, to kick my heels waiting for trains now running to a no-man's timetable. Eventually, one came late in the afternoon, by when, had I known, I could have finished both prayers and lunch and been well in time to catch it. However, priorities fed my pride, if not my stomach. Nothing moved on the pavements of Kingsway that evening. The emptiness was the token of a curfew self-imposed by the good citizens of a war-tense Metropolis. Inside the Air Ministry the light, the air and the walls were grey, in gloomy contrast to the sunlit church of the morning. Walking about sipping cold tea and with an air of aimless monotony were three officials, known as the 'three Fs' – Fish, Fife and Finch – and a fellow called Bush. All were to be senior colleagues of mine, bright professionals turned colourless bureaucrats. Like Anthony Smith's blind white fish in the *quanāts* of Tehran, their facilities of open vision and ready action had atrophied from serving too long in the confines of 'the usual channels' of official practice.

Fish was saturnine, dark and heavy, his face deep grooved, as if cut from raw granite with a cold chisel and sledgehammer. Fife, in contrast, was a cheerful character; a ready smile made him approachable even behind never-dwindling piles of dirty buff files. There was a touch of the music of life about him. He could as well have been christened Lenny Lute as Freddie Fife. Finch also lived up to his name. Birdlike with drooping tail-feathers, he pecked his way between the 'in' and the 'out' trays. Bush was different again, a rotund body, keener of wit, and with a tooth-brush moustache. He should have been in uniform; indeed he soon was, for he went over to the Land Officers at the War Office, there to be kitted out in khaki with three pips to the shoulder. We counterparts at the Air Ministry were never so caparisoned. Uniform! To be or not to be? That was, forsooth, the question pointlessly debated over the cold tea.

I waited in dutiful patience. One of the 'three Fs' looked up.

'Ah! Denman,' he said. 'Thank you for coming. You can go home now.'

I looked perplexed.

'Oh! Sorry, your posting? Yes, of course, you want to know your posting. Ah! Yes, let me see.'

After some fumbling, he muttered without looking up, 'Yes, here it is. You are to report to Headquarters at Cambridge on Monday.'

His words were freighted with a consequence neither he nor I could have imagined. At the time they neutralised my euphoria. So that was it. Why all the fuss? A telephone call would have been ten times cheaper and a hundred times quicker. These were exceptional times, however, and the lack of what seemed to be common sense and expedition could doubtlessly be blamed on the 'usual channels'. Those accustomed paths were, however, a little risky in the chaos of the new cloak and dagger activity. Herr Goering, it seemed, across in Berlin, had to be kept guessing on whether so inconspicuous a mortal as I, had been dispatched to Timbuctoo or Salisbury. A posting to Cambridge was top secret, hence no telephoning, but a whisper and a nod over the now empty teacups.

For too long, the Prime Minister and his henchmen, Rab Butler and Lord Dundass (now Lord Home of the Hirsel) among them, had tried to talk to Hitler as if he were a business man of integrity, a man whose word was his bond from upper-class Birmingham. Public opinion, in large measure, was behind these politicians at the time and some heed had to be paid to the ambiguous sentiments of the Labour Party trumpeted in the accents of Aneurin Bevan. Appeasement had ended in time to justify Neville Chamberlain's 'We are ready'.

Because the Government had introduced the conscription of 20-year olds, albeit as late as April, Bevan claimed that Britain had already lost to Hitler. And in the interests of democracy there was the niggling of Vera Brittain and the Peace Pledge Union to cope with. By the autumn of 1938, however, a change to *Realpolitik* was showing through in the increasing number of new official appointments announced day by day. Production of aircraft had multiplied fifty-fold but little thought had been given to the provision of suitable airfields.

To catch up, the Air Ministry were calling for qualified professionals to be recruited as Lands Officers in a regiment of gnomes, grounded to serve the airmen in the skies. Prospects in Eastbourne had become more and more bleak in the gathering war-clouds. Brother Roland was away building factories for munitions in Wales and Sidney had left Eastbourne for the north. Circumstances left me to sell out and join the gnomish Land Service of the Air Ministry. My aspirations were backed by a friend, Air Commodore Wiseman, whose signature I deemed of sounder worth to support my credentials than the bland assurances of Dr. John Bodkin Adams.

War and its consequences are not all damnation. The conflict then visited upon the civilised nations changed a number of things for the better. Under my very eyes, the exigencies of the emergency were already freeing the bureaucracy in the Air Ministry from the ineffi-ciency of its 'usual channels'. The nature of those deadly ducts was unknown to me when at the end of 1938 the red-tape brigade recruited my services. Their baleful existence, however, was discovered later when, in the bliss of total ignorance, I disregarded them. *Ignoti nulla cupido*!

My colleagues and comrade Bush had welcomed me, as a new recruit, on a dark landing many floors up in the dismal Air Ministry building behind grimy windows whose opaque panes gazed upon the fairer face of Bush House across Kingsway. Unlike Little Jack Horner, I sat in a corner *sans* pie and purpose for some days wondering how

I was to be employed. After a while a file or two were handed over. Each file was a folio of correspondence embroidered over the inner cover with a lacework of messages and memoranda and a picoté of red tabs. The tabs marked the place where the story in the file had got to and were, at the same time, signals to take the action further. These files, some fat with age, others mere beginners like myself, were conveyed from desk to desk and from room to room by messengers endlessly on the trot. Use of a telephone was rare. Even more infrequently did memo writers visit one another to discuss the business in hand. Only tea, served recurrently throughout the day in white elephantine china, would break the tedium and, on occasion, slop over from a tired hand to leave its umber stain on the lacework and letters.

Now and again there were excursions into the bright world beyond the grimy windows. Fife, it was, who from among the Fs one morning when the hawthorn was white in the Vale of Avon, called me to his desk and waving a clean, slim file asked if I would care for a trip to Bristol.

'Read it up and report back', he said and passed the papers duly flagged by red tabs into my eager paw.

A production works at Filton and its appurtenant airfield had to be put on a war-footing. Wider boundaries opening up the flying space well beyond the present cramped perimeters were to be set and the land acquired between the old and the new. Of immediate urgency was the need to find out who owned the land. By the evening of the following day, the facts were in my keeping. It was clear to me that the speed of acquisition would turn on what was wanted and when and where. So I set about finding out. The Chairman of the works company, Sir Graham White, was sealed up like a prisoner.

The best part of next day was spent using my Air Ministry Pass to pick the locks of officialdom which barred access to the great man. Perseverence won. By late afternoon when, back at the Air Ministry, the Fs would be sipping tea, I was being introduced to Sir Graham. His response was genuine surprise and pleasure. Here was someone he could talk to, face to face, a veritable Hermes from the Gods. Our impromptu agenda had covered facilities and priorities, when his face fell somewhat:

'Of course, all this must be approved by the Air Officer Command-ing Bomber Command at Farnborough,' he said, as he pondered what to do.

'Let me go and see him tomorrow, Sir,' I offered.

Sir Graham reached for a telephone, spoke to a voice which rang with the authority of command at the other end of the line. He nodded and turned to assure me of a waiting open passage through the steel gates, wire and sentries at Farnborough Air Base. Another day of travel, persuasion and vetting of credentials went by ere I could, in person, present the facts to the Chief of Bomber Command. That brisk man of action was even more overjoyed at seeing me than Sir Graham had been. More telephone calls, this time to Bristol, and a plan of acquisition was prepared and agreed. A weekend intervened before reporting back to Kingsway on the following Monday. Understandably, I was elated with what I had accomplished. Imagine my fury and chagrin when the Fs, fulminating with querulous indignation and pitiless reprimand, demanded an explanation of my long absence.

'Where have you been?' When I told them, their anger sizzled and spat; I was in the frying pan and the fire!

'Surely you know better. You should never have acted as you did, taking matters into your own hands, using initiative. Don't you know there are the usual channels?'

Weeks behind the grimy windows passed into later summer before they trusted me out again. Someone must have had second thoughts. Perhaps Bomber Command had sent a dispatch of thanks for services rendered. Whatever the explanation, a four-days sea trip to the Isle of Man came my way, there to pin-point on a bleak headland looking seawards a site for a radio mast. Caught up in the sun and saline breezes, I revelled among the bracken and saw no higher life than Swaledales and sea-gulls. I returned to base to feed a postage stamp of fact finding into the 'usual channels' and accept the gracious thanks of the Fs for my pains and pleasure. So it came about that by September 1939 I knew how those doleful ducts the 'usual channels' ran. By then, however, the war-time spirit, helped maybe by my own impertinence, was spreading a kind of in-house openness among the bureaucrats at the Air Ministry. The 'usual channels' were doomed to be an early and welcome war casualty.

❧ ❧

Cambridge once had the longest railway platform known to man, until a large-pouched Yankie boasted through rings of cigar smoke to have beaten the record by a foot when building some railroad in the States. His hands were probably free to do so. When the Great Eastern Railway Co. brought the line to Cambridge they had not reckoned with the medieval conflict of 'town' and 'gown' which still beset what

Trevelyan has called its 'double public'. 'Gown' had prevailed over 'town' to banish the station and the sidings well beyond the majesty of the collegiate city. Later, three-storeyed Victorian dwellings of upper middle-class pretensions flanked the main approach way to the city from the railway station. Two of these in 1939 housed the regional Headquarters of the Air Ministry. The HQ had little logistic merit, apart from easy access to rail connections. Its region of command was a triangle with Cambridge at one angle, Oxford at the other and Stamford at the apex. Centuries ago, these three mediaeval cities had been linked by common academic aspirations, as Oxford University wanted to move to Stamford. Maybe an Air Vice Marshall who had read the History Tripos at Cambridge took a hand in planning the location of this Air Ministry Command.

Although aliens among the 'double public' of Cambridge, the Air Ministry personnel would in the conflict uphold the 'town'. With the onset of what was to be one of the coldest winters on recent record, an historical confrontation opened negotiations between the airmen and the University people. Terms had to be arranged for the Initial Training Wing of the RAF to occupy selected colleges.

The mandarins in Kingsway had sent their man, selected not from the Fs, alas, but a fellow called Palmer. His job was to lead the bureaucrats alongside the airmen who were headed up by an Air Vice Marshal. The University proved to be seductive hosts with unsuspected props at their disposal. Within the ample recesses of twin Jacobean fireplaces logs had blazed from early morning to warm the Combination Room of St John's College, one of the most handsome chambers of Cantabrigian Gothic to survive the centuries. The ambience was an opiate. Long refectory tables, placed end to end, were polished to give back, together with the displayed silver upon them, light from a colonnade of candles and the blazing logs of the open fires. The autumn afternoon outside wept in bleak contrast to the mediaeval comfort within. Opposite the airmen and the civilian officials, among whom I sat as a junior, the table length was occupied by Heads of Houses and Bursars of colleges presided over by the wit and wisdom of Sir Monty Butler, erstwhile Governor of the Central Provinces of India and now Master of Pembroke College.

Sir Monty's opening gambit evoked the spirit of centuries past as Lands Officer Palmer passed across the table a clutch of compensation forms.

'What are these?' croaked Sir Monty, as if he didn't know.

'Forms, Master,' came the reply. 'On which you must claim payment for the rooms to be occupied by the RAF.'

'And who are you to say so?' The question set the company back on their seats.

'Oh! I'm a Civil Servant in charge of the acquisition proceedings,' came the bewildered reply.

'Civil Servant. Eh?' said Sir Monty in feigned surprise. 'I've been a Civil Servant. I know what you get up to. Who gave you those forms?'

'They are authorised under the Compensation (Defence) Act and Defence Regulations.' The luckless Palmer, shaken and confused, was manfully trying to hold his own unaware that the erstwhile Governor was testing the vitality of his humour.

'What's the Compensation (Defence) Act?' Sir Monty, now thoroughly enjoying himself, went on relentlessly.

'An Act which Parliament has just passed as a wartime measure. All is in order, Master, I assure you'.

'Parliament!' exclaimed the jovial tormentor, tossing the forms back to Palmer through the candlelight. 'Parliament! I'll have you know, young man, that this University is older than Parliament. You come here under a Gentleman's Agreement, or you don't come at all.'

The exchange of dialogue later passed to the Air Vice Marshall whose note-pad had remained an innocent white for an hour or so. By the time the candles had burnt low in a forest of descending shadows and agreement was almost in sight, the Bursar of Trinity Hall inclined his head and was heard to lisp,

'May I enter a caveat, Master?'

After a whispered exchange, Sir Monty's eyebrows levelled and he addressed the opposite side.

'Ah! Air Vice Marshal, there's one more question. What about the hobnail boot?'

'The hobnail boot? I don't understand.' The Air Chief was tired and not a little irritated.

'In the rooms you will be taking,' the Master's voice was as *pukka* as he could make it, 'are expensive carpets. They wear well because our young men don't have hobnail boots. Your young men will have hobnail boots. Would you like us to take the carpets up or will you pay for them?'

By now the level of the rent payments had already been agreed. So it transpired that with the removal of the carpets, and much else, the Air Ministry was paying fully-furnished apartment rents for nearly bare carpetless rooms.

Ultimately, the RAF were revenged. One evening, smoking dog-ends left over after an illicit party in the now-occupied Pembroke College set light to some curtains. New Court was ablaze. Fear rose with the flames, lest the conflagration should attract Luftwaffe bombers. The Dean in dismay and near panic banned all photography. Where then, we asked next morning, had the *Daily Mail* got its front page pictures? No one knew. The only certainty was that the carpets had been stored elsewhere.

Enemy intelligence was sufficiently well-informed not to mistake a college fire for the flames of a planned bombing raid. About this time when the 'phoney war' tempered the fury of the Western Front, the Luftwaffe gave proof of how surely they could put a finger on what we were up to. Airfields, old and new, were fairly well strung out at more or less regular intervals up the eastern side of the regional triangle. Wiseheads up at Kingsway and among the brass hats of the fighting squadrons, on the assumption that Jerry couldn't tell one flare path from another, hit on the idea of devising a decoy. Some fifteen miles south of Stamford much time, ingenuity and cash were put into constructing a dummy airfield, complete with runways, flare paths and dispersal points to house replica aircraft. A few farmhouses would be in danger when the enemy spilled his bombs and these had to be evacuated. Only old cow pastures would then suffer the pasting. Operational activity was simulated to attract night attacks. Jerry apparently took the bait. He came out of the east in impressive numbers while authority on this side watched, waited and laughed. When it came, the bombing was amazingly accurate, straight down the runway and on every dispersal point. The performance, however, was an eerie bathos: thud, thud, thud, but no explosions. Jerry was dropping *ersatz* bombs pinpointing their targets. The only benefit on our side was our knowledge of his knowledge. He was given a 'thumbs up' for good humour. The British taxpayer picked up the tab and the cows went back to pasture.

The compensation legislation on which Sir Monty Butler had whimsically quizzed the hapless Palmer was, in the main, fair, even generous. It needed to be. As the war hotted up after Dunkirk and the skies and the seas became the country's only defence shield, new airfields grew like plantains and old sites devoured adjacent farmlands and crops to extend their boundaries. Normal courtesies, the 'usual channels', Notices to Treat and other paraphernalia of peacetime acquisition procedure were brushed aside. At the height of harvest, bulldozers would go into standing corn, without leave or notice and while the farmer was at his breakfast; by lunchtime a grey raft of concrete had replaced the golden corn.

The Lands Section in the Cambridge HQ was housed in a hut down the garden. A buzzer rang: I was wanted in the main building. To cross that threshold was never a happy experience – a summons spelt trouble ahead.

The Finance Officer had sent it. He thumped a pile of forms and demanded, 'Denman. What the devil do you mean by authorising payment of compensation to these farmers above what they have claimed?'

'Munns,' I replied. The little man was jumping up and down in blind fury. 'I was appointed here as a Civil Servant. So I regard myself as one who should wait on his fellow citizens and help them understand what the Government, authorised by Parliament, has provided as compensation for damage inflicted upon them. Many do not know the law or the mind of Parliament. It is therefore plainly my duty to tell them. If their claims are wrong, too little or too much, I require them to be adjusted, up or down. You have before you some of those I advised should be amended upwards.'

'Rubbish!' The finance wizard screeched in indignation. 'Your job is to get the claims as low as possible.'

By this time my hackles, so easily roused by pigheaded officialdom, were bristling. Smarting under what I regard as insults to my professional integrity, I declaimed,

'I am a surveyor, not a pig-dealer,' and walked out of the interview. A stormy meeting with the Superintending Engineer followed and for me a period of being confined to barracks. In this and other ways, I had discovered that crack which so flawed the Civil Service from top to bottom: the dominance of a laicised administrative directorate over highly-competent professionals of various kinds. Administrative officers, however, are probably no more to blame than are the professionals for the strait-jackets which cramp judgment and stifle all sense of personal responsibility and practical service. The system is the culprit, the dead hand of a modern mortmain.

The more jovial the soul, the freer the spirit and the warmer the breast, the more deadly are felt the clutches of impersonal officialdom. Harry, the Lands Officer who presided over the small team of land professionals at Cambridge, was a sad and sorry victim of the implacable system. He was a pudding-like fellow with two large brown Welsh eyes that peered at the world with immutable wonder. We all pulled together, we, two officers and a splendid clerical assistant, Gerry, who had stepped straight out of one of John Buchan's better novels. He and I had a problem on our hands. It was Harry. With regular intemperance Harry would return from lunch about four o'clock each day incapable of going further. Necessity required he be incarcerated in the loo – a solution of great annoyance to others – until steady enough to drive home.

As a rabid teetotaller who had a soft spot for the inebriate, I took up the task of keeping the barque steady while the Captain slept off his stupors. Life was all the more difficult owing to Harry's habit of accumulating old files and stacking them in odd corners. In the February of 1942, he went sick. Here was an opportunity. For two whole days the dog-legged smoke-stack which served as a chimney vent to the stove in the 'Lands Hut' never ceased to belch forth dense smoke and flakes of irradiant cardboard. Assisted by Gerry, I was burning every file and pile of paper judged to be disposable, thereby, in our view, streamlining the office for efficient action.

Expecting thanks, the reward was, again, reprimands. Harry had lost his bearings, the Air Ministry had lost its archives and I was about to lose my job. A Mr Clark, a doleful fellow whose religious outlook matched that of old Stanley Woodcock, was dispatched from Kingsway to make a full, analytical inquiry into the black-smoke catharsis of the Lands Section at Cambridge. This solemn-faced inquisitor warned me that he intended to pronounce a doom of summary dismissal. Happily, he was too late. The Cumberland War Agricultural Executive Committee had just appointed me Deputy Executive Officer.

᪥

War drives a man and a maid to enter the marriage stakes regardless of form and, with odds stacked against them, to outrun, if possible, the pale Horseman of the Apocalypse. Within a year of the outbreak of hostilities my two brothers were married; Roland to a distant cousin, Mary Rogerson, in February 1940 and Sidney, five months later, to Elaine Graham-Brown. Hope Prior, the girl for me, had sat in church with sister Peggy two rows behind on the memorable morning in September 1939. My heel-clicking departure made little impression on them. Their only concern was how to get home, my car would no longer be outside.

There were however plenty of rivals in reserve. With the Dresden china factor still a regulator, my wooing was conducted from a hide-out in Cambridge with courtesies a hundred years out of date, including a letter to her widowed mother. However reserved and proper, all went well regardless of the war. We were formally engaged in May 1940 as the German armies swept towards the French coast opposite Eastbourne.

Jessie Hope, known to the family as 'Twinks', was a lovely meadow nymph best dressed in greens to offset her flowing Titian tresses. An

Left Jessie Hope Prior   *Right* Graduation

exceptional clothes-sense and natural elegance saw to it that her turn-out was always flawless, be it muslin or mink. She married me under a straw boater slightly tilted over her left eye, at the familiar altar of Holy Trinity Church, Eastbourne in the following April. Hoarded in my pocket were precious petrol coupons collected for a honeymoon. We set out westward in the early afternoon and wended over country now as a wartime precaution bereft of all signposts and pub signs. The terrain became more and more unfamiliar as the evening shadows lengthened. Town after town were puzzles only solved by waylaying strangers to get our bearings. Some places we knew. The broad expanse of Marlborough High Street, for example, was indisputable. But an hour later on the shoulder of the Cotswold escarpment darkness had fallen. By then the planned time-table was in tatters and petrol was perilously short. Headlights were forbidden and having once taken a wrong turning, map-reading was pointless with no signposts to check by.

Journey's end was to be the Hop Pole Inn at Worcester, a destination recommended but quite unknown. Another two hours navigating by the North Star and we sensed success. A blue-black outline, predominately of Tudor architecture standing against a navy star-lit sky suggested the approaches to Worcester. No one walked the streets. All was dark except for the occasional swish of a searchlight across the night. As the car crawled up what looked like a main throughfare, a slit of light from a half-closed door shone obliquely on a pole sign jutting over the pavement. *Mirabile visu*, The Hop Pole! And so it proved to be.

'I have a double room booked here, name of Denman.' The words pant out, as my heavy, throbbing head droops over the Register.

An awful pause; the lass at the desk pulls the tell-tale register closer and runs a hurried finger up and down the scribbled entries.

'No. There's nothing booked here in that name. We are full to the roof. When did you say you booked?'

This was too much. 'Booked? What do you mean? I booked weeks and weeks ago in writing.'

'Have you our confirmation, Sir?' She is trying to be as helpful as possible.

'Of course I have. It's in my baggage in the car.' The words hiss out between my clenched teeth.

'Would you mind checking for me? There are private quarters. We will have to do something if we have made a mistake.'

Dropping with fatigue, I rummage among the socks and shirts; toss aside the new pyjamas and pounce on the vital correspondence; then rush in triumph back to the desk.

'This is for the Hop Pole, Worcester,' she exclaims.

'Yes, I know. So what! You are the Hop Pole aren't you.' What is the matter with the woman?

'Sure, but this is Tewkesbury.'

Everything falls to pieces as she tries to explain the way to Worcester, a maze of 'lefts' and 'rights' into the darkness.

Towards midnight, we collapsed on the doorstep of the intended Worcester hostelry. Fatigued not only from tiredness but at the sight of its uninviting, gaunt, filthy exterior. Fortunately, it has since disappeared, another unmourned wartime casualty. Once inside, dejection mounted to despair. The bedroom was barren and bare. An old iron bedstead poked its skeleton limbs through a shroud of thin, dirty bed linen and a dwarf pitcher of yesterday's water mocked the expanse of a huge white Victorian washing bowl, its dust-laden, fissured hemisphere as repellent and forboding as Antarctic ice. Served by a trickle of tepid water, the bath occupied a landing at the top of a rickety stair and was screened from view by a trellis of wire and cardboard.

A solitary light switched on for our benefit to reveal a dingy dining room, heavy in gloom. Out of its tenebrous distance, following a clatter of falling crockery, shuffled a slippered waiter. He thumped two portions of last week's ham, dried, crinkled and curly onto a stained tablecloth.

This unsavoury offal at the unwashed, tobacco-stained hands of the *soi-disant* waiter snuffed out all appetite. Only humour kept us going and an impromptu parody of Byron's pentameter:

'The waiter came forth with a menu of old;
His shirt-front all glistening in gravy gone cold.'

It was all we could bear. Leaving the horrors to the night and the cold meats where they belonged, we traipsed leaden-footed up the rickety stair, pulled the paper-thin sheets over our half-dressed bodies and prayed for the dawn.

With the morning came Easter Day, the Festival of new life, and with it golden April sunshine, lanes aflame with Heart of England daffodils, the first morning of life together, the first of thousands yet to be. However, the war was still with us, an ever-rumbling trouble like the faulty brakes on our decrepit Morris; nagging, gnawing things we could do nothing about. So they were banished from our newly-wed minds by the swelling 'Alleluias!' in the red-stone church of Leominster. Across the far horizon on that beguiling afternoon, the undulating purple of the Welsh hills told of mountain contours, steep inclines and the bleat of new-born mountain lambs. The frail but valiant Morris 8 was determined to get there, if the petrol held out.

Within those mountains as the sun went down beyond Snowdonia, our day-old marriage nearly came to an abrupt end. To the north of Lake Vryrnwy, a loose-stone mountain track struggles upwards to within a few feet of a cliff edge and a sheer drop of 1500 feet to the valley below. The panting Morris 8, slipping back at times over the rough surface, finally made the top, sank down on its haunches a few feet from the precipice and blew a mushroom of belching steam over the lingering snows – an offering of thankfulness and relief. To staunch the gusher, I went in search of a water-hole, leaving my newly-wed, oblivious of her fate, to watch the crows circling over the abyss. Half way back, I saw the car move. It was on a slope open to the cliff edge. Minutes later it would have gone over, wife and all.

Grabbing the vehicle made little difference against its momentum. My wife kicked open the car door, tumbled to the grass, picked herself up and ran for a nearby loose boulder. I made to jump clear and leave the car to its fate, when the boulder gripped the back wheel. We drew breath, tempered the steam, filled the radiator and trembling with relief jerked upwards to the crest of the pass. Once over the top, however, the brakes gave way altogether. The car was uncheckable. Rocks and boulders flashed by either side of the precipitous, hair-pin sheep track. We hurtled down, down. Sheep-gate after sheep-gate met head on were smashed to splinters. Flocks of sheep, maddened as Gadarene swine, raced through the jagged gaps pell-mell to forbidden mea-dowlands below. Three miles were covered in as many minutes. Straining against the hand-brake but to no avail and hysterical between tears and laughter, we held on till the car lost momentum and came to a stop a mile along the water edge where Bala Lake empties into the Dee.

Ideally, a honeymoon should be a prelude to a life freshly opened on all sides and from all angles, with the backdoor shut on the past. Destiny for us was to mean hills, sheep, conifers and silent waters.

As it was, Cambridge lay ahead. For a year we dipped into unpacked luggage like marooned passengers in an airport departure lounge and never settled into the second-hand furnished house that had been my home. We waited there for April to come, our month of change and challenge. With it, in 1942, came the 'call', no uncertain sound, an echo from the North to other hills, to mountains which Norsemen centuries ago had occupied as 'fells'.

The road north from Cambridge was a ribbon of expectancy. All we had left behind of value were the assurance of a welcome back from good friends and, from the 'gown' side of that 'two-public' city, a strong hint of a future among the cloisters of learning by the Cam. Happy with these promises cashable in the future, there were no present regrets. So, with its anatomy more trussed up and wired in than ever, the faithful Morris 8 at an average of 20 mph. carried our blithe spirits up the Great North Road. That journey, though unrealised in the joy of it, was nothing less than an uprooting and a replanting. Emotions of adventure compounded with wonder and the awe of a tomorrow that had no shape deepened as the distance lengthened. As so often in my life, the outlook was westwards. In the evening light, the journey ended over the crest of Cold Fell, downwards to the floor of the arresting beauty of the Eden Valley tinged crimson in the sunset beyond Blencathra and the distant peak of Skiddaw. Later that evening reality came back. In the County Station Hotel, the voice of Winston Churchill broadcast live to the nation. The beauty in life was yet to be shared with days and nights of blitzkrieg.

The city of Carlisle has a rough maleness that stands in contrast to the fair femininity of Cambridge. In red madder and grey stone, from the Courts at the southern edge to the Cathedral and Castle in the north, the city is a romantic memorial to the long-ago Border warfare with the Scots. On arrival, a local monsoon had been blowing for weeks and now whirled the city's dust, debris and litter of wartime neglect into our faces.

Before a week was out, however, there were warm-hearted new friends to embrace us, friends, whose bonds of love were to last a lifetime. As prayer is answered before it is uttered, so George Duncan, a most handsome prelate and rising 'star' among Evangelicals, knew of us before we arrived. More Presbyterian than Anglican, he was a true 'bishop' to the flock at St James', among whom we were to find a northern home. Miss Rogerson, a lady whom the etiquette of the day would count too senior in years to be called Margaret and whom St Luke would have numbered among his 'devout women not a few', offered us lodging in her inner-city residence where The Dam, a flowing mill-race, gurgled under the bedroom window.

Fire-watching through the long night soon absorbed the after-hours. This pastime was a hive of gossip, salacious, hurtful and helpful. But for the helpful 'have you heards', we could have lodged at The Dam throughout the winter. The Transport Officer of the 'War Ag' was a regular among the firewatchers. One night, soon after our coming, he disclosed over the midnight coffee, knowledge of Rose Cottage, elegant, traditional, well-furnished and nigh the sparkling waters of the Petteril River. The owner wanted a tenant.

The only snag was the rent, so excessive that only a man of blind faith or imprudence would face it. My earnings were our only source of income. Slashed by war-time income tax, they contributed between £30 to £50 a month. Miss Fairhurst, the landlord of Rose Cottage, wanted more than a third of that sum as rent. Rationed food, however, set by the standard of one egg per person per week, barred the way to any prodigal extravagance. We lived on vegetables and faith and, sustained by these resources, a deal was struck with the Fairhurst landlord.

Rose Cottage meant settlement, in the Anglo-Saxon sense of the word. We were seated there, impregnable against all comers – our very first real home. The front entrance traditionally faced sideways off the road, to open upon a terraced garden, herbaceous and neat, its Cumbrian turf close cropped by rabbits of all ages. Roof-rafters, broken only by an enormous chimney stack, formed low, sloping ceilings to the bedrooms. An open fire cheered the sitting room.

The bathroom, an after-thought of modern days, snuggled for warmth behind an ancient kitchen stove installed when Prince Albert was courting Queen Victoria. Beyond these agreeable quarters stretched an endless scullery, a veritable oven in summer and a deep-freeze in winter where the pipes burst every December.

Downhill, in the nether garden, *phlox paniculata* carpeted over a fertile cesspit. Sometimes of a summer's evening, the music of the Petteril would lull us to sleep. The only drawback to the place was the guilt complex its idyllic setting aroused, as it would in anyone who had the privilege of enjoying its peace in wartime.

No one will ever see Cumberland as we came to know and love it, free of tourists, motor cars and Sellafield. More intimate, more seductive than the grandeur of the Scottish Highlands, its poetic beauty holds a peculiar loveliness, a fascinating allurement. Tiny mountain tarns, when the moon is bright, the water calm and the dragonflies asleep, reflect in their oval mirrors the breasts of proud mountains which rounded by the sculpture of time and nipple-crested by pointed cairns plunge in shameless décolleté to the deep cleft valleys below.

≈§ §≈

The guilt complex which beset us as bombers, making for Clydeside, flew over Rose Cottage was an inevitable consequence of wartime self-analysis; particularly if one was not a conscientious objector but a conscientious participant.

At the back of my mind was the thought, irrational maybe and certainly ill-digested, that the death, destruction and danger of the war should, as far as possible, be shared equally between everybody. The wonderful peace and loveliness of the Petteril Valley and the balm and comfort of Rose Cottage, on the surface of things, placed us in an exceptional position where our portion of the bruises of war was minimal and, therefore, vaguely unjust. Unlike members of the Peace Pledge Union and of other similar movements, I could not espouse pacifism while brave men and women, on my behalf, fought and died. The pacifists' position appeared immoral and their outlook doubly indictable when they denounced Churchill and the British nation for going to war to oppose the evil of Nazism.

At issue in the War was the very existence of the democratic ideal. By condemning the majority for holding to its view and expecting it to adhere to the 'righteous' claims of a minority smacked of the negation of democracy. Hitler opposed democracy, and so did the pacifists. The Government, after all, was a National Government and, therefore, as representative as it could be constituted in the circumstances. That my personal lot in serving the Government had fallen in a fair ground was none of my seeking; any more than was the enviable position of an army officer friend of mine who spent the better part of the war hunting wild boar in the forests of Iran while on army service. Someone had to carry the executive task of running the War Agricultural Committees in Cumberland's land of Goshen. Why not me? The boom and blessing of Rose Cottage were accepted with thanks together with a resolve to make my service more efficient than it otherwise might have been. While helping, on a tight time-schedule, beleaguered Britain to grow her daily bread and feeling happier about living in rural quiet, there was, nevertheless, ever within me something of Paul's disquiet; 'Woe is me if I preach not the Gospel'.

From a lack of allegiance to the denominational order, some wit had dubbed me a 'roaming catholic'. So it was; happy enough to preach the Gospel among the Anglicans of Southwaite; or among the Methodist chapels of the East Fell-side; or with the Presbyterians north of the Border at Gretna and Moffat. The Southwaite Meeting was an unlicenced offshoot of the parish church at Wreay and Southwaite a boundary village the vicar seldom visited. A few enthusiasts and other friends had installed me in the village hall to run a Sunday evening, well-attended and lively unorthodox service. The Methodists, on the other side of the Eden valley, were homely folk, well endowed with the Bread of Life and, also, with wartime victuals from the abundant fertility of their farms.

The dour Presbyterians north of the Border had little respect of persons. The Rev. McDonald 'ministered the Word' in Moffat to those whom Robbie Burns would have called 'the unco guide'. One Sabbath morn after church, two elderly ladies of the congregation, sitting side by side, occupied the back seat of my car. The pulpit and the Word had been put in my care that day by Mr McDonald. The elder of the two worshippers who had accepted a lift home was a lady in her eighties with a stentorian voice, like a fog-horn with a Borders accent. To the embarrassment of her companion, she had not recognised me. They swopped impressions. Had she, like her companion, enjoyed the sermon? The quieter of the two in a whisper put the question to the other, expecting a diplomatic affirmation. Instead, the raucous voice with no uncertain conviction declaimed an emphatic contradiction:

'Nay, nay. I prefer Mr McDonald!'

Her judgment, thereafter, became a ready weapon in the armoury of family criticisms. Should I or anyone else for that matter, slip up, the rest would chorus – 'We prefer Mr McDonald.'

Once the kirk at Gretna, not to be confused with the Parish Church of Gretna Green, had been in my exclusive care for three weeks on the trot. It was the final evening, the last hymn had been sung and I was making my getaway when a lady of vintage years arrested me by the arm.

'Before ye'r awa,' she said, 'I want to wish ye weel and to thank ye for your meesiges.'

Wishing to ascribe any merit she thought my words possessed to the inspiration of the Holy Spirit, I with a hypocritical display of humility assured her the messages were not mine.

'Och! Weel,' she insisted, 'ye read them verra weel.'

I vowed never to be humble again and departed in the comfort that she, for one, had not preferred Mr McDonald.

◦§ §◦

Farming in the war was regimented into the Fourth Service. All land resources, those who tilled them and those who owned them were under the direct command of an authoritarian Minister of Agriculture and Fisheries. Power devolved from Whitehall to Liaison Officers in the Regions, to Land Commissioners in counties and to the 'infantry' – the County War Agricultural Executive Committees (WAECs). Among the county committees, the Cumberland WAEC had the fortunate responsibility for setting the pace of food production over a landscape of outstanding natural beauty and in a county with an East-West divide. The East Fellside differed in sheep, contour and language from the West Fellside with its lakes, high peaks and Hardwick shepherds. Between them ran the lush valley of the Eden.

My job was to shadow the Executive Officer as his Deputy in all his doings (except his domestic divertissements which fed rife rumour) within a personal brief which distinguished land activities from farming proper. We were 'of' the war but not 'in' the war. Every night German bombers flew overhead molesting neither us nor, even, the lakeland reservoirs which supplied water to the great cities of the North. Some of the best farmers would spend all day on the WAEC. They directed State farming policies under the Chairmanship of Charles Roberts, a Border dignitary, landowner, farmer, University don, former Liberal MP and teetotaller. At night the same worthy committeemen, masters of modern scientific know-how and ancient folk lore, would return to their homesteads to 'stitch' the land, till, sow and reap under the creeping lights of the hooded tractors. Sleep was at a high premium.

The near absolute power which this wartime authoritarianism put into puny hands (including my own) was alarming. Its misuse was a formidable forge which has moulded my political outlook for life and which consolidated in me the opposition, already kindled by Oswald

Mosley, against State domination over the affairs of free citizens, whether of a democracy or other polity.

Local subcommittees scrutinised the farming of every acre and made recommendations for the dispossession of farmers to the main Executive. Those classified as 'C3 farmers' were put under judgment. Once a damning denunciation of a farm at the foothills of Skiddaw reached Carlisle urging instant action. Aided by the light of a long summer's evening, I surveyed the place. A lone farmer, single-handed had driven his soul into the soil to produce the best which old age and peat-bog could yield. This was no case for dispossession. I urged the Land Commissioner to adjudicate. He upheld my plea for a reprieve. It transpired later that the unfortunate husbandman was the butt of a local vendetta and his foes held the ear of the local committee. The farmer, if dispossessed, would have lost his livelihood and good name and the nation his contribution to a depleted larder. Of such is the iniquity of State control without appeal.

The power of dispossession was absolute, relentless and merciless; and at the same time necessary while U-boats were demolishing our shipping to the verge of national starvation. The Minister, Lord Hudson, brooked neither sentiment nor slackness. His ruthlessness made unbiased judgment all the more essential. On another occasion, a local committee with jurisdiction over the fertile lands of Solway had, wittingly or unwittingly, passed adverse judgment on the farming standards of one who, it transpired, was Chairman of the Cumberland County Agricultural Committee (a vestigal remain of World War I) and contemporaneously sitting as Vice Chairman of the WAEC. The disturbing news reached the Minister. Soon afterwards, in the opening stages of an actual Executive Committee session, the telephone rang for the Land Commissioner. Ten minutes later, ashen grey, he returned, whispered something to the Chairman and left the room. The Committee meeting was forthwith adjourned amid a welter of speculation. Lord Hudson, the Minister, had been in Westmorland across the border staying at Lother Castle and had called his Land Commissioner.

'In half an hour,' he said, his voice angry and hard, 'I shall be in Cumberland. By then I want "that man" (the Vice Chairman) off the War Agricultural Executive Committee and out of his farm.'

There was no reprieve. When the Committee reconvened in the afternoon, the Vice Chairman's seat was vacant. *C'est la guerre.*

Lord Hudson, when not a Minister of State, was Chairman of Knights Castile Soap Co. Soft soap, however, was never his line. His method with farmers and land girls was rough towels and carbolic.

Hardwick Willie and the wild sheepmen off the West Fellside tried to corner him once in a public meeting in Carlisle.

'Minister. How come with ratching yows in whin bush?'

Hudson, almost certainly ignorant of what the fellsiders were getting at, shot straight from the hip. 'Don't waste my time,' he countered, 'if you can't answer that question yourself, you have no right to be farming the fells. We'll have you out before the week's gone.'

Only once to my knowledge were his defences breached and, then, not by any sleight of espionage but inadvertently, in a way spy-catchers could well heed. In the New Year of 1943, Britain's losses at sea were nearing four million tons of shipping. The fatal figures were under wraps although a desperate Government had authorised Hudson to take the farming authorities into his confidence. Every risk had to be run to boost winter ploughing and spring sowing campaigns. Hudson called a top-secret conference at the Queens Hotel, Leeds. On the evening before, when the carefully-vetted delegates were assembling, my twin brother, who lived in Leeds, was by chance leaning over the bar in the hotel. The Liaison Officer for the Northern Region came in with the Minister.

'Ah! Denman,' he called, 'You've beaten us to it.' He proceeded to introduce the Minister to my twin, who kept mum although fully aware of his mistaken identity. The three retired to a confiding drinking nook. At a point in the conversation beyond which boded acute embarrassment, even danger, my brother interrupted the soap Baron with the caveat:

'Minister, I must explain that I am not the man you take me to be.' Next morning, my arrival confirmed the slip-up, by which time the Minister, in a spirit of sporting good humour, had sworn my brother to secrecy and seen him off the premises.

While this comedy of errors provided a light relief, it was discovered later that sinister attempts had been made to penetrate the Conference and make off with the secrets. Alien agents had been in the wings watching the movements of delegates; all officials and anyone who consorted with them were marked men; no unguarded papers or briefcases were safe.

Reggie Lofthouse, a young friend and colleague of mine who had been a District Officer with the Cumberland WAEC and had recently been transferred to the Ministry of Agriculture and Fisheries, had met me at Leeds Station. Together we went to the Conference in his official car. Late the previous evening, I had put the final touches to an 800-page manuscript, to be submitted to London University as a doctoral thesis. These precious pages were a monument to years of research and writing conducted through countless night hours, on fire-watch duties and in air raids, the sum of the 'own time' gaps in my schedule of wartime obligations. They were the joint effort of an over-loaded official and a lonely housewife. Now all was finished! With a light heart, I had tossed them into a special carrier to be read through on the journey to Leeds. Although a legal analysis of wartime agricultural policy, the thesis, in itself was of little consequence to the Conference agenda, so it had been left secure in a brief case on the back seat of the car.

Some hours later, as we came from the Conference, my mind seized of gruesome tales of merchantmen, marines and submarine onslaughts, was shattered in blind incredulity. The car had been rifled. The back seat was empty and the manuscript had vanished; never to be seen again. At that time I was unaware of sharing a like fate with Lawrence of Arabia whose newly-finished manuscript of *The Seven Pillars of Wisdom* reposing in a black Gladstone bag had been stolen from a seat in Paddington Station while he went to buy an evening paper. What most came to my aid, however, was the memory of my Belgian riding-master and his fierce insistence 'si vous tombez – montez!' Fate had given me a nasty toss. I heeded the riding lesson, remounted and charged forward. Near Leeds Station was a stationer's shop where I bought a ream of wartime, rough yellow paper. By the time the train had reached Carlisle, three pages had been written of a new thesis. These were handed with pathetic bravado to my incredulous wife standing at the kitchen door on my return. The new version was new in every way for neither copy nor notes of the stolen work remained. The loss was never counted – the sum was too great to calculate. What is certain is the merit of the second attempt over the first.

Doubtless, the raiders who filched the manuscript regarded themselves as war heroes intercepting enemy intelligence. Fair enough; our lot were doing similar things in Germany. Such acts are within the ethos, within, even, the ethics of war. They are a long remove from those petty offences which, apart from their inherent wrong, in wartime trespass against the prevailing selflessness that binds society together in a common cause.

Alas, they occurred. There was, for example, the case of bribery among the bilberries. A hill farm near Hartside where Swaledales nibbled the bilberry buds was 'in hand' to the CWAEC. Tenders were out for repairing and extending the homestead. Because petrol was hard to come by, I had motored a local builder to the site where he made a survey and estimate of the work. One comes down Hartside, hairpin after hairpin, both hands gripping the steering wheel. Seeing my hands preoccupied, the builder chose to stuff my pockets with £5 notes and dump a gigantic box of chocolates (where he got it from only heaven knew) on my lap with a 'for the wife'. He explained that the presents would help me recognise his tender when it was opened. It was not possible to get at my pockets to resist the attack and thrust the money back at him. To abandon the briber on the roadside would put me in jeopardy; his word against mine to make me the venal party offering privileges for graft. As it was, he crumbled scared. Bribery was not his accustomed way of business. So I retorted with as much reserve as I could command:

'Take that money out of my pockets at once; and the chocolates from my lap; and I will undertake not to drive you to the police but drive you home and no more will be heard of this lapse. By the way,' I added, 'don't bother to put in your tender.'

Putting secondary causes and narrow party politics before the nation's war effort rankled with me. In the very bowels of the Cumberland WAEC a Communist zealot probably 'planted' by the Party was an example of this sort of thing. When Britain's fortunes were low and the conflict high and hot, this humourless, dour female, bemoaning the conditions of child labour in the eighteenth-century mines, tried to call the entire secretarial staff of the WAEC out on strike. The bosses (they included me) had been insistent that because youths were answering the nation's needs in the fighting forces, our secretaries would be required to stick stamps on all letters themselves as they wrote them. The Communist bigot and false citizen genuinely thought this an exploitation of labour. Fortunately, she had called the local representative of the TGWU to support her. He came to give advice and proceeded to squash the Communist hornet flat and to urge instant and effective dismissal as the proper reponse to disobedience. All, with the exception of the sour-faced Communist, and most of all the secretaries themselves, were encouraged by this commonsense, patriotism and the spirit of Ernest Bevin which flamed within him.

Another frenzy involved the Spiritualist Church of Carlisle and brought it into conflict with the Cumberland WAEC over what the locals call 'bogles'. The Committee had acquired Aspatria Hall to house girls employed from the Women's Land Army. Unknown to us, a 'bogle' was already in possession of the Hall. The girls only endured one night. Weak-kneed and white, they walked out before breakfast vowing never to return. Bombs, fire, shells, death and destruction they were prepared to face, but not another night at Aspatria Hall! A strutting Commandant of the Women's Land Army, short kilted in thumping jack-boots, pooh-poohed the fuss. She'd sleep there, bogle or no bogle – just to prove her point. Next morning her knees within the jack-boots had turned to jelly. It happened in the wee hours. The haunted room had become a deep-freezer pervaded with the stink of purtrefying flesh and death.

A day or two later, in the bustle of a busy afternoon, my office door in Carlisle was flung open without notice or courtesy. Scent of another odour entered, overpowering and soporific. The Head of the local Spiritualist Church had flounced into the room.

'I've come for the keys of Aspatria Hall,' she announced. The office felt constricted and cramped. The room was overcrowded anyway. A secretary sat squashed against the window and the chief clerk against the opposite wall, with his back to me.

'Madam,' I pleaded, 'please sit down and stop swaying about'.

She poised herself in front of me and bored into my soul with gimlet eyes. 'We have arranged to hold a séance in Aspatria Hall tonight to release an earth-bound spirit. The worshippers are on the train from Euston'.

Flaming fury possessed me.

'That you certainly cannot do,' I flung at her. 'I cannot allow Government premises to be used for occult experiments.'

There was a pause while the tension mounted.

'However,' I went on. 'it is my duty to report your request to the Executive Officer.'

She rose in a shower of perfume and powder to cry: 'Have you ever seen an earth-bound spirit leave the body? Have you? Do you know what it is like to be earth-bound? Do you? I knew you would resist me. I felt it. Your vibrations hit my bosom as I came into the room. And another's hit me from behind.'

She swung round waving a parasol at the chief clerk. At that point the secretary fled from the room shrieking that I'd been bewitched.

'You do not know what you do,' the persistent spiritualist was threatening. 'I shall have you reported. Duty! Duty! It's the coldest word I know. Why don't you try love?'

Without giving me an opportunity to comply she left us as discourteously as she had come. Only the perfume hung about for a day or two.

Unknown to me, another leading lady of the Women's Land Army, wealthy and without jack-boots, had taken her chauffeur (yes, a chauffeur in wartime) to Aspatria Hall. Later, he told of an encounter in the haunted bedroom between his lady employer and her Ouija and planchette which, at her request, he had brought and put on the bedroom table.

'If,' said the lady inquisitor, speaking through the planchette, 'you inhabit this room, who are you?'

Immediately the index pointer swung and counter-swung across the board. Letter after letter spelt out: 'My name is Stinking Lilly.'

Aspatria Hall was decommissioned after that. The Land Army sought other pastures. My indignant spiritualist gate-crasher got what she wanted from a compliant Chairman of the local Carlisle committee. Nothing more was heard of Stinking Lilly.

Bribery and the unscrupulous degrading of defenceless farmers were examples among many of the temptations into which the wartime centralised *dirigiste* government could lead simple citizens. The scarcity of those who yielded to these and similar temptations speaks volumes for the common decency and fair-mindedness of wartime society. Indeed, the record stands to the credit of all concerned – Parliament who authorised the King in Council; the Privy Council who authorised Ministers of State; the Ministers who authorised the Committees; and the Committees who authorised the farmers and landowners – that, despite the lapses, the imposition of universal central planning directives was not found more intolerable. Explanation lay in the absence of tension. Farmers and landowners at best (which meant the bulk of them all) wanted to be told what was expected of them, how to use themselves and their stock to the most efficient advantage of the nation – they wanted directions. There is a form of service which is truly free; wartime response sampled it. Lack of friction, however, lead many otherwise farseeing statesmen after the War to make the profound and mistaken assumption that because authoritarian directives achieved so much in wartime, similar means could be as effective and welcome in peacetime. Wartime planning had a purpose – to overcome this enemy. In peacetime there are as many objects motivating national plans as there are planners to make and direct them – what Professor Hayek has called the *Fatal Conceit*.

By coincidence, Hayek's *Road to Serfdom* had a prominent place among my books in those early Cumberland days. The memoranda, minutes and meetings which made up the daily chores of the WAEC vividly endorsed the tenets of that profound book. So much so, that when in 1945 a misguided public threw out Churchill and elected Attlee as the first peacetime Prime Minister with a programme to re-enact the substance of the Defence Regulations as traditional Acts of Parliament, I resigned from the CWAC the next morning. The Chairman, Charles Roberts, in a voice like water running through a sluice-gate, tried to still my prancing nerves. As I was an over-zealous individualist and he an erstwhile Liberal MP destined to beget a Labour Party progeny, his counsel had little effect. I stayed a little longer. But my days among the bureaucrats were numbered.

4

EXPECTANT YEARS

Ask why God made the gem so small,
An' why so huge the granite?
Because God meant mankind should set
That higher value on it.

Lines in a Moffat Inn, Robert Burns

Pastoral ease at Cambridge, its gilded and silvered youth, lectures at leisure and the temptations of intellectual 'homosocial' coteries never came my way. Grind without discharge was the road I plodded to graduation. Those laurels, once won, were in consequence probably too highly valued. Even so, they were no more than foliage adorning the foot-hills of my ambition. Climbing further to higher degrees had to take third place behind earning a living and fighting a war. Opportunities for study were snatched interludes, strap-hanging in overcrowded tube trains, lunch breaks and war-weary nights behind blackout curtains. Only once did I attend a lecture. It taught nothing I could not have gained from books. Books, indeed, were my mentors. Tomes that baffled on first reading were read ten times if necessary; none was discarded and many advanced works would be used as directories pointing the way to the rudiments of a subject yet to be mastered. While at the Air Ministry in Kingsway, the Convocation Library of London University came in very handy. Listed titles gathered there were later bought in David's Bookshop, the renowned second-hand dealer in Cambridge, and among the nearby market stalls when 'up' at Cambridge on war service.

My academic path was a narrow track with a fixed destination – a degree in estate management and a surveyor's professional qualification attached. The track was hardly discernible on the academic landscape. For twenty years both London and Cambridge Universities had offered degrees in the subject but no one had, so far, pursued post-graduate studies to the point where the subject was secure on the academic map. There was no academic literature, no scholars jealous of high academic credentials, no Socrates to guide the development of advanced thinking. When, in 1939, I applied to London University for admission to read for the MSc. Degree in estate management no one knew what to do. The authorities left it to me to frame the course, pursue it and write a thesis to cap my auto-suggested procedure. The examiner, when asked for a criterion by which to judge my perform-ance, suggested my ability to demonstrate a width of knowledge broader than his own – no very onerous task! Later, to the incredulity and annoyance of the University, I aimed to go further and to stand among the doctors of philosophy in their carmine gowns. Polite irritation and dithering bordering on despair in official quarters merely strengthened my resolve. What was to be done? No one seemed to know. Again I was launched upon an uncharted sea but permitted to fly the flag of the Faculty of Economics and Politics. The first attempt in 1943 was a failure – data insufficient to support formulae. Second time around, when in the process my thesis was stolen, brought success. Whatever London University thought of this achievement, the Univer-sity people at Cambridge opened their eyes and, although careful not to show the slightest trace of enthusiasm, suggested a talk *après la guerre fini* with a view to future years.

Thus it was that Cambridge called me; I did not call Cambridge. The logical assumption was that the University wanted me. The door was held wide open; neither a knock nor a push were needed. The open door signalled obviously a warm welcome on the other side; so I thought. The signal belied the facts. There was no cloistered ease. On the contrary, the door opened upon a curious struggle for survival and upon a University in dispute with itself. As an 'alien', my ignorance gave no inkling of what was happening, nor had anyone warned me, least of all the posse of dons who had given the interview and invited me to come. Disputes at Cambridge are well ventilated because the University prides itself on a Cantabrigian species of democracy. Voting takes place in the Regent House, the University's parliament where proposals are often supported on slender majorities. Minorities who *non placet* Graces in the Regent House can often champion majorities throughout the Colleges whose independence can thereby effect a *de facto* reversal of University intentions. At the time, the Regent House had given the go-ahead to a resolution of a Syndicate to strengthen and develop teaching in estate management. Many colleges were against the policy. Consequently, the job the University wanted me to do met a barrage of criticism which bewildered me and evoked a sense of betrayal. The mood did not last long. Impersonal, genuine criticism helped me find my bearings and, at least, was couched in a language germane to the dispute but found only in short supply among my 'professional' colleagues at the intended Department of Estate Management.

When Stanley Woodcock met me off the Ingatestone train to accompany him to a farm valuation on that never-to-be forgotten autumn morning in 1928, he could have had no idea that his initiation of an ignorant youth was to sow a seed from which grew a tree of knowledge of international proportions. Lawyers speak of 'tenant right' to denote the right to claim compensation for items of benefit to the landlord which a farming tenant on quitting his farm cannot take with him. Valuers assess the claim along with the landlord's counter claim for dilapidations which is set off against it. Behind this simple 'farm valuation' process lies a long, complicated history of custom and usage, of common law and Statute and a critical relationship of parties with reciprocal proprietary interests. It was this relationship which provided me with a key to a calculus of proprietary structures by which analyses of landholding systems could be made. Some years were to pass before this phenomenon developed and the ability to demonstrate the use of it in fundamental research.

Fascination with tenant right, however, had set me going. Claims could only arise at the interface between proprietary interests limited in duration (e.g. leaseholds) and the superior proprietary interests from which they are derived. The history of English land right from Norman times to the present day is a story of a hierarchy of derivative interests (estates), changing, expanding and contracting, under a paramount lordship. My first serious research traced the changing features of tenant right in that moving story. By the time war-service sent me to Cambridge, my book on the subject, *Tenant-Right Valuation in History and Modern Practice*, had not yet been published.

Private and public interests in land as the evolved land law of England had framed them were rudely disturbed by the rigorous directives of the war-time Government. Going for a doctorate, the enabling legislation controlling wartime agricultural policy which provoked this disturbance was exceptional and worthy of analysis.

The job was done before going to Cambridge in 1946. These research studies equipped me with a tool to analyse the power base which, in any economy, determines the ownership, use and development of land and they gave me unique experience in handling it. Although new to Cambridge itself and its mysteries, when appointed in 1946, I was by then no tyro in academical research and impatient of the limited horizons of knowledge which sufficed to satisfy the qualified practitioner.

Disillusion followed disillusion. To help erect a new teaching and research edifice was the prospect which allured me. Much preliminary spadework was inevitable and was a happily accepted chore. To conduct it, however, I had expected an academic position, if only a junior one. The post given me was a let-down, a contradiction and a paradox. The university appointed me as a 'land agent', a job wholly out of context in a prodigious seat of learning. Horror at the appointment was mixed with inequity. The staff of the new Department were expected to earn fees running a professional service for the University, colleges and cognate bodies. Fee-earning was used to finance the lecturing. This practice of 'taking in washing' meant doing two jobs for the pay of one. To the two-fold imposition was added, in my case, the extra burden of being unpaid Secretary to the newly established Board of Estate Management. The Head of the Department was a chartered surveyor who was doing a good full-time professional job. So to my hand was left the responsibility for drafting academic policy memoranda, the text of the new Handbook of the Department, new teaching courses and so on. Howbeit, I knuckled down to try and make bricks with little straw, although 'the Pharaohs' seemed to have small notion of how their monuments should be built. What rancoured most was carrying creative academic responsibility as a 'land agent' in such a place. Treason raised its beguiling head more than once as I listened to the siren voices of critics. My heart, bruised but not broken, was however truly wedded to the opportunity before me and was strengthened by a silent vow never to quit.

Even so, other irritations darkened the skies of those early Cambridge days. Like the peace of God, the Cambridge ethos passeth understanding. A particular annoyance was the convention which denied me the courtesy of my doctor's title. Had the doctorate been awarded by Oxford or Durham Universities, or Trinity College, Dublin, all would have been well. London University, it transpired, was beyond the pale of recognition. This prudery still mystifies me, even after forty-five years of trying to acclimatise to the Cambridge 'way'. The fuss about democracy also seemed hypocritical in a society which condoned the arbitrary bestowal by colleges of the privileges of college fellowships. Fortunately, time and humour work solace and the sheer grandeur of Cambridge overcomes the pettyfogging anomalies of the place.

≈§ §≈

These tangible woes were accompanied by a pervading sense of desolation. Back in Cumberland, fate had cast me in the role of a petty Caesar, second in command over a hundred or so staff, waited upon by Peggy McAlindon my personal secretary and possessed of a private office. At the Cambridge end, in the Easter of 1946, my destination turned out to be an empty room in a half-occupied building on the corner of a by-road, Mill Lane, which dropped the unwary into the River Cam. Only a caretaker noted my arrival. For many days, the bare room was used as a dump, housing all our wordly belongings in two closely packed tea-chests. Life revived somewhat with the return to duty of the Head of Department, Noel Dean; but a kind of inner demise gripped me on being introduced to his secretary. Speaking from a great height of condescension, she would offer to 'do a letter' for me on occasion. She was a small, grey bundle of femininity quivering with the anxiety of her own importance. My heart yearned for Peggy whom fate had handed over to a Commissioner of the Duchy of Lancaster. Desolation and despair moved me to threaten resignation if Peggy were not wrenched from the Duchy, recruited to the staff of the Department and appointed my aide. Noel Dean showed a remarkable sympathy; so she came – the first shaft of light in a dark place.

Noel Dean was the most unlikely person one would have expected to find heading up a University teaching and research establishment. A pragmatist, innocent of academic obsession, he saw life in black and white images devoid of half tones. From early manhood in Nottingham, he had been a surveyor of buildings and public works. Single-mindedness, one of his simple virtues, hampered the execution of the dual role of surveyor and university teacher into which events had cast him. An efficient and growing professional service over which he presided absorbed the larger share of his abounding energy. A man of short vision, whether thinking or writing, long-term intellectual prospecting and academic strategy disinterested him. Until convinced by results of the practical value of research, he mistrusted it and begrudged time and money expended on it. As Head of Department, he encouraged teaching activities providing no teacher wanted to play first violin in the departmental orchestra. He would accept ideas he little understood; but woe to the promoters of them if promises were not fulfilled. Bodily a large man, laconic and growly, Noel Dean and I struck up a workable rapport over the years. He became a useful baffle to my unorthodoxy, far more so than a more querulous debater would have been.

At the other end of King's Parade, Sam Weller, the clinically precise Bursar of Gonville and Caius College, presided over the fortunes of Estate Management like Moses, the lawgiver of old. Weller was the first Chairman of the Board of Estate Management and to him all issues of consequence were referred. His were a visionless, razor-sharp mind, sparse frame, finely-drawn features and hair besilvered before its years to invest him with an air of asceticism exonerated somewhat by a subtle, inner humour. He was an all-sided contrast to his friend and colleague Noel Dean. Had Sam never switched from academic teaching to administration and the stewardship of College affairs, he might have given a much-needed intellectual lead to the new ventures in the academic development of estate management. But he preferred the practitioner to the sophist. As Secretary to his Chairmanship, my lot was to see a great deal of him. Later, in session after session, I would bridle with scorn as his poised, fussy pencil (he never used a pen) was used as a stiletto to stab and niggle at my draft minutes.

The mists of uncertainty gradually cleared as the summer months went by to reveal a small Department in the flux of parturition between its own ultimate independence and the parentage of the Faculty of Agriculture. Besides Noel Dean, two other formal University Lecturers made up the established staff. All new-comers were merely 'advisory' auxiliaries. There were no introductions. My two 'university' colleagues were encountered, either deliberately or inadvertently on the staircase. One affectionately known as 'Timber' (C.H. Thompson) was a lugubrious fellow who taught forestry. Soon after, I met him coming down stairs repleat in full morning dress and silver buckles. 'Ah' I called, 'Off to a wedding?' His hauteur towered above me like the Mattherhorn. 'I am the Esquire Bedell. The Vice Chancellor is waiting for me to precede him in Procession to the Senate House.' This little incident was typical of the contretemps that beset my blundering way. R.P.F. Roberts was the other University Lecturer, urbane, taut and lonely, a sad man obviously unhappy with the new transition. My mission puzzled him. It was simple enough in my own mind: to get his suggestions for a detailed text of the law syllabus. 'Quite unnecessary', he said with no little heat. 'Just put down Land Law and Landlord and Tenant'. I wandered off to compile my own version.

Alas, by the turn of the year Roberts had committed suicide. Cambridge seemed more ominous than ever. Happily, there was promise of a sunnier future when a new Advisory Officer came to share my room. Because no one was ever given proper guidance, C.W.N. Miles (Charles to us all) was, like me, expected to exercise extra-sensory perception in understanding what was wanted of him. His all-round commonsense, however, never disclosed that he was due at his humble desk alongside me in the bare room two weeks before he turned up. Charles Miles is one of those rare people to whom kindliness, courtesy and ready hospitality jointly with 'Dickie', his wife, are precepts of natural law to be followed without question. He could read traffic lights better than I could and so when, some eight years later the lights ahead turned amber presaging the red, Charles resigned his University Lectureship for green pastures in Hampshire and Anglesey. Later, after a successful professional career, he came back to the realm of academics as a Professor at Reading University – a man of renown and greatly beloved.

❦

In the days before town planners and developers spoilt the centre of Cambridge, a yard led from St. Andrew's Street through a low Tudor archway into a cobbled mediaeval sinus under the lee of the golden sandstone of Christ's College. There, in Bradwells Court, scented and festooned by the early wallflowers of 1946, my wife and I discovered a poetic *pied-à-terre*. Every twist of its winding stairs, the closet cupboard with the hip-bath inside and the asymmetrical oak panelling of the living room, were elfin delights and testimoney to craftsmanship long gone. They echoed the laughter that accompanied one to the loo, housed in a shack of cubicles at the further side of the courtyard. History was built into the place and was often sealed there with a royal signet. Queen Mary, the Queen Mother, would be seen walking the cobbles and peering through her lorgnette in search of 'finds' in the antique shops under the archway. Under overhanging eves, our open windows, no more than six feet above the cobbles, dispensed a vapour of morning coffee and biscuits to entice friends in for a happy ending to a morning's shopping.

The balm of that cosy niche was too good to last. Life was still nomadic. We had flitted quickly through hotel luxury and backroom lodgings to the delectable courtyard but always in search of an abiding haven. Friends of friends from Eastbourne days were moving from the top flat of a tall, chimney-like dwelling house overlooking Parker's Piece. Different in every way from Bradwells Court, this watch-tower had a subtle attraction about it and, besides which, it was unfurnished and promised the prospect of a first real home of our own. We took it, No 7 Regent Terrace, and set about the furnishings. Hugh Dalton, when Minister of Trade, had continued to impose on the nation rations, coupons and utility furniture. These impediments had still to be reckoned with. Our coupons ran out after buying the front window curtains. All the others were makeshifts, including bedsheets dyed a hideous olive green. Up to my arms in a tub of green liquid and besplattered from head to foot, decidedly no sight for visitors, I was caught when the door-bell rang from the street below. Freddy Grahame, a friend, had called, the immaculate Freddy! 'Come up', my wife yelled over the banisters. 'Donald's up here dying'. When Fred saw me, he reckoned that if *rigor mortis* had not yet set in, it soon would do so. With jokes at every turn to liven our step, we darted in and out of auction rooms, bric-à-brac shops, secondhand dealers and the more serious antique vendors. My first buy was unintended; an auctioneer knocked down a dustbin to 'the gentleman over there' pointing at me. Days after, we were seen struggling over Parkers Piece, bent double with an armchair apiece over our shoulders. Thus the flat high up among the Cambridge crows slowly became a home.

The depth of our purse, no deeper than the cash from the sale of the valiant Morris 8 after it had brought us to Cambridge, and the narrowness of the three flights of stairs from the street to the living room, contrived to limit in a practical way the range of hard furnishings possible. Knowing nothing of antiques, our criterion was appearance. This empirical satisfaction landed us an early 17th Century mahogany and oak supper table, Hepplewhite chairs and a lowboy. Shrewder judgment could not have done better. Over the forty years of Keynesian inflation that followed, these foundation pieces multiplied in value by a corresponding factor. As to the rest, it was Dalton utility beds, tables and chairs, home carpentry, embroidery and the dyed curtains. When complete and Peggy McAlindon, in desperation for a roof, came to stay with us, the available floor space was reduced to very slender margins. The predominant feature of this high flat was the staircase with closets and bedrooms leading off it like the branches of a conifer trunk. In those days, a university lecturer (even if he were a land agent) who had no room of his own, either in college or in a University Department, where he could supervise, had no choice but to use his home. Our staircase rang with the activity of this imperative 'cottage industry'. One undergraduate, Michael Barnes, would beat out a war march in rising crescendo as he ascended it. Michael had an advantage. As a serving officer in the King's Own Scottish Borderers, Hitler's war had deprived him of a leg. Nothing daunting, he became the pipe-major of the University Pipe Band and with his metal leg would drum out the pibroch on the steps and risers of the staircase. Michael's habit was to sign his name 'McDoogan' after a character he had played in a departmental mock arbitration. This 'McDoogan' pressed his disability to make it serve his every need. Petrol was still rationed; coupons were needed and could only be had from girls dispensing them in a Government office at the top of another towering stair. Thumping out *The Campbells are Coming* McDoogan would stump up the stairs to the ladies in charge. On reaching the top landing, he would twist the squeaking screws in his artificial limb, unstrap and decouple it and fling the artefact over his shoulder. Hopping across the floor to where the maidens were in hysterics, half way between tears and laughter, he would bang it down on the counter and plead, as only a lowland Scot can, for compassion and coupons. He got both.

Within a year our domestic borders were extended to take in the flat below. Twelve months later we had bought the freehold of the whole facinatingly ugly building. By the next year, things were different again; the family had changed from a family of two to a family of three. Lugging a pram up and down those dreadful stairs was intolerable. Change was essential. Mrs. Prior, my mother-in-law, came over from Oxford and took on the flat. Twinks and I bought an ancient converted inn – The Bells – by the river Ouse in Hemingford Grey. Purchase was possible because the family company in Eastbourne had been wound up.

When Queen Anne reigned, The Bells was a newly-built coaching house in a commanding position where the village high street ran out at a wharf-side by the Ouse.

Time had contrived to provide an architectural harmony of polished flagstone floors and fitted carpets, spacious inglenooks and central heating, and a garage block from converted stables.

The happy mix was secluded behind an elongated facade of brick and stucco, built hard up against the highway. Generations of care had wrought an old-world garden, its rosebeds, lawns, pathways and beds traced out by low, six-inch *buxus* borders, a mature arbour at peace with itself within high brick walls.

The place, so it soon proved to be, was an escape from Cambridge, where we should have been. Cambridge was too far away. The 'cottage industry' languished, restricted to the few pupils who had motor-cars and wanted an excuse for using them – petrol coupons were still *de rigueur*. The facilities of Regent Terrace were sorely missed. Make-do arrangements kept me away from home, late of an evening and absorbed the best part of weekends. Besides the logistic problems, the river turned hostile. Its damp miasma brought pleurisy and pneumonia and, in the late winter and early spring something even more sinister. Behind an insecure door which opened on to the flagstones of the dining room was a cellar. When the river rose, the cellar became a treacherous pool of dank, slimy water ten feet deep which lipped the dining room flagstones at its doorway – no place for a lively eighteen-month old boy or his mother.

Christmas, nevertheless, at The Bells was comfy and intact. Among the guests was a handsome ebony Ghanian – Alex Kwapong, of Kings College, a classical scholar destined to dominate higher education in the developing world as first Vice Chancellor of Ghana University and later on as Vice Rector of the UN University of Tokyo. Alex was a close friend of the Obo of Benin, a royal chief who, aided and abetted by Kwapong, was in hiding in various 'holes' off King's Parade in Cambridge, keeping out of the way of my friend Canon Cecil Bewes of the Church Missionary Society. As we understood the story: the Obo had infuriated his subjects in Benin by refusing to honour his kingly obligations and marry some thirty or so wives and, thus, give a traditional lead to the marital order of his people, as his royal forebears had done. On the contrary, the Obo and his fiancée were awaiting marriage according to the Christian marital rite. The chase was on because the royal chief had, unexpectedly, changed his mind and was now prepared to accept the kingly burden of polygamy. Cecil was after him to push for a second recantation. The outcome we never knew.

Resolve to sell The Bells was not fully confirmed until another winter had deepened our fears. A few discrete advertisements in the national newspapers tested the house market. Among others who showed interest was P. Dudley Ward, the Fabian socialite, another King's man. At the time, he lived at the Old Vicarage, Grantchester, so his enquiry much intrigued me. The asking price was deliberately high because, at heart, we were reluctant vendors and the prospect of trying to find a place in Cambridge a chilling one. It was early spring in 1951 before a suitable offer was made. Acceptance was made in the faith that a new home unknown to us was waiting somewhere. When the morning came for exchanging contracts, I set out for the lawyers in Cambridge like Abraham leaving Ur of the Chaldees, not knowing where lay the promised land.

Back at The Bells, Twinks, having begged me not to clinch the deal and so leave us homeless, had gone upstairs to make the beds. There was a knock at the front door. A neighbour from across the road brought an invitation to coffee later that morning. The coffee hostess was one of two sisters who lived in a mansion by the river bank with their elderly brother, Dr Wilson, lately Master of Clare College. Our homelessness not only kept the chatter going at the coffee party, it also, galvanised an exchange between the two elderly sisters. 'Why, yes, now we come to think of it, it's just what SC is looking for.' 'I suppose you are right.' 'It's worth trying, you know'. 'Such a nice family, such a darling little boy, couldn't be better'.

Justifying this dialogue was a very close friendship between Dr.

Number 12, Chaucer Road, Cambridge, circa 1961

Wilson, his sisters and Sir Sidney Roberts who had recently been made Master of Pembroke College. 'S.C', as they called him, had a large house in Chaucer Road which he had cut in two to solve the post-war housing problem of his daughter-in-law, a war-widow. She was to have the upper floor and the Roberts the garden flat to which he had added living quarters for servants. After all the turmoil and upset, SC and his wife Marjorie never lived in the new flat. Before it was ready, they moved into the Master's Lodge at Pembroke, leaving the commodious ground floor flat to be temporarily occupied by Kingsford, the Director of the Syndics of the University Press.

Thus it was that the new Master of Pembroke was on the look out for someone to replace the Kingsfords. Back at Hemingford Grey, the Wilson sisters were convinced they had found 'just the family SC was looking for'. A telephone call was put through to catch me at the lawyer's in the act of depriving my family of house and home. Dr Wilson's name was the only credential wanted. Before the morning was out and at his suggestion, I had reconnoitred the precincts of No. 12 Chaucer Road and was aghast at the dimensions of the place, especially the spacious rolling garden. SC was cautious about the rent. He reckoned £200 high for a lecturer with a rising family. The hidden vacillation in my belly was not revealed as I assured his trusting, genial soul that all would be well and the rent covenant honoured. If there was ever any doubt that the Divine Hand rules in the affairs of men, here was proof enough to dispel it. The Wilson sisters reckoned their cup of coffee was central to the divine plan.

The freehold tenure of The Bells, the character of the place, its riverside, its room for expansion had all the promises of a 'soil' where the roots of a family country home would grow and run deep. In less than two years, however, we had come and gone, to enter 12 Chaucer Road as little more than caretaker tenants. The owner fully intended to return there after his days in the Master's Lodge at Pembroke. Our tenancy was precarious and gave no warranty of an abiding dwelling place. We have been there over forty years!

After cutting an access way through an inner wall and bricking up a doorway, the ground floor flat became virtually a fully self-contained five-bedroom dwelling with two bathrooms. The exceptionally spacious rooms, parquet floors and corniced ceilings had built into them the solid confidence and aplomb of Edwardian England. The garden, then an expert gardener's domain, matched in character. Lavish with flower-beds, a mixed orchard of apples, pears, peaches and plums, soft fruit, terraces, strawberry beds and asparagus clamps, it spread beneath the majesty of great trees, willow, maple and ash and a high-wired tennis court provided a social completeness. Here was *rus in urbe*, an uncommon acre of inner Cambridge. Somewhat awe-struck at first, we wondered how with only meagre belongings we could furnish the rooms, let alone fit into this heartland of professorial and professional Cambridge. For we had landed virtually unknowingly on the notorious Latham and Chaucer Estate surrounding Southacre Park where, shortly after 'our house' had been built in 1906, Prince Albert (later King George V) and Prince Henry were content to live while up at Cambridge. Needless to say, the 'cottage industry' revived and flourished mightily. The Vandyke brown of the panelled study in Chaucer Road had nothing a college parlour could better.

◄❧ ❧►

College supervision is a luxury for undergraduates provided at Oxbridge. Its drinks and fireside approach is incomparably more rewarding than formal lectures. Ideally, supervision should be conducted in college by college teachers. However it is done, leading scholars and lesser teachers are personally in touch with undergraduates in a warmth of informality. Friendships are forged that can last a lifetime. Bonds for me were tied which after forty and more years still hold.

Memory of those first supervisions and lectures at Cambridge is peopled with faces which stand out from the misty hosts not only in lasting friendship but because of some conversation, some word, a quip, a mannerism, an incident; or because time and circumstances have invested unknown youth with a public image.

There is Richard Orde-Powlett facing an essay on *The consolidation of the gentry in the Sixteenth Century*. By way of gentle briefing, I remind him that the life styles of the flamboyant Elizabethan gentry could camouflage plebian origins and instance the family of the Marquis of Exeter and the grandeur of Burghley House. 'I know', says the grinning pupil, 'He's my uncle. But the blood of Orde-Powlett has enriched the Exeter veins, Sir. You needn't worry unduly.'

Michael Foljambe's tale is a similar one. In his best Etonian diction, he is making comparison between Anglo-Saxon and Anglo-Norman land tenures. Repetition of the second person plural is becoming tiresome, so I interject to ask why he is using the royal 'we'. 'We Normans, Sir', and here he bows with a nice blend of politeness and disdain. 'We Normans must distinguish ourselves from the mere Englishry. My spindle legged forebears were butlers to the Plantagenet kings. To make their calves more comely their hose were surreptitiously stuffed with straw. When a Prince's rapier in jest exposed the artifice, the family was dubbed "Fool's Leg"; and so we Foljambe's have remained ever since. In name and ancestry we are truly Norman.'

Many undergraduates of the immediate post-war years were contemporaries of the younger dons; and a few – the Air Commodores, Major Generals and Naval Commanders among them – were senior in years. The iron in their souls forged by years of war was annealed with genuine laughter. They were free of the mockery which twenty years later was mistaken for gaiety. War, however, had wasted their years. Intent on catching up, they were men of purpose and knew what they wanted. Many were impatient among this altogether delightful breed.

Typical of them was a six foot six naval officer. Standing in the doorway one morning, blocking all light from the entrance passage and bending over me like a coy dinosaur, he whispers 'Is this estate management?'

'Yes,' I reply. 'How can we help?' 'My name is Pettit', he drawls with a whimsical smile. 'I have ridden the seas for four years giving chase to the Nazi vermin. Now, drenched with water and reeking of salt, I want to know something about the land.'

He took his time and taught us more about rugger than we taught him about land. Unlike Geoffrey Pettit, there were men in a hurry, content to skip terms and sit examinations in June and October of the same year. For these fliers, coaching and study went on throughout the summer. After the Long Vacation Term, all college responsibility for them stopped. Supervisions were arranged privately and fees levied not without hazard. A young Baronet played me up for months. On being relentlessly pressed, this defaulter defended himself on the grounds that his family were 'beerage' not peerage and so did not recognise *noblesse oblige* and the courtesies that it implied, including trivial things like paying debts!

Another man in a hurry was J.M.L. Prior. He sat the examination in October 1950, sent me a cheque for supervision fees and expressed the hope that he would not have to see me the next June! All was well. Jim Prior went off with a First Class degree. We did not meet again until he was Minister of Agriculture. Unlike the defaulter, he became a most worthy and outstanding member of the Peerage, as Baron Prior of Brampton. Jim Prior had shared his supervision hours with James Crowden, an oarsman pre-eminent in his generation among the Captains of the Cambridge Boat. James was another who took his time. He had to do so: one of the more profound laws of nature states that it is impossible to sit a Cambridge examination if, at the same time, rowing obligations want you in the USA with the Cambridge Boat! By a similar law, it is difficult to sit an examination if, at the same hour you are rehearsing the Queen's Coronation ceremonies in West-minster Abbey. Such was the plight of Michael, the Earl of Bective. As one of the few remaining traditional peers of Ireland, he held some office at Court. By royal command, Bective was due to attend rehearsals for the Coronation in the Abbey each morning at 8.00am throughout the very week when the University wanted him seated by 9.00am behind closed doors for the Degree Examinations. As a compromise, he was met off the London train each lunchtime, imprisoned in a kind of black-maria, taken to a windowless dungeon and subjected to six hours of written papers. With a wink and a gesture, Michael Bective told me how at the Coronation he had suffered crossed loyalties. Were it not for his English title alongside the Irish one, he would have been constrained (so he said) when the Archbishop turned to him with the proclamation 'I present unto you your undoubted Queen' to draw his sword with the riposte 'Not of Southern Ireland'. As it was, he failed that high moment of history and the Examination.

Lecturing is a performance in cousinly relationship with the histrionic arts. There is an audience to be held, a communication to make and criticism, inevitable criticism, to be encountered. A remarkable commonplace is the indifference of University authorities to the quality of lecturing. University lecturers are appointed and no questions asked, no standards or tests applied to discover whether the candidates can open their mouths in public. Erudition and scholarship are nothing to go by. Alas! they can go hand in hand with a bad delivery and diction. Once it fell to me to deal with a lecturer whose academic credentials were of the highest order but as he stood up to speak, could never make himself heard. He was admonished to make a special effort. Next lecture at the edge of the podium, he opened his mouth and made a sound like a muffled fog-horn, paused, then standing on tip-toe was actually heard to gulp, 'Oh! I see you can't then.'

As the terms ticked by, time sorted out my 'does' and 'don'ts'. Lecturing suited my temperament. There was no difficulty with delivery and enunciation. A voice trained in elocution and tried out in the pulpit served me well. When a lecture failed to raise a laugh, it was counted a failure. Foremost among the taboos was 'chewing the cud', regurgitating another's text, from a manuscript, notes, paper or book. To require an audience to take 'notes' always seemed to me to be the surest way to distract attention from what was being said. After a while my stratagem was to hand out copies of notes before a lecture, followed by an impromptu rendition. Interruptions were, in the main, amusing and were meant to be so. The warrior undergraduates were always good humoured. Flogging black-market eggs and other rationed produce in the back row was tolerated in the hard-up days; indeed I regarded the practice as a demonstration of the free market in operation. Once when lecturing on Shaftesbury's Housing of the Working Classes legislation, a booming voice from Bethnal Green called out: 'Wot Hi wanter know his oo har the workin' classes?'

Chalk battles especially when waged by the Earl of Bective had to be stamped out. Two Ghanian fellows, Fred Owosu and John Agbettoe, had to put up with a certain show of good-humoured banter. On the day of Ghana's Independence, Fred Owosu came to lectures dressed in full Ashanti robes. He was cheered to the ceiling. Ten minutes later John Agbettoe arrived in collar and tie, to be booed from the room and debarred re-entry. These two would each carry a pair of light brown kid gloves which were religiously laid on the desk. Michael Foljambe or some other leader of ribaldry was seen holding up the gloves against the ebony cheeks of Owosu with the comment, 'You are bit off colour today, Freddy.'

Invigilations were by contrast straight-faced and undisturbed, except when Elstob was among the candidates.

He was another demobbed navy man, thick-set, whose hands and arms moved with the deliberate thrust of an articulated excavator. In examinations he would sit toying with pen and paper waiting for one of the candidates either side of him to finish a folio and go up for more paper or, better still, indulge in making preliminary notes on loose sheets. The excavator arm would then go into action to grab whatever was going. Later, the arm moved into reverse would restore the cribbed material to its rightful owner. Needless to say, Elstob was disqualified.

Much was forgiven him, for this jolly tar spread a rapport of amusement far and wide and livened it with stories against himself. One in particular must not be forgotten. Before Cambridge, Elstob was, for a spell, the youngest midshipman in the Mediterranean Fleet. He tells this story. The Admiral and his Flagship are approaching Valletta Harbour, where the whole fleet is in attendance, a guard of honour, dressed over-all, beflagged and festooned with ratings and officers lining the decks and the yardarms. Slowly the Flagship is brought alongside and docked with mathematical precision. The final act is to shackle the towering bow to a ten foot iron orb that rises and falls in the harbour swell. The honour is assigned to Elstob. He is lowered over the side in a pinnace with a brace of ratings to do the strong stuff. All eyes are focussed on the operation. Elstob and the pinnace make for the anchor chain as the mighty ship rising on the swell converges on the buoy. There is no retreat. Battleship and buoy kiss each other and between them squeeze the tiny boat and its valiant crew high out of the water like a lemon pip between impatient fingers. The ebb moves the ship off the buoy. Elstob, pinnace and ratings together drop like a stone to sink out of sight in a whirl-pool of bubbles. Later, the Fleet still at attention, a sodden oozy wreck of a midshipman climbs aboard the Flagship to hear the Tannoy blaring, 'Elstob, to Captain.' The Captain, his Flagship still unshackled, glares at this awful apparition from the deeps of the harbour mud and at the greasy puddles tracking his progress over the immaculate carpet. Both men are in agony; one from suppressed shame, the other from suppressed laughter. 'Elstob,' the Captain is merciful, 'I think you have had your punishment. Dismiss.'

The only point and purpose of getting married is to have a family. Such was my wife's deeply-held conviction. For eight years all efforts failed. There were many beginnings but no full terms. Lactation on lactation led to a state of physical congestion which had alarming suggestions of malignancy. All was well, however, and with patience the accumulated milk dispersed. We never marked time deliberately, as did many couples over the war period. On the contrary, in the thick of it in Carlisle, tests and consultations were had on either side of the sexual divide. Not until Cambridge, under the benign eye of a caring family GP and clever men at Papworth Hospital, was the root of the impediment discovered. It ran back to the days of the First World War when, in 1916, a small four-years old Jessie Hope went down with a serious bout of pneumonia. She recovered but the illness left her a victim of a chronic racking cough and feverish rigours whose recurrent paroxysms would oscillate and tremble through every organ and bone of her body. Papworth eventually diagnosed Bronchiectasis. Nowadays, antibiotics would cure the primary infection before permanent damage was inflicted. In its advanced stage, afflicting both lungs, the damage had to be lived with.

Reaction was a two-pronged defiance. From the early summer of 1948, we went in for long-distance hiking. Walking was fun. Calculating nightfall however could be a problem. South to the Chilterns and westwards through the Hampdens took us from Cambridge to Oxford in three days. Later in the summer, after running a childrens' mission at Croyde, we walked the Cornish coastal path round Lizard to Penzance. As we walked, plans were made for adopting a baby girl pre-named 'Holly'. Archie Hanton, the family physician and a wiser psychologist than he knew himself to be, backed this bid to beat the system and even found a suitable 'Holly'. It was not to be. Just in time, defiance ended in victory. The golden days of late autumn turned into days of promise, followed by a winter of calculated care. Lugging over-laden shopping baskets up the dreadful stairs at Regent Terrace and similar risks were strictly prohibited.

Winter made way for spring. By April, the family's month of destiny, the pregnancy was still holding but a new anxiety bestrode. The baby, ready for turning, was obdurate, fixed and the wrong way up. While obstetricians pummelled and pushed and the saints among us prayed, 'Dick Whittington' as we now called the child in the womb refused to turn. Critical weeks passed into days, to the point when Ossie Lloyd, the brilliant specialist in charge, appealed unto Caesar. Not only was the babe hewn from its mother's side, but the operation was placarded among the staff and students of Addenbrookes, the famous Cambridge teaching hospital, as a spectacle of prime consequence – it was to be the very first Caesarian Section performed in the hospital under a local anaesthetic. No West End box-office could have wished for a greater draw. With standing room only in the operating theatre of Old Addenbrookes, Jonathan Donald Denman, in a blaze of publicity, was yanked screaming into the world. No wonder he has assiduously avoided the limelight thereafter. The manner of his birth set a precedent for two more dramatic deliveries to come.

Nature never forgave us for defying the processes. The milk supply which previously had exceeded demand was now cut short and an early decline set in. The precious babe lost bloom, bounce and baby-fat, till it resembled a famine waif from the Ethiopian deserts. Nurse, eking out the essential Farex, had gone by the book and not by the baby. 'Give him all he wants', our doctor was emphatic and exploded in anger at the pathetic sight. Within a fortnight, instead of surveying a lifeless, shrivelled scrap of starving humanity, the pram was occupied by a cherubic image of Winston Churchill, imperiously demanding Farex by the bagful.

For the next month or two while at The Bells, Jonathan Denman waxed fat. His parents waned for want of sleep and worse. At his most vociferous, at 3.00am every morning, neither lullabies nor lavish helpings of Farex would dampen him down. One morning, the father, due at a lecture, was found crumpled up on the floor under his office desk. Peggy had left to get married and my new secretary mistaking the pallor or tiredness for *rigor mortis* had raised the alarm, supposing I had shot myself. Twinks some weeks before Christmas went down with pleurisy and pneumonia and in a toss-up between Papworth Hospital and the Evelyn Nursing Home in Cambridge landed up in the latter for Christmas and the New Year. On recovery, our burdens were lightened by the kind heart and hands of Margaret Sandford, a friend and qualified nurse who came to stay at The Bells for a while. She was destined, although she knew it not, to become a lady of the highest distinction in church and state as Mrs Maurice Wood, wife of the Bishop of Norwich. We greatly loved and missed her when she went.

Distress at Christmas was followed by unexpected calamity at Easter. The telephone rang at The Bells. My wife heard the blurred voice of her mother speaking from Regent Terrace. Although garbled, the gist of the message was clear – she was desperately ill. Maggie, the *au pair* girl, was dispatched to Cambridge to meet me who, in the meantime, had rallied the police to force the front door of No 7 Regent Terrace as we had no keys. My mother-in-law had suffered a severe stroke.

By June there was a tolerable recovery. We had moved into Chaucer Road by then and had achieved the unbelievable – number two baby was on the way. It was July 1951 and I, having beeen admitted a Diocesan Lay Reader in the April, was conducting a church service. The church-warden passed me a note – reminiscent of 3 September 1939 – I was to go home at once. Great preparations were afoot for an undergradaute tea party and my mother-in-law had been helping at cake-making. She had run to answer a potty-summons from Jonathan and had collapsed in the doorway of his nursery. Death was instantaneous. By the time I arrived, the doctor was on the premises and neighbours had moved the body to a spare bedroom.

Although we had lost, so unexpectedly, someone very precious, we experienced an inexplicable transcendant elevation of spirit. A deep, deep peace possessed the home. Outside, the benison of a beautiful summer morning enfolded the place. Within and without, a vivid sense of the presence of our mother and an invisible company who, in the Spirit, had brought the assurance that 'in thy Presence is fullness of joy' pervaded everything. Until eventide this palpable *parousia* abode with us, too intense to allow either grief or sadness.

The funeral committed physical dust to Chichester clay – an historical event in the dimension of space-time. Margaret Prior, our experience assured us, was sentient in a happier, higher dimension of being and must have been amused at earth's puny litanies. The critical factor, at that moment, was the effect of these events on the newly expectant life on earth. In a lodging house at Bosham where the sea comes up to the edge of Chichester, the old familiar signs of miscarriage signalled action. Jessie Hope was rushed to St Richard's Hospital to be under the able eye of cousin Dr Everal Prior. Jonathan and I returned to Cambridge with Aunty Peggy; the little boy meeting for the first time life's uncertainty was heard murmuring to himself with self-induced assurance, 'Daddy's not gonin away'.

It was November before the curtain rose on the full drama. With a bout of influenza and a fever raging at 103°, I was groaning beneath the sheets when my wife appeared in the doorway to say she was pouring blood over the bathroom floor. Her plight froze the fever out of me. It literally drained away as I rushed to the telephone. Previously, Ossie Lloyd, the specialist, had cautioned me to look out for what he called symptoms of *placenta praevia*. This was it. Ossie Lloyd was in Cardiff watching a rugby match. Archie Hanton, the ambulance and the hospital took charge. At what speed Ossie Lloyd drove back from Cardiff must never be revealed to the police. But by midnight, with I useless in the outer wind-swept corridors of the hospital, all was over. Richard Martyn, like his brother before him, had been hewn from his mother's side.

How a double Caesarian wound can heal is beyond my imagination. But it did. So much so, that by the autumn of 1952 Number Three was *in utero*. The coming drama was to leave the other abnormalities far behind. Pneumonia accompanied by extreme, feverish hallucinations was a prelude. Ultimately, in the spring, Ossie Lloyd diagnosed *polyhydramios* and warned me to anticipate very serious consequences. In April the swollen waters burst. A snap third Caesarian Section delivered a new life, to flicker out after only a few days. Margaret, a tiny, wee heiress of Downs Syndrome and congenital heart malformation, was baptised in clinical intensive care by our dear friend Rev Kenneth Hooker; his attention an endorsement of faith in a life yet to be. Her mother, meanwhile, suffering from acute paralysis of the lower abdomen was close to the borders of death herself. To live was imperative. The new baby apart, Richard and Jonathan were there to be cared for. Without further to-do she demanded from the long-faced physicians, surgeons and nurses ranged round her bedside, a bottle of Lucozade! The intuitive wisdom saved her life, much to the astonishment of all but Ossie Lloyd. He sent the nurse running for the bottle and sensed where the borders of science met the mysteries of life. Ossie also pronounced with great solemnity that this conception must mark the end of our perilous attempts at proliferation of life. Nature has her limits and we had reached them. The good man more or less demanded my signature authorising his taking appropriate steps to ensure that no further like events occurred.

∾∾

Life at the University was what one made it. The politics, promises and opportunities never fully absorbed all my energies. They were not meant to. If they had done so, destiny in one way or another would have been double-crossed. As the War drew to an end, many friends for various reasons and from different quarters urged me to take up holy orders. The prospect of becoming the Bishop of Pimlico or the Rural Dean of Potters Bar was a fascinating fancy to be toyed with in lighter moments. It could never, however, dispel the deeper 'direction within' which relentlessly brought me back to the unsought summons from

Cambridge University. Why had it come? Faith could provide justification. Either presumption or faith could conclude that Divine ordination had written my passport and stamped my visa to Cambridge. Maybe the posting, from a heavenly viewpoint, was ambassadorial; to be on the spot to commend the Kingdom of God among my donnish duties. Time would test my inner convictions.

The *tout ensemble* of existence at Cambridge embraces much more than the University itself. 'Cambridge' is an adjective qualifying many appendices and appurtenances, as variegated and curious as humanity itself. Some take themselves seriously. Others are there to take what they can get. Among the graver, other-worldly and deep-rooted Societies is the Cambridge Inter-Collegiate Christian Union (CICCU). George Goyder once called them a bunch of Christian logical positivists. His sardonic badinage was getting at CICCU's four-square literalism of biblical text and meaning. As Senior Treasurer of the Union, my heart went with them nearly all the way. Their concern for heavenly citizenship, however, could all too easily let the world with its autogenous political and social muddles swan along to damnation. Such exclusiveness made me unhappy and I tried to provide ways and means to wider envangelical activity.

The outcome was the founding of the William Temple Society. Dr Fisher, then Archbishop of Canterbury, willingly became the Patron. Finding Vice Patrons, however, was difficult. The dark divisions of worldly politics were all too clearly manifest in that province of Christian witness. Sir Henry Willink, the right-wing Master of Magdalene, declared roundly his adversion to being Vice Patron if it meant sharing the office with Canon Charles Raven, the left-wing Master of Christ's. Charles Raven, equally contentious, rejected any risk of meeting Willink on a joint platform. Both Masters avoided letting the sun go down on their anger, recanted and happily with Quintin Hogg MP made a trio of most worthy Vice Patrons. The William Temple Society maintained a tolerable political balance for some years. From the outset, I confess, my hope had been to uphold individualism, to see reflected in the body politic that autonomy of personal choice which the grace of God allows each soul at the cross-ways of spiritual destiny. Movements from the left in secular politics and from liberalism in Christian thinking, however, prevailed, over time, to extol collectivism. It was the 1960s, the era of the Left.

Other calls mainly extramural claimed my attention. Through the letter-box and over the telephone with mounting frequency came requests to lecture, preach, conduct meetings and assemblies, from schools, Universities, professional bodies and business houses. Non-conformist churches and chapels were regularly on the list; only the Church of England banned my unlicensed feet from ascending its pulpit steps. The one-sidedness was a petty question of convention. The Rural Dean of Cambridge promised to help remedy this inconsistency. The process of assuring Edward 'by Divine Permission Lord Bishop of Ely' that he should 'make it known unto all men' that he had admitted me to the Office of Diocesan Reader in the Church took over two years. His licence in my pocket and a chain of office round my neck equipped me to serve him anywhere in the Diocese of Ely. Pitfalls were many, rewards abounding and here and there came a touch of pantomine.

One thundery December night blowing up to a snow-blizzard, a summons came to help at Burwell parish church, some twenty miles in the Fens. The cosy vestry of the parish church had a huge fire blazing in the grate in merry contrast to the vast nave. Burwell's high-vaulted, mediaeval sanctuary was heated only by two-bar electric stoves ranged down the aisles. By the time the service began, a handful of wor-shippers had gathered in the farthest corner of the stark expanse of empty pews. At *Nunc Dimittis* lightning struck. Every light was extinguished, the electric organ came to a whining halt and the avenue of vermilion red flares from the electric stoves faded into the blackness. After some panicky hesitation, I decided to continue the service in the dark. The bravado of faith was rewarded. A speck of light moved from the back of the church slowly round the perimeter of the outer dark. The lady verger shielding a tiny candle with her hand was making towards the vestry door. The Creed and the Litany were accomplished and we had arrived at the Third Collect. At that very moment, to the prayer 'Lighten our darkness we beseech thee, O Lord; and by thy great mercy defend us from all perils and dangers of this night', the vestry door opened to admit into the Chancel the lady verger carrying two four-foot candles each ablaze with a Pentecostal flame. The candles, with determined aplomb, were raised into the ample iron candlesticks at the foot of the Chancel steps. Everyone rose to the occasion. The shadowy worshippers surged out of the remote gloom to the Chancel and the candle-light. There, unaccompanied by the organ, which had died with the electrical power, hymns and praise were lifted to the Lord and to the lady verger, to round off one of the most memorable services ever known in the Parish of Burwell.

Until recent times with their Synod, governance of the Church of England was committed to Church Assembly, a legislative chamber responsible to the Queen in Parliament for Measures controlling the secular affairs of the Church and for framing and sustaining Canon Law over its ecclesiastical life. Only one of the Houses, the House of Laity, was an elected body. The other two, the House of Clergy and the House of Bishops had their own peculiar constitutions. Goaded by friends, dared even by some, I agreed to contest the Ely Diocesan election to the House of Laity in 1955 and confounded my own incredulity by carrying the day second in preference among those elected. Top of the list came Evelyn Garth-Moore QC, an eminent ecclesiastical lawyer and Fellow of Corpus Christi College. Elections were held every five years and confirmed me through five quinquennia.

On my first election to Church Assembly, Edward Ely the Bishop wrote to welcome me and, in an aside, admitted that 'there are times when that august body somewhat irritates me'. Certainly members would sit hour after hour carrying democracy's burden, as other 'parliaments' do, over business which an autocrat would dispatch in ten minutes. A great deal depends on the ability of the Chairman to hold a debate together, dismiss the long-winded and accomplish business. Archbishop Fisher was a virtuoso, second to none. Only once in my days did he side-step a Chairman's decorum and then to deliberate purpose. For two years the Assembly had debated Canon Law to the point where the only outstanding issue was the nature of the final appellate court. Choice lay between the Privy Council and a new tribunal – the Court of Ecclesiastical Causes Reserved. As with all such problems, a Commission had been set up to deal with it. The Archbishop, however, was impatient. He called for a 'straw' vote, to test the direction of the voting wind. Bishops and clergy voted for the new court and the laity for the Privy Council. The Archbishop wryly thanked the voters and hardly suppressing his irritation added, as became a retired Headmaster of Repton: 'I now know where the votes lie and where the merit lies; and they don't lie in the same place.'

Committees and commisssions soaked up a large part of my time. One of these met for days and nights in Lambeth Palace, lovingly called by the sojourners 'The Lambeth Arms'. As a member, they made me responsible for outlining a scheme of Diocesan land management. It was based on methods used by breweries to manage pubs. Specifically, it proposed setting up Diocesan land offices to look after glebe and parsonages and was designed to prevent the further growth of centralised bureaucracy. The proposal was not socialist enough and was stamped on by the collectivist-minded of the Church Assembly. The attack was lead by the First Church Estates Commissioner, Sir Malcolm Trustham-Eve, who would have all glebe siezed of his fiefdom.

There were no party politics only private feuds and factions aplenty, highlighted by much comic relief. Ivor Bulmer-Thomas MP sought every opportunity to tilt at the Archbishop. At the end of a tedious debate on the Qualification of Church Wardens Measure, he challenged the Chair with: 'Your Grace, do you know what you are doing? This draft Measure specifies the qualifications for aspirants to the office of church-warden by cross-reference to the Qualification of Clergy Measure. Consequently, it lays down that no one shall be eligible to stand for appointment who "has been in prison or committed adultery." By so doing, Your Grace, you first preclude both St Peter and St. Paul. And as for adultery, if you have sat on as many Parochial Church Councils as I have you will know that the risk is imperceptible. I move we remove the cross-reference.'

&9 8&

Notes for a speech have a habit of growing into texts, texts into published articles and published articles can become books. Articles, let it be known to those who would write for money, reward more quickly and far more handsomely, effort for effort, than books ever do. Even best-sellers have to woo the public. Money from writing was a most welcome tributary feeding the meagre stream of my earnings which since the advent of a family had to fund chronically inflated overdrafts. Even so, money was a low voltage impulse. A subliminal 'thrusting' which could only find its catharsis in writing, however puerile, seemed to possess me.

Sandwiched between prosy pieces for professional journals and prissy contributions to the religious press ran a mercurial vein of children's stories. Peter Pan still occupied a Wendy House somewhere in my heart. And there, unlike St Paul, I thought as a child although 'grown up'. Albeit the Lord had said 'Except you become as little children, you cannot enter the kingdom of heaven.' Perhaps I was not far off. Anyway, while running childrens' services on the sands of Croyde the urge came. Within a year a publisher's contract had been signed up with the SPCK for a book of children's stories and poems – *First Footprints*. Sadly, wartime impoverishment was still about and the eventual publication had the look of Hugh Dalton's utility furniture about it. Some one, however, sent it to the BBC. And from then on, my story-telling was to reach a radio audience. Richard Tatlock, producer of the BBC's *Five to Ten* programmes commissioned short stories off and on over the next three years. Some of these appeared in a BBC anthology based on the programme.

First Footprints and *Five to Ten* had religious themes. About this time the Editor of *The Field*, with an eye to something else, included me among the regular contributors to his well-known journal. Again the BBC took note and I was offered a contract to speak live on the *Farming Today* evening programme. My initiation to that ordeal was hair-raising. Seated behind a plate-glass wall in Broadcasting House in a cell of utter silence I was conscious of being, for the whole of six minutes, *the* voice of the BBC from London. Half way through my talk, the operator on the other side of the glass started jumping up and down, gesticulating and waving his arms at the producer sitting at my side. Shaken by what even to him was a quite novel experience and despite the perspiration dripping from my face and neck, he got up, shot out of the room and left me with the BBC on my hands. Later, amid bouquets of thanks and congratulations for not sinking the ship, the truth was revealed. With four minutes of my time still to go, the announcer in another part of Broadcasting House suddenly realised that immediately after I had finished he would have to announce the name of the next week's speaker. Only the producer sitting in with me knew it. There was only one course open: to get him out and leave me to get on with it.

5

THE VISION AND THE FALLACY

Was not my fate to mix with earthly vanity,
Learn the inane, and then impart inanity?
And when I ventured what I could of sense
Dislike and protest grew the more intense.

Faust, Part Two, Goethe

Sir William Cecil Dampier-Whetham, landowner, farmer and Cambridge don had, in the west, inherited an estate of fine farms and fair acres in the neighbourhood of Piddletrenthide under the Dorset downs and, in the east, had built himself a worthy house, Upwater Lodge, where the city fringes of Cambridge ran out in riverside meadows.

Around Candlemas 1917, in the throes of the ploughing-up campaign promoted to hold the food-line of the nation, help victual its larder and make good the ravages of the German U-boat onslaught, William Dampier-Whetham had a vision, a future 'call to arms' in a peacetime yet to be. As Senior Tutor of Trinity College and propelled by his revelation, he used his privileged position to circulate a Flysheet among Heads of Houses, Bursars and such other memberrs of a truncated Regent House as were still in residence. Dated 7 February 1917, the Flysheet, although somewhat quixotic, came from a man with as much practical purpose in his pen as in his plough and who that Candlemas had broken permanent grasslands in preparation for the spring-sown wheat campaign.

As a literary piece, the Flysheet was a symphony of logic. The prelude, vital to the argument, was emphatic and read:

'After the war it will be necessary to place on a permanent footing the efforts now being made to increase to the utmost the resources of the agricultural land of England. In this movement, education must play a great part. Not only must scientific training be given to farmers and land agents, but it is hoped that English landowners, more than in the immediate past, will set an example of scientific management by farming large areas of their own land on economic lines and regard their career as a definite profession needing a professional education not in agriculture and forestry only, but in all other departments of rural economy'.

After a questionable assumption that 'in the education of land-owners, land agents and large farmers, the University of Cambridge has a great responsibility, and, by a natural development of work already well begun, a unique opportunity for usefulness and success,' the William Dampier-Whetham text suggested that 'a great School of Rural Economy should be built up at Cambridge. It would include the existing Schools of Agriculture and Forestry, a central Land Office, to which the University and Colleges should be invited to entrust the technical part of the management of their estates, and one side of the existing school of Architecture. . . . In the Land Office, either concurrently or successively with their course in the School of Agriculture, future landowners and land agents would gain a practical acquaintance with all details of estate management'.

Dampier-Whetham's challenge was straight-flung. It bore all the hallmarks of original thinking and yet, timing and circumstances suggest there could have been other men's thoughts reflected in it. Coincident with Sir William's war-time appointment to the Food Production Department of the Ministry of Agriculture was Sir William Henry Wells's position as Chief Livestock Commissioner. The two men might well have bumped into each other and exchanged views in the corridors of the Ministry. William Wells, when out of Government office, was head of Chestertons, the leading firm of surveyors and brother of Sir Sydney Russell Wells, Vice-Chancellor of London University. It is remarkable that within a year of the Dampier-Whetham Fly-sheet fluttering round Cambridge, Sydney Wells had introduced and established an external Degree in Estate Management in London University. Even if the two Williams had never met and cross-fertilised each other's minds, it is yet possible that pollination was carried in the speech of mutual, like-minded friends, particularly the two well-known land agents, Charles Bidwell and George Fordham who, alongside Dampier-Whetham, were at that time on the Board of Agricultural Studies of Cambridge University as external members.

The Dampier-Whetham gospel never found a St Paul, an apostle whose mind, wisdom and selflessness were adequate to its propagation. The Schools of Agriculture and Forestry at Cambridge were not content to lose their souls to the higher grace of some grand palace of Rural Economy. So long as such new ideas could foster the fortunes of the current vested interests, guarded as they were under the twin-command of Professors Wood and Biffen, they were welcome. Cambridge, perhaps too earthy at the time, never aspired to a 'Great

School of Rural Economy'. The idea of its prophet was, however, thoroughly raked over for leads which could advantage the existing Schools. Agriculture and Forestry remained the name of the game. Only the rules were altered to allow Estate Management and Horticulture in among the players. These subjects were included 'to widen the educational basis of the course', to use the comfortable words of the Board of Agricultural Studies when in 1919, some two years after the famous Fly-sheet, the desired changes were proposed to the University.

No sooner had the new options been accepted than the formidable Wood-cum-Biffen duo pressed opportunism further, put a spotlight on Estate Management and lit a fuse of controversy which sparked and spluttered into the future to be encountered by myself some thirty years later. The Professors wanted a Readership for whomsoever should be appointed to nurture the new fledgling. A Readership in Estate Management and right away; what audacity! Opposition would have been more strident had not the Government of the day offered to foot the bill. As it was, antagonists in a formal Discussion on the proposal denounced the professorial impudence as a blatant waste of public money, a misuse of the high academic distinction of Reader and a near insult to 'the most competent men in the world, the College Bursars who certainly did not require the assistance of a Reader in Estate Management'. Protest, however, was to no avail. The Dampier-Whetham advocacy worked its magic: Col Frank Braybrooke Smith CMG of Downing College was appointed Reader in Estate Management from 1 October 1919 for a five years tenure at an annual stipend of £500. The critics may well have been right for there is no evidence of the new Reader giving in return anything worthy of much academic acclaim.

Reading history backwards from the Second World War, there is weighty evidence to suggest that Agriculture at Cambridge became hampered with parasites because the authorities failed to follow the letter of the Dampier-Whetham doctrine and to build the grand School of Rural Economy. Unlike the scriptual grain of wheat, the School of Agriculture hadn't the nerve to let go, fall into the soil of self-abandonment and rise to rebirth and a life of grandeur.

Instead, the main stem of the plant was encumbered by alien growths grafted into it to produce an ugly mongrel. No School of Rural Economy given to the academic understanding of the management, use and development of rural resources replaced the old School of Agriculture. The latter retained its form and the old denomination over its portals through which were admitted a medley of horticulturalists, foresters and estate managers to join hands with agriculturalists proper in a *sans souci* Ring-a-Ring-of-Roses. The mix was not a happy one. From the start in 1919 to the nemesis some seventy years later, changes, twists and turns, innovations and rescissions marred an excellence to which the School aspired and deserved to attain.

The many heads of Agriculture were a hydra of trouble, only to be countered by chopping them off one by one and reverting to a purer discipline, a tortuous and prolonged process which was to have a profound bearing on the Cambridge scene. Suppressing university posts that have once been occupied can inflict damaging lesions on the university corpus unless appeasing and delicate surgery is employed. The Readerships in Estate Management and Forestry and the University Lectureship in Horticulture were firmly prehensile, especially after 1923 when each was anchored in the newly-established Special Board of Agriculture and Forestry set up to give executive backbone to the jelly-like, amorphous Board of Agricultural Studies. The Special Board was the forerunner of the later Faculty Board of Agriculture. When established at the end of the decade, it cleverly avoided having in its constitution any formal representative of the Estate Management side. Jealous of its teaching territory, each head of the agricultural hydra wanted more room among the curricula than, in sum, could be found. As a result, the study loads for the degrees in the agricultural subjects were tight-packed and far more demanding of time and effort than was required for other Ordinary Degrees. Consequently it came to pass in 1924 that the agricultural establishment by sleight of hand tried to introduce a new-fangled degree which would have dropped the 'Ordinary' from the degree title. The President of Queens' exploded and crying out in his anguish warned the Regent House that 'they will be wanting a Degree in Engineering next'.

Not to be outdone by the slap-down, the agriculturalists in pursuit of an enhanced academic currency erected a fence which virtually divided the dirty-boot farmers from the scientists. On one side were to be diploma-holders in agricultural science who were not permitted to enter for the Diploma Examination unless they had graduated by sitting the Natural Science Tripos. On the other side, were the graduates in Agriculture, the sons-of-the-soil, who had read the Ordinary Degree. These changes were part of a surgical process. Horticulture did not survive even the knife-sharpening stage and was summarily decapitated in 1927. Forestry lost its claims on the Ordinary Degree and lingered for some time as the core of the Diploma in Forestry. So by the end of the decade immediately preceding the outbreak of war, only Estate Management remained of the adventitious growths which had irritated Agriculture for the preceding twenty years; and a particularly virulent irritant it was.

Although fully-orbed in its general purpose, the Dampier-Whetham vision was remarkably opaque towards the existence of the land profession, its institutions, practitioners and qualifying examinations and the service it offered to Government and public alike. In this, the Cambridge prophet's outlook differed much from that of Sir William Wells. Contemporary academic developments at London University were focussed directly and of purpose on the profession, busy with the managment, marketing and development of land. Even so, one of the leading professional bodies, the Surveyors' Institution, ever myopic towards academic horizons, was more obstructive than helpful. With the Surveyors lukewarm towards the London University initiative what chance was there of a ready Hurrah for Cambridge! Besides which, a Land Office on the Dampier-Whetham scale could in professional eyes spell competition, especially among practising land agents.

However, a modicum of bursarial support encouraged the Board of Agricultural Studies of Cambridge to provide and equip an embryonic Estate Management Branch. No records remain as minutes, files or folios of its activities and transactions. Occasional laconic references to rising demands and growing services appear among the yearly reported activities of the School of Agriculture. Between the lines, such progress as was made appears as a one-legged march. In the mid-1920s, the Reader in Estate Management is calling for an assistant. There is recruited to his side neither land agent nor valuer, but a young building surveyor, one Noel Dean out of Nottingham, to be 'available for the supervision of repairs to buildings, roads etc. on College and University estates'. Dean's special attributes, skills, training and inelasticity tilted further development one way. First and foremost, the professional work became a building service while rural land management languished for want of matching vigour and skill; besides which professional competition was not far from the doorstep. When the Readership lapsed, E.P. Weller an agricultural economist and surveyor was appointed as a University Lecturer to replace it. With him came the hope of better balanced development. Within a year, however, Weller had become the Bursar of a College. The growing building service became progressively more and more ill-fitting within the agricultural household to which it belonged. In 1938, a Syndicate, set up for the purpose, justified its recommendations for a separate *ad hoc* command and special committee over the affairs of the Estate Management Branch by recording publicly that the professional work is 'at present mainly on the urban side'; and 'during the long period in which the future of the Branch had been under discussion the development of the Agricultural Section had been hampered'. This Syndicate felt inadequate to deal with the problems of total separation and future teaching and called for a better qualified successor than itself to do so.

In the last analysis, the Branch and its activities were more of a hindrance than a help to the academic credentials of Estate Management as a subject. Teaching in Estate Management became in substance more and more detached from the guidance of the Drapers Professor of Agriculture and more aligned to narrow professional criteria. The Agricultural parent was only too happy to untie the apron strings. When, by linking its postgraduate Diplomas to the Natural Science Tripos, Agriculture bettered itself, Estate Management shied away, noticeably in 1930 when E.P. Weller advised those responsible against having a postgraduate Diploma in Estate Management contingent upon a Tripos examination. Three years later, the professional validity of the degree course was written in indelible ink by the University awarding a Certificate of Proficiency along with the Degree. And, as if to put the question beyond all doubt and by way of a belated acknowledgement of Rural Economy, a Paper in Rural Economy including Estate Management was included among the Special Studies for the Ordinary BA Degree. Graduates who selected the Rural Economy option stood over against the professionally qualified holders of the Ordinary Degree in Estate Management. Estate Management was, thus, gradually drawn away from Agriculture to seek justification on its own merits yet displaying a rural-urban dichotomy. It was, nevertheless, labelled with a reputation of being insensitive to academic aspirations beyond the qualifying Ordinary Degree. Such was the state I found it in, bereft of parents and encompassed by critics.

The Gilbey Lectures on the history of Agriculture were the one staple which for twenty-five years made a recognisable collage of the bits and pieces and of the choppings and changes in the courses and examination schedules of the Department of Agriculture before 1939. Dr Venn who eventually became President of Queens' College was appointed Gilbey Lecturer in 1924. His informed scholarship and liberal pen provided a kind of metaphysical perspective of the Dampier-Whetham vision. History justified itself when, after his death in 1957, Edith Whetham, herself the daughter of Sir William Dampier-Whetham, became the Gilbey Lecturer.

Testimonies to the memory of Dr Venn were worldwide. Jeffrey Switzer my friend and colleague and I were crashing through the Iron Curtain in that year to keep a rendezvous at the University of Wrocław, in Poland. International tension made us somewhat apprehensive. Our fears, however, were groundless. We were welcomed into the Vice Chancellor's study by Professor Stys who, taking our hands in his own, lamented the death of Dr Venn with whom he had studied while at King's College in 1926.

Stys, an academic replica of Lech Walesa, hastened to inform us that he was also a 'praarphet' and enquired whether we would care to hear his latest 'praarpheesy'. Jeffrey and I nodded a polite assent. A posse of professors were ranged round the table in due histrionic solemnity. 'My praarpheesy,' says Stys, 'is a praarpheesy of twenty-five yeese. Twelve and a half yeese have passed. In zoze twelve and a half yeese, zee Americans rattled zee sabre; and zee Russians rattled zee sabre. For zee next twelve and half yeese, the Americans will rattle zee sabre; and the Russians will rattle zee sabre. Zen zee Russians will say to zee Americans: "Loooke wot we've got! Zeeze are our terms: General Eisenhower and zee Queen of England, to Siberia – to comb zee hairs of zee white bears!" Zen it will be peace in all zee world – the Pax Sovietica.'

We looked dumb. 'Wot's zee matter?' Stys' bright blue eyes were smiling from under a dense thatch of dark Slavic hair. 'Don't you like my praarpheesy? Ah', he continued, 'I should have told you that I am no ordinary praarphet. I am a praarphet who hopes his praarpheesy never comes true!'

☙ ❧

It is 10.25am on the morning of 10 July 1946 in the Financial Board Room of the Old Schools of Cambridge University. A morbid silence rises from the soft pile of the Cambridge-blue carpet, like marsh mist over a soundless fen. The noble door swings on muffled hinges to admit the Bursar of Trinity Hall followed at silent intervals by his bursarial colleagues from Gonville and Caius, Downing and Christ's. Muted smiles are exchanged as each arrival pivots the blue leather seat of a noiseless chair from under the polished mirror-surface of the expansive centre table. Ronald Ede, the Secretary of the Faculty of Agriculture, has led in his contingent of University Teachers when finally, as if on tip-toe, the University Treasurer assumes his chair. Seated in silence, they wait for the last two minutes to pass before the 10.30am deadline. Like the stranger in the parable of the wedding feast, I am present but unrobed, a mere graduate from London University, called to serve in academic nakedness these begowned 'Guardians' of a Platonist Cambridge.

A Westminster chime from somewhere in the bowels of the old stone building breaks the silence. And precisely to the minute, the first meeting of the Board of Estate Management opens. There are eighty more yet to be before the Minute Book will be closed some fifteen years later. With very few exceptions, the Board in its lifetime would meet either in this solemn chamber or in the Council Room where the *gravitas* could be even more oppressive, if that were possible. These elegant, silent, high-ceilinged chambers generate a remoteness which can distance speech and thought from the realities outside. Had the Board met elsewhere, the history of Estate Management and my own story may well have been different. Later, in a new age when the successor to the Board would meet on home ground, speech would be freer and the clamour of the real world more truly vocal in our ears.

At that moment, the posturing and proceedings of that first meeting of the Board of Estate Management, on the warm July morning, could give an impression to the unknowing that the entire University beheld the subject in unclouded unanimity. Within ten minutes, E.P. Weller had been put in the Chair; Noel Dean proposed as Head of Department and I installed as unpaid general factotum over the administrative, financial and policy-making business of the Board. Non-payment was not an oversight; it was meticulously minuted. The decision had some logic behind it. After all, when the Creator made the universe, He had to wait until He had finished the job before He was paid in kind – a holiday on the Seventh Day. So what had I to complain about? A Secretary of a Board in a state of genesis – new curricula, new lecture lists, new examination schedules, new budgets and annual and quinquennial estimates without precedent. Someone had to produce first thoughts and promulgate them 'in house' and 'out house'. For me the handicap of having no formal University *locus standi* was truly inhibiting. Here I was a 'land agent' on a year's tenure expected to discuss long-term, novel academic programmes and expenditures with Secretaries of Faculty Boards, tutors and University officials. Such University Teachers as there were to hand belonged to the Faculty of Agriculture; one a forester, another a lawyer and the third a building surveyor who, as Head of the new Department, was fully absorbed regulating the University building service. Each, in his own way, was deeply grooved into the old order. What could they know of a new one? Or, in some cases, want to know? And yet the Board was there, ponderous and purposeful, because two Syndicates, in war and peace, had unreservedly agreed that 'the establishment in the University of a Department for teaching and practical work in urban and rural estate management was educationally desirable'. If the Board had not met in those beguiling chambers they might have heard more clearly the hostile notes which pervaded, in no small measure, general University opinion and which I, as a kind of emissary at large, was soon to encounter.

Over the years, I came to recognise the trouble we were in as being in part due to the mistaken belief that we preached what we practised. The mistake was understandable. The University had been advised by its Vice Chancellor, by Sir William Dampier-Whetham and other members of a special Syndicate that 'there should be a close relation between teaching and the professional work in Estate Management'. That the relationship had been mooted so formally, clearly and consistently justified the supposition that it was so. Moreover, the evidence ran backwards to the very beginning, to the Fly-sheet and the analogy it drew between a Lands Office and the teaching laboratories of the applied science faculties. Nevertheless, the stunted growth of Estate Management research at Cambridge over the first fifteen years of teaching from 1920 belies the presumption and exposes the fallacy that professional work and academic teaching are complementary and mutually beneficial. Professional work in a Lands Office is limited to the demand for it, is subject to clients' deadlines and other conditions on which fees are based, is ill-adapted to teaching methods, and is ingenuous, conservative and often deliberately cryptic lest rivals should be privy to a practitioner's secrets. And yet it was its professional countenance which greeted Colleges and University Departments day by day and by which the new Department of Estate Management was recognised. Hitherto, teaching in Estate Management had been an *ad hoc* derivative from generalised agricultural instruction. That placental support had in 1946 been severed. The comfy womb of agriculture had been exchanged for a crib of criticism. Naked and newly born, the subject had now to find its academic soul. The search was on. Those who looked no further saw only the workshop and scorned it as an unworthy academy. Seekers after truth, however, were not dismayed and asked only that such false images should be cleared from the judgements of the genuine-hearted.

The time would come when we would find a way of drawing together what then was an array of discrete subjects, of holding them in a coherent unit which the touchstone of a disciplined mind can recognise as the embodiment of universally relevant truth. Now, in the year 1946/47, the old form on its rural side is void of any trace of the spirit of new birth which the hour requires, a spirit so new as to harmonise with a virgin course on the urban side. Opportunity is there, together with circumstances which demand rethinking, innovation and daring. My masters on the Board wait for a blueprint of ideas, something from the hand of the humble Secretary they can knock about. We are a long way from finding the touchstone. Separate threads, however, even then are brought together and a pattern of sorts becomes discernible. So much is accomplished, that by 1949 our scheme of interrelating economic theory and structure is in place, with the philosophy of value and the study of the historical evolution of juridical concepts essential to the administration of land resources. There is constituted a body of knowledge which, it is argued, can slot into social studies which at that time are enlivening interest within the University. The Board's proposals for the first stages of the new look were debated and accepted in October 1947. Controversy, however, was not stilled. A mission of enlightenment to College tutors was necessary and a campaign to destroy the false image. Among the tutors of the time, three prominent critics stand out whose strictures contributed to ultimate success.

Among the genuine-hearted with an outlook influenced by the false image was Noel Annan, now Lord Annan, then but recently returned to King's College from war service in Berlin with the British Control Commission. Along with other tutors, he was unstinting in condemnation of what he supposed we had in mind. Over the road from King's flourished The Copper Kettle, a well-packed eating-house. On the impulse of a genuine gesture, Annan invited me to meet him there for a bite one lunchtime. As a host he was urbane and generous; as a critic he allowed a vexed tongue to baste me and to level the accusation of teaching undergraduates to aim no higher than comprehension of the anatomy of drains and to study the sinuous relief of sewers. He was of course tearing at the false image. It was, nevertheless, a helpful denunciation so far as it warned me where the false trails ran. My own thoughts were too insubstantial to trade them with one so irate. Annan picked up the bill, accepted my thanks and nodded an admonitive but not unfriendly farewell.

Censure also came from the Senior Tutor of Emmanuel College, Edward Wellbourne who boasted a rustic Lincolnshire ancestry and was at the time living in a rural Victorian mansion on the outskirts of Cambridge. True product of the Danelaw, he resembled a land-based Viking and had the propensity to bless and curse with equal vehemence. Again, I was invited to lunch; this time in the Tutor's rooms at 12.55pm precisely. Duly seated at his fireside, I counted every book and surveyed all the prints and pictures on the walls wondering the while what had happened to my host. Just off the stroke of two o'clock, he burst into the room. 'God!' he cried .'I'd forgotten all about you.' Rushing to the bell cord he summoned a tray of 'whatever you've got left' to be sent up at once. I, protesting that his words and wisdom would be as good as a feast, was told to shut up and ordered to consume a mess of cold chops, sprouts and potatoes, white in mutton fat and congealed in frozen gravy. My thoughts went back to the honeymoon night at the Hop Pole, Worcester.

Speech came at irregular intervals. Much weight, I explained, was being put on tracing and teaching the antecedents of land systems and tenures. Wellbourne grunted, then swung round on his heel with 'What are your sources? To whom do you go? What are the titles?' He catapulted out the questions. Confident of my authors but ignorant of the Senior Tutor, I offered reliance on a number of scholars but in particular F.W. Maitland, Downing Professor of the Laws of England and an acknowledged authority on the origins of land law beyond Domesday. Wellbourne leaped in the air like a sprung mouse-trap. 'Maitland!' he yelled. 'Maitland! God damn his soul! Maitland! Maitland! He's a nihilist, a nihilist!' We got no further.

If the late Downing Professor had been a nihilist my knowledge of his views was no help in facing the reality of the cold mutton. The kitchens, duly thanked, received back their own untouched. The Senior Tutor stopped flaying the air and growled something about dirty boots, empirical evidence and the beauties of the Lincolnshire countryside. A remembered appointment and the cynical thought that to forget it would be tantamount to nihilism came to my aid. With the excuse on my lips, I hastened back to Pollock and Maitland.

Over at Trinity, from whence thirty-three years earlier the famous Dampier-Whetham Fly-sheet had emanated, sat John Morrison, just back from Durham University where he had been Professor of Greek, to become Senior Tutor of his old college. Our interview was sharp, short and from my standpoint satisfactory. He had taken care to read through and cogitate upon the various academic menus from Faculties, Departments and Schools so as to be able to choose the best diets for Trinity men. The new courses from the two sides of Estate Management, rural and urban, were included in his assessments.

He was ready for me. There was no false image in the way. The subject, he acknowledged, could be worthwhile, even appealing to some, but the schedules as prescribed might do for butterflies but not for Trinity men. We were, he commented, fluttering in an academic garden, sampling a flower here, a flower there, pursuing fleeting visits of curiosity. Where was the consistency, the cohesion, the unity of concept and core which could round a pupil's thinking and discipline the mind, he wanted to know. It was a googly which stumped me and frightened the butterflies off the pitch. His cool amending wisdom helped me, was taken to heart and used to shape and sharpen the touchstone we were seeking. Not only understood, his criticism was welcomed because to a small degree it had been anticipated in the closer-knit pattern of the reformed and new courses. Trinity was ever critical of us but always in the acceptable spirit of *patria potestas*.

⋐ ⋑

Because it was conceived by a Syndicate who passed away with the event, the Department of Estate Management and its Board had no normal parentage. At birth, on 18 May 1946, the infant Department, regarded by some as an untouchable mongrel, was entrusted to the frosty care of the General Board of the Faculties. Fortunately the Department's natal day came just within the Zodiacal ambit of Taurus. The child would need every bullish quality it possessed to confront the beady eye of its fosterparent who, over the first six years, tried to starve the infant to death.

The struggle for survival as it unfolded did nothing to strengthen the morale of 'nursemaids', like myself, recruited to help nurture the child. We would have been better placed if fully briefed on what had gone before. Nobody seemed to know or to bother or to deem it necessary to do so. Ignorance on one side and cunning on the other were at the root of the tussle. When the progenitors made their Report to the University early in 1945, their thoughts on finance were positive but carefully laced with caution and conditional observations.

Future independence was advocated but was foreseen to depend upon an 'if' here and a 'maybe' there. More specifically the Report insisted upon the continuation of the financial support then given by the Government for Estate Management to the Faculty of Agriculture; upon the transfer from that Faculty of all relevant endowments; upon the free use of fees earned from teaching and advisory work; upon the *ad hoc* provision of external funds to help with any urban courses that might be introduced; and upon donations or provision by the University of finance for any unforseeable future developments. By 1951, the new Department was facing bankruptcy. Over the past six years, at the instance of the General Board its foster-parent, the University had established within the Department urban courses without first seeking the outside support forecast as needful; had introduced unexpectedly higher stipend scales retrospectively; had let go all the Government earmarked finance previously available, except a small support grant for Forestry; and had deliberately, by specific footnote, excluded the Department of Estate Management from the official claims of the University made quinquennially upon the University Grants Committee (UGC).

Yearly budgets submitted by the Department, despite sanguine forecasts of student numbers and the start of a modest flow of external donations, showed inescapable deficits as the new and successive stipend scales took hold. Pleas for help from the University were rejected on the grounds that the UGC had made no provision for the Department of Estate Management! Mitigation by tricks and devices was haltingly offered and accepted from time to time. They provided momentarily parsimonious relief on a wholly inadequate scale. By a curious transmutation Noel Dean was made Director of Estate Management so as to claw his stipend out of the University Chest; all very unorthodox, all very unsettling, especially when the authorities used what to the Estate Management Board looked like cloak and dagger tactics.

Certain University Teaching Officers within the Department (I had been appointed a University Lecturer in October 1947) had, besides the uncertainty, other irritants to vex them. Although members of the Regent House, they were not members of a Faculty and hence were disenfranchised from electing members to the General Board. So deprived, they were denied even that remote curb over its make-up, attitudes and machinations. Deprivation bred remoteness, and remoteness suspicion. The covert and inexplicable actions of the General Board reinforced the misgivings.

A typical example of the General Board's method, insensitivity and discourtesy is found in the run-up to the 1952–57 Quinquenium. At the time, it was perplexing to account for the Board's behaviour; in retrospect it is even more cryptic. No sooner had the Lent Term 1951 begun than the General Board, with no prior consultation, intimation or warning of any kind, curtly informed the Board of Estate Management that a committee had been set up 'to consider the future of the teaching of Estate Management in the University'. For six months, staff and Board were kept in the dark. Unease set in, morale plummeted and rumours ran rife. A subtle leak made things worse: the General Board let it be known that they were taking a poll of College opinion and expectations concerning Estate Managment. By the summer our patience ran out. Discourtesy had degenerated to a calculated snub and bordered on insult. Lord Caldecote, the General Board's representative on the Board of Estate Management and a champion in sympathy with our grievance, was appointed a plenipotentiary to the General Board to make known to them, in person, the following unprecedented Minute:

'The Chairman informed the Board that he had asked the Secretary General of the Faculties for the terms of reference of the General Board's committee, but they have not yet been received. The Head of the Department and the representatives of the teaching staff on the Board said that lack of knowledge as to the scope of the Committee's enquiry was having a very unsettling effect upon the staff. The Board recorded its grave concern at the effect of this uncertainty on the work and on the staff of the Department, and instructed the Chairman to inform the General Board accordingly.'

The intelligence opened hostilities. A counter-attack mounted in June was answered by a spirited defensive the following month. Behind the General Board's tetchiness was the tension which had always accompanied the five-yearly negotiation with the UGC. On this occasion, it was heightened by memories of the shabby treatment of Estate Management five years earlier and by our hunch that behind the scenes was evidence of the General Board's hope of justifying a repetition of it. Brought to bay, the General Board alleged there was evidence of disquiet among certain colleges. They exposed their own bias or maybe ignorance by referring to previous grants from the Education Fund which the Department in fact had never had. In addition, general statistics showing rising costs and declining numbers were cited as if they applied exclusively to the Department of Estate Management. Resolute opposition to these tactics paid off. The cloak was never cast off; but the dagger was never unsheathed. The child had grown too big and muscular for its foster parent again to attempt infanticide.

Incompatibility of temperament and its associated misunderstandings continued to dog the unhappy life with the General Board. The original fallacy of supposing teaching and research could run harmoniously with active professional service was, in some measure, to blame; but not so as fully to exculpate the General Board. Officials in the central offices never understood the demands his professional service made upon the time and administrative energies of the Head of the Department. Noel Dean, unlike other Heads, had a double load aggravated exponentially by the very success it achieved. In practice, Dean's priorities lent towards the professional. The executive oversight of teaching and eventually of research fell into my lap. Representation after representation was made to the Financial Board begging a modest £50 a year in recognition of my services. All to no avail: 'the Head of the Department handles all administration and is paid to do so'. And that was that! QED, Cambridge University! Attempts to find a formula for 'seniority' followed and flopped. C.W. Turner of Trinity Hall, a great-heart among the bursarial fraternity, and now Chairman of the Board of Estate Management, tried in vain to resurrect the Readership in Estate Management.

Fallout from the explosive impact with the General Board was not entirely noxious. Some of it was manna from heaven sent to feed the promise of a better time to come. Life, for me, became easier. Proposals to change the curricula and widen the research field were listened to now; a response so different from the frowns of yesterday which used to meet my warnings and such impertinances as my early suggestion to change 'Estate Management' to a designation which would make sense abroad and would not cast us with estate agency at home.

Critics among the Examiners sometimes helped to radiate redemptive enlightenment in the narrow minds of my superiors. The Smuts Reader, Peter Bauer, (now Lord Bauer) a leading Hayekian scholar and intellectual magpie given to pecking at socialists and other weaklings, once examined in Economics. His Austrian brogue tinctured with caustic wit would deliver merciless condemnations. On one occasion, it opposed the Chairman who was inclined to champion the efforts of two candidates, Higgs and Hewett, who were sitting another Paper – *Sanitation and Buildings.*

'Zay are no good', declared Peter.

'Do you wish to fail them?' The Chairman sought a clear decision.

'Zay can't write the English – vorse than me'.

'Their Paper could be better. Should we not, however, give them the benefit of the doubt.' The defence was a soliloquy murmured to the table.

'I do not understand, Chairman'. Peter squirmed in his seat. 'You say to 'Iggs and 'Ewett – "Vare do you put zee lavatory?" They say, "On zee stairs". You say, "Rrrright, a Degree from Cambridge".'

Later, when asked to sign the Pass List, Peter Bauer snatched it, waved it in the air as a signal to the world of his disfavour and groaned.

◦§ §◦

A restive spirit within me had from youth continuously whispered of 'something hidden – go and find it'. The spirit drove me out of the money-round which contemporaries, including my twin, had unquestionably joined. The whisper 'as bad as conscience' hinted of wider knowledge somewhere in the unknown. Heeding it, as I did, took me to far horizons. Once there, I saw in retrospect the treadmills of practice illuminated in the light of deep perspectives hitherto unknown. I saw the need for research as relevant to the prospect and how the trackways ran out into a compelling expanse of unexplored country. It was a lonely outlook. Sought for and expected companions were yet to be found.

How could the lack of research in Estate Management be explained throughout all the inter-war years at Cambridge when the subject had been a component of the hybrid degree courses in the School of Agriculture? Neither want of money, nor want of time could account for the sterility. Penury might be an obstacle but never a deterrent. I knew also that at the expense of sleep, work into the small hours can earn handsome dividends. Did the cause lie with the old mistake, the old fallacy of trying to combine teaching and research with professional practice? Such evidence as existed pointed that way. What time had a busy practitioner in the Estate Management Branch to pursue independent research, even if he had the ability and inclination to do so?

When in 1946, the Branch from the School of Agriculture set out on an independent life, its staff possessed ingrained habits and hardened attitudes. Not being of the company, I did only as much professional work as my 'other conscience' required. Nights at Regent Terrace were, from the start, shared between sleeping and burrowing into stacks of books – archaeology, classical and mediaeval history, law and economics to educe the story of the origins of landownership in Britain. Ten years on, my book *Origins of Ownership* appeared. As the land-ownership and tenure patterns of the past took shape and found their place in the historical relativism of the unfolding story, they threw revealing light on the character of modern institutions.

Back in the Department, Noel Dean's criticism accused me of 'sitting on my ass doing nothing'. Such resentment was perhaps excusable in Noel Dean who had no money to spend save what he and his staff earned from fees. Hostility to research, however, was inexcusable. Research was the sure if not the only way of making progress towards the academic credulity implicit in the University's policy of establishing the independent Department. Slowly, antipathy subsided. Yet, in 1950 when the Board of Estate Management took the first step to appoint specific research staff, the obligations of those essentially academic officers were split between research and giving advisory service. Professionalism still hung like a tightened halter round the neck of the academic. Ironically, as the research work of the Department found increasing favour with Government and private sector opinion, the inherent impracticality of an independent academic body giving professional service inflamed resentment within the profession itself. The RICS in 1956 received complaints from its members against the Department.

These took exception to the Department acting as a firm of land agents and surveyors for College and University properties and to individuals and members of the Department taking outside work. It would allay unease, somewhat, the complainant suggested, if the Department and individual members of it did not continue activities as private practitioners. At last the incongruity which had troubled me for so long was forced upon the attention of the die-hard old guard and from a quarter, perhaps the only quarter, to whom they would pay heed.

By the time the complaint hardened in 1956, priorities and placings in the Departmental staffing pattern had swung far in favour of research. The Board of Estate Management were even asking the General Board to create an additional University Lectureship so as to lighten my teaching load and allow me more time to devote to orchestrating the rapidly expanding research programme.

For some years the only serious research had been my own into the early structures of landowning and tenures. Concurrently with these efforts, Hugh Dalton as Chancellor of the Exchequer and Tom Williams as Minister of Agriculture and Fisheries, one of the best in the business in the post-war years, were each engaged in the destruction of each other's policies. Dalton's neurotic obsession to break up family fortunes and distribute the fragments on the winds of socialist prejudice had provoked him to impose crushing inheritance taxes. Williams, to counter the damage, was paying back the money as capital expenditure allowances on farm and forestry improvements, deductible against income tax. Between them, they had managed to pin-point the critical place of landed wealth in society and in the national economy. Where had the landowners come from? What was their *status* (estate) and function in the modern world? Here were questions which only persons who could stand back with time and patience to hold them at a distance and without prejudice could answer with equanimity.

The first object of a Departmental research programme should surely be to search out evidence to help answer in part, if not wholly, these fundamental questions.

And so it transpired. Our extensive series of pioneering studies led on to a major empirical investigation into the manner in which property rights and the institution of property transform the economists' 'land' as a factor of production alongside capital and labour, into a species of capital itself which we designated 'estate capital'. The results widely disseminated in articles and more substantial publications, including my own book *Estate Capital*, demonstrated how estate capital was affected by taxation of all kinds, was generated by investment and was complementary to other factors in the production equation. The Duke of Northumberland in his foreword to *Estate Capital* reminds the readers that 'the economics of landownership are new and therefore require new definitions of which "estate capital" is one'. The new research work was acclaimed and financially supported by Government Departments, by Conditional Aid (financed via Marshall Aid), and by landowners, farmers and the professional bodies. Academically, although at first not recognised as such, we had espoused law and economics in a particular union and by so doing had moved a long way forward in the direction of the integrated discipline of thought on which a fertile future depended.

By the time the major project had been completed, data had been gathered from four field surveys covering one and a half million acres of Great Britain. Little reliance was placed on the postman to carry questionnaires and stamped addressed envelopes in his bag. Personal interviews wherever possible were conducted with landowners or their legal advisers and agents. Without their cooperation and readiness to entrust confidential intelligence to our keeping, the work could never have been done. Family anecdotes were distinguished from family secrets but on one occasion the trust of the confidant was shaken badly. The shock was momentary but traumatic.

One evening in Edinburgh, I was the guest at dinner of a senior partner in a leading firm of Writers to the Signet. Lord Glentana had agreed to my meeting his lawyer and the generous Writer had me as his guest at the New Club. He had seen to it that the Keeper of the Archives of Scotland should be about in the club that evening. After dinner, with the port warm and red, the Archivist regaled us with stories from the late 17th century. 'Land speculators from Yorkshire would invade the Highlands and buy up land, after the Clearances,' he said. 'Ah!' commented my host, 'Just like Margrave today'.

The worthy lawyer apparently knew a great deal about the ramifications of Margrave, the successful Yorkshire property company. He was, however, ignorant of the identity of its managing director, otherwise he would have been more canny, for it was my twin brother. Our physical conformities were at the time so nearly alike that it was almost impossible for the unsuspecting to tell the difference between us. Three weeks later came the contretemps. In the office of my host of the earlier evening, I had closed my notebook on the matters of the Glentana estate and taking his good-hearted co-operation further, enquired if he could help me on similar lines with the estates of the Farquharson of Invercauld. 'He is certainly our client', he nodded, 'but my colleague, Norman, deals with his affairs. I'll call him down'.

Five minutes later a spruce military figure stood in the doorway, his moustache twitching in agitation and astonishment.

'Margrave!', he thundered. 'What the devil are you doing here?'

My host turned crimson and looked at me with incredulous alarm. Panic quickly subsided as the two lawyers compared what each knew of our respective identities and doings to assure themselves that I was not Margrave the property man.

Glentana's confidences were respected. Some gin and tonics later, so were those of the Farquharson, safe in my keeping.

Among the family tales of long ago which helped to quicken interest on the research trail was a story which, if I had not known the raconteur, I'd have dismissed as apocryphal nonsense. It smacked of fiction and high fancy. In the Civil War, Charles the Monarch had stabled his horses in the farmsteads of the royal physician whose estate was nigh the battlefield of Naseby. After the Restoration the same landed family continued to be physicians royal. Doubt had arisen concerning the whereabouts of the interred body of Charles King and Martyr. Some historians wondered whether it had been fittingly laid in the royal vault at Windsor. Consequently, a private midnight disinterment ceremony, attended by members of the Royal Family, church dignatories and the royal physician, was solemnised in the Royal Chapel. The Stuart coffin once identified was opened, the contents verified, the lid secured and with due reverence returned to its resting place. Dispatch alas had bettered watchfulness. As the company made to leave the Chapel, the severed vertebra was observed to have fallen out and to be lodged in a crevice in the flagstones. Dawn was whitening the sky, so the sad relic was hastily entrusted to the royal physician. Thus it transpired that for generations, encradled in silk within a small velvet casket, the treasured relic passed from heir to heir. Late in the reign of Queen Victoria, the last of the line of former royal physicians then well-stricken in years was concerned for the fate of this rare family possession. Following a letter of explanation which he had sent to the Palace, the old man was summoned there by the then Prince of Wales and commanded to bring with him the casket and its contents. On arrival, this conscientious action received no acknowledgement of its virtue and, indeed, went unrewarded by thanks of any kind; on the contrary the bearer encountered a display of Princely anger. The landowner who gave me this tale was a cousin some degrees removed from the gallant and loyal last descendant of the direct line of former royal physicians. 'Rumour has it' he said at the end of his narrative and with an enigmatic smile 'that later on another ceremony was held in the darkness before the dawn in the Royal Chapel to return what should never have left it. Once opened, however, the royal vault revealed the coffins to be inaccessible. So a sling of royal handkerchiefs, gently swaying from a shaky hand, lovingly lowered the royal vestige to rest on a lid which the company present supposed to be its proper place of repose'.

◦§ ◊◦

Empirical research into the finances of landownership, especially the sifting of evidence through the annals of heritages long founded, needs money and is no work for penniless scholars travelling with the gypsies. Funds were wanted for the new research programme both intrinsically and because the sanguine forecasts of the advisory Syndicates twenty years back which had proposed teaching courses in urban estate management had encouraged the University to reckon on external sources being available to finance them. Slowly, far more slowly than the prophets had forecast, a formal Appeal gained momentum. Small down-payments in cheques and cash and a dozen or so more generous, but by any standards, skimpy, covenanted donations made up the beginnings of a purse. No one was reassured. Noel Dean, however, came into his own. Widely scattered and useful contacts among the professionals, a bull-dog determination not to let them off and forceful persuasion did not go unrewarded. Noel in this attitude displayed a curious ambivalence; almost as if he approved of the fund-raising but was less than enthusiastic about its purpose. His was another example of what Emmanuel Kant had once called 'disinterested approval'.

Stanley Edgson of Berkeley Square handled the Appeal to the property world and became the first Chairman of the optimistically designated Development Fund Committee. The hesitant response from the ruck of the profession reflected the temper of the average taxpayer of the day. Universities, through the UGC and local government, milked him for money to subsidise higher education, so why should he be expected to contribute twice. What's more it was Cambridge! Wealthy folk who wanted to send sons and daughters to Oxbridge to idle away youth and years should be made to pay appropriate fees to prevent the Universities having to turn mendicants. As to research, the professional man saw little call for it. Even where goodwill prevailed, Stanley Edgson faced a tricky chicken and egg impasse. What was Stanley selling? What had he to show in exchange for the money asked for? Until even the right-minded would part with their shekels, they wanted to see the merchandise. For want of money, alas, there was little or no merchandise to show. Blind faith, the staple by which the just shall live, was expected. Big men, here and there, men accustomed to risky ventures and the professional institutes took the plunge out of a fellow feeling for the universities. The aristocracy of leading firms and the RICS itself occupied the vanguard. First discrete trickles flowed to feed bolder streams and these in time became riverlets. Three years were to pass before the flow was ample enough to be tapped.

Like Jehovah's Witnesses, we found button-holing in pairs less embarrassing and more rewarding than begging one by one. When Stanley Edgson took over, Noel Dean would make the contact, knock at the door so to speak and bring me alongside as the patter-man with the sales talk. One day, Noel's friend Graham Saunders of John D. Wood & Co. walked the two of us over to the Grosvenor Estates Office in Davies Street. There for the first time, I met George Ridley, the field marshall in command of the Grosvenor Estate battalions at home and abroad. George's penetrating intelligence hit one before his introductory handshake had made contact. He was master of consummate professional skills, the wizard of financial finesse and there shone from his person an *Aurora Borealis* of goodwill which no difference of opinion could ever dim. His anger was always just and shortlived. We spoke of financial strategy, of the benefits of comparative research studies and discussed the outline specifications of research programmes. Our minds had met. Noel Dean and Saunders left us to it. Suddenly George Ridley picked up the phone and called in his chief accountant. Immaculate and smooth, the man effected a matey introduction. In the event it was unnecessary. I had recognised him at once.

'Hello! Martin', I said. 'It's half a lifetime since we met. How's your mother?'

A sardonic smile flitted across his face to register a moment of surprise and perceptible unease. Ridley plugged into our electricity, cut short the interview and summarily dismissed his man of finance. Turning to me, he asked with some show of suspense where we had met. 'His family were tenants of ours on the Finchley estate', I explained.

Nothing more was said. That evening at home there was a call from Graham Saunders. George Ridley had asked him to find out if I could 'help with enquiries' concerning Martin, the chief accountant. I did so. Some weeks later it all came out. Martin was no longer on the pay-roll of the Grosvenor Estate – he had fled the country rather than be arrested. My statements, given over the telephone in all innocence to Saunders, in no way matched what George Ridley and his colleagues had been led to accept. Those disclosures were acted on, the blood-hounds turned loose, first tracking the top-hatted young man down to the Ascot races. On the negative side, they discovered he was not a Brigadier in His Majesty's forces; nor a MA of Cambridge University; nor a Harrovian; nor a chartered accountant; nor a scion of a landed family as he had led them all to believe. On the positive side, he had practised on a grand scale, deceit, forgery, misappropriation of funds, falsification of the books, embezzlement and much else. The meeting in Davies Street that morning had turned to everyone's advantage, except that of the chief accountant; the office had laid hands on a trickster, the Development Fund had gained in magnitude and I had found a life-long friend in George Ridley. Fate could have made George Ridley Chancellor of the Exchequer and he would have been second to none. As it was, he remained at his post to build the fortunes of the not undeserving Grosvenors.

Fortune turned a fairer face upon us after the momentous summer of 1951. The General Board was less provocative and had written a supportive note on behalf of the Department for the UGC to honour. There was, of course, nothing in it for research; but by meeting part of the teaching costs it left more in reserve for research. The Development Fund, renamed the Endowment Fund, was under a new Chairman, B.G.K. Allsop and was burgeoning with springtime growth fertilised by a new, active well-wisher, the Ministry of Agriculture and Fisheries. The inflow of new money was sufficient to pay the newly recruited research pioneers. The Ministry had saluted the Department in a formal communication in which an offer had been made to finance research on the ground that 'the Department is one of the very few appropriate bodies available for doing research into land ownership economics'. Good to their word, over the ensuing five years, the Ministry met 76% of our research budget.

The trick of the trade in those days was to keep away from long-term research staff appointments. Such luxuries would have to wait for firmer finances. The pioneers were teamed up from the eager-hearted and good-looking, especially the women, and taken on within the confines of specific project budgets. Personable qualities counted more than technical expertise when selecting the most suitable candidates. Nothing was stereotyped. The confidence of landowners had to be won at the dinner table and on the hunting field. Neil Elliott, a pace-maker among forerunners, would, with notebook and camera, pack a John Peel hunting coat and black tie as he set out for the shires. Without Hervina Roberts' Grecian knot, outstanding natural beauty and Celtic lineage, the farm rent survey of Wales would have withered on a fruitless stalk. Robin Stallard, another pioneer, before taking over his own farm in Devonshire, developed the knack of trading valuable farming tips for confidential costings.

The University bureaucracy was ill-adapted to such free enterprise research. A rigid, parsimonious and short-sighted outlook on travelling and subsistence expenses stood in the way of fair treatment. Salaries of the researchers were low by any standards. When the University expected the investigators to meet out of their own meagre pockets a high proportion of the costs of essential travelling, it was time to take serious issue with myopic officialdom. Inappropriate claim forms, ambiguous and only with difficulty capable of completion, never accounted for more than 40% of reasonable costs. One of these brain-teasers had to be filled in and submitted for each trip. Waste of time, annoyance and ultimate indignation at the folly and unfairness hampered work and crippled progress. The absurdity had to be opposed on behalf of the staff as well as myself.

Fortunately the Assistant Treasurer of the University at the time was a MacDonald, a squat fiery man, red of face, with a keg of bottled-up fury in his breast and much hard-headed sense above it. He understood, when confronted with the evidence, that our research laboratory stretched from Lands End to John O'Groats. Travel was imperative using hotels and lodgings for nights on end. Prices at all hotels, the cheap, the modest and the five-star places varied with the seasons; how, then, on that count alone, could a fixed subsistence rate throughout the year be right and fair? Maximum allowances far below the price of the most modest hotel, even in winter, piled indignity upon injustice in the summertime. Travelling costs also were far below what the car-users had to bear. MacDonald understood these points. On my assured word to limit all claims to the actual costs incurred in running inexpensive cars and using only modest accommodation and to keep all costs within the budgets acceptable to the external research funders, he agreed to tearing up the pitiable, absurd formal claim-forms and to our living as free men and women in a world of reality and fairplay. MacDonald probably remembered Glencoe, where the MacDonalds of yore suffered their own grave injustice.

In the late summer of 1955, Neil Elliott and I set off for the Highlands, to visit the Earl of Moray, Lord of the ancient castles of Darnaway and Doon. Castle Doon is remembered in Scotland's ballads for its tears and its tragedy. A valiant former Earl had espoused the Queen's cause by fire and sword against rebellious clansmen only to be worsted by these *sans-culotte* among the heather on the braeside. The ballad tells of tears of lamentation flowing down the grey walls of

Castle Doon for 'The bonny Earl of Moray who lay deed upon the plain. He was the Queen's ain love'.

Darnaway, sister Castle to Doon, overlooking the fair waters of the Moray Firth was sunlit in splendour from drawbridge to turrets when we espied it. A clarion call responding to a pull on the ample bell-rope brought to its ancient portal an immaculate footman who apparently could only bend from the hips at an angle of three degrees. Before stepping aside to permit entry, he enquired whether it was 'Dr Denman for his Lordship?' With an almost imperceptable inclination of the head he purred the question and to my nodded assent replied, 'You are wanted, Sir, on the telephone this instant. Please follow me.'

The 'phone call was Noel Dean in a state of exceptional excitement. 'Can you return at once? Ben Allsop has a financier in tow who has promised to find all the money we need. Ben wants to know what the sum should be. We must meet immediately to give him the answer: the day after tomorrow. You must be there'.

The incumbent Earl, descendant of the luckless Queen's ain love, accepted my apologies – as became the grace and civility of the Stuarts. Neil Elliott was left alone to get into his dinner jacket and conduct the evening's 'research'. His acquaintence with Scottish ballads stood him in better stead than any insights he might have had into taxation policies. On the way to dinner, Neil was introduced by the Earl to the Fiery Cross which his ill-fated ancestor had sent through the glens to rally loyal kinsmen to the Queen's cause. As the port circulated after dinner, the Lord Moray suddenly downed his wine and turned to Neil with the amazing query, 'Elliott, do you like peanuts? Go and get 'em, there's a good fellow. Over there on the sideboard'. (The butler had withdrawn). Being well-groomed not to answer flummoxing questions, Neil completed the errand in silence. After sampling the peanuts, the Earl, smiling half in languor and half in pity, asks for a verdict.

'Fine'. What else could be said while wondering what in heaven's name his Lordship was getting at.

'Do you like the dish?' Another awkward question which diplomacy met with a grin. The receptacle was a small tripod standing on wrought silver legs and carrying a white bone-coloured bowl.

'Don't you know what it is?' enquires the noble host allowing his smile to widen.

'No idea'. Elliott shakes his head.

'That is the cranium of the bonny Earl of Moray who lay deep upon the plain. Take care! He was the Queen's ain love.'

6

TIDE AND PREJUDICE

They that dig foundations deep,
Fit for realms to rise upon,
Little honour do they reap
Of their generation,
Any more than mountains gain
Stature till we reach the plain.

The Pro-Consuls, Rudyard Kipling

At a press conference in November 1955, Noel Dean told the world at large of the new wealth of the Department of Estate Management at Cambridge and of the consequential decision to wind up the seven-years' Appeal. It all happened over the port at a lunch and within hours of my returning south from Darnaway Castle. Harold Samuel, Chairman of Land Securities, twirled his wineglass and with calculated insouciance asked what the sum was 'which would stop his friend here, Ben Allsop, from pestering the City of London with his begging bowl'. Matching incredulity with daring, Noel Dean came out with – '£250,000' (a few millions in present day money). Before the glasses were drained, Harold Samuel had promised the sum, a benefaction among the biggest ever to be bestowed on a small Department in the University – hence the euphoria and haste to tell the world.

The tide of fortune was full and the horizon fair but Noel Dean, sanguine fellow, had overlooked the enemy, amazed, indignant and infuriated lurking in hidden creeks. No sooner had Harold Samuel's most generous gesture been made known to the University than letters expressing horror, dismay and alarm were piling on the Vice-Chancellor's desk. The good souls wanted to say 'Thank You' and to do so at once but the narrow-minded, the ill-informed, the bigots, in the name of 'right learning' and the prudent use of resources and motivated by subconscious jealousy, abhorred the prospect of Estate Management receiving a benefaction of such unprecedented magnitude. These attitudes of irrational prejudice then, and at other times in the past, had kindled towards the Universities, in the minds of industry and commerce, suspicion, mistrust and indifference. The mistrust has left an echo not entirely imperceptible today, although now a different spirit and a universal, urgent need pervades the academic world. When Harold Samuel gave his Benefaction, three months were to pass before the University promoted a Grace accepting it. In Samuel's own words, 'It took the University longer to say "Thank you" than it took me to make the money'.

The adversaries who had insisted on acting so discourteously towards Harold Samuel seemed incapable of accepting Estate Management as a serious field of academic study. Money spent on it would be frittered away on superficialities, in semblance if not in fact, on a routine of champagne lunches, tight-skirted secretaries and fast sports cars. To imagine anything but wanton waste was beyond them. Chagrin was too deeply rooted in wrong motives to be mollified by the Vice-Chancellor's publicly expressed gratitude. The University authorities expressed delight, relief, shared with our own, at the bestowal of an endowment which from henceforth would provide for an impecunious infant adopted ten years ago in faith and charity and whose upkeep was becoming more burdensome by the year. The child was hungry and had a claim upon its foster-parent, the General Board. Although in infancy, it had given no premonitory sign of being successful and was now manifesting evidence of promise and an unexpected comeliness which justified care and understanding. There was, indeed, a long-term research programme still waiting acceptance and sufficient money to conduct it. Contrary to the ignorant misconceptions, there was nothing wanton, whimsical nor worthless about it. The programme called for a permanent research staff and the planned exploitation of a virgin field of research in a way which would eventually warrant the establishment of an honours school in the subject. All that was now possible, *Laus Deo*.

Resentment off-set by jovial dismissal of the carping disapproval prevailed in the Estate Management camp. There was resentment because the adversaries never came out of hiding but sneaked behind our backs to the Vice-Chancellor with their caveats and half-informed lamentations. They were never clean fighters.

From the earliest days at Cambridge, I was aware of the disapproval of the subject in certain corners and was not unsympathetic, even, in a measure, grateful for it. I was aware also that the critique was not peculiar to Cambridge discontents, for the Loveday Committee on Higher Agricultural Education advising the Government, while recognising Estate Management as a respectable pursuit in the halls of higher learning, rejected surveying, building construction and valuations, then current items on the Cambridge menu. In the Committee's view they were indigestibles in an otherwise acceptable diet of higher learning for undergraduates. The Committee therefore counselled the assignment of the subject to the post-graduate schools at Oxford!

Some among the nobler antagonists, when sought out, talked to and asked to listen, would nod slowly in pontifical assent as if they were giving an aspiring freshman the benefit of the doubt over some point which their superior intellects deliberately found difficult to grasp. My own research work from pre-Cambridge days had led my thinking far away from the pragmatics encountered at Cambridge. 'Estate Management' was an unhappy and ambiguous designation. Either it implied the study of the historical derivation of some of the more arcane aspects of feudal land law; or it suggested gumboot jaunts over downland, fen and the lusher meadowlands by Piddletrenthide. The latter view was not acceptable to me.

The former was a pathway which led behind the title 'Estate Management' to landownership, an entire, self-contained realm of decision-making; to a realm critical to economic structure and social order, where lawyers held sway, economists seldom trod and where the only academics in sight were discursive historians who, having described the scene, left it at that. Here was a universal domain recognisable wherever man, in the name of human settlement, drew litigious bounds between mine and thine. Here was a intellectual field of promise.

On 'landownership', stood the whole economy of civilised man – his home, his farmstead, his factory.

Step by step, it was possible to bend the inflexible minds of my colleagues at Cambridge. The first revision of the syllabus after the foundation of the Department was accepted by the University in 1947. It deflected from the old order towards some of the principles advocated by the Loveday Report. But this was only a short step forward towards a real beginning and the new realm of understanding which Loveday had failed to identify, let alone fathom. At the next syllabus upheaval, immediately after the fracas with the General Board and on the eve of the great Benefaction, changes were radical and transforming. The ambiguous title which was causing so much mischief hung on but behind its hindrance teaching was confined, consolidated and compressed under three subject heads – Land Ownership; Land Economy; and Land Resources. New knowledge, new thinking and an incipient literature were the scaffolding upholding the new edifice. Literary output took all forms – articles, booklets and hardbacks. The hardbacks were either edited by me or came solely or in part from my pen and included – *Estate Capital; Origins of Ownership; Land Ownership and Resources* and a *Bibliography of Rural Land Economy and Landownership 1900–1957*. Funds were secure. Scholars were now needed, learned men and women, graduates with vision. Harold Samuel's Benefaction had come at the propitious moment. Assistants in Research from then on augmented the staff and were closely followed by the first postgraduate Studentships.

Apart from helping shape, organise and cost the research programmes which absorbed its money, the Benefaction was no business of mine just then. Jeffrey Switzer of Sydney Sussex College who had succeeded me as Secretary of the Board of Estate Management executed the wishes of the Managers of the newly-created Estate Management Development Fund. There were four managers – Ben Allsop, the Chairman; Sam Weller, Bursar of Gonville and Caius College; Noel Dean, Head of the Department and Trevor Thomas, Bursar of St John's College. They were authorised to 'manage the Development Fund at their own discretion upon trust for the furtherance of research in and study of Estate Management.' This somewhat ambiguous charge was premonitory of misunderstandings and disputes to come. While the Fund assured seven years of plenty, the whole gift was not available as a single cornucopia. There were to be seven instalments. Prudently invested these paid for Assistants in Research and Studentships and supplemented teaching finance. Projects from outside financed *ad hoc* were welcome and, indeed, came more frequently as money from the Benefaction was made available to make good shortfalls between costs and grants.

⋘ ⋙

Leonard Elmshirst had used Dartington Hall, his country seat in Devonshire, with open-handed generosity to gather about him in 1929 a band of agricultural economists from home and abroad to form a Standing Conference. It suffered oblivion during the War but rose again at Dartington in 1947. Eight years later, the Department of Estate Management was invited to the Ninth of a series of conferences, to be held at Otaniemi, an ambitious rendezvous in Finland. By the nature of things, estate management was not in the bosom of this family of economists. Items on the agenda, however had name-tabs and affinities with our research projects close enough to suggest to our Board that they might, without too severely twisting conscience, allow us to tap the research grants to meet the cost of attendance. So off I went.

Finland, the enchanted land of a thousand lakes, came to me through the golden mist of a sea dawn. The night mail-boat from Stockholm flying the blue and white Suomalainen Lippu over her stern against a receding emerald west slipped among the first islets of welcome off the Bothnian shore. At breakfast, the boat took in the majesty of Helsinki Harbour, crested by the pure white Citadel against a Baltic Blue. Otaniemi, some fifteen miles from the city, was a new Conference Centre built to emotionally attractive Finnish architecture and lost in sea-laped woodlands where pine-needles snapped underfoot and the air was fragrant with the scent of resins.

As with other international get-togethers where simultaneous translation make tedious Papers even more fatiguing, the lasting benefits of the Otaniemi Conference were not the formal contributions but personal exchanges and new friendships. Philip Raup of Illinois traded draft entries of an annotated bibliography of land tenure for some of my better but thinner lists of authors and titles. And there for the first time in my ken was Constantin von Dietze of Freiburg. His School of *Agrarwirtschaft* came nearest to Land Economy in the German. In a moment of bravado, swords were crossed in debate with Professor Bolgov from the USSR who had told the delegates that the collective farms in Russia were held on free tenure. 'Was there then', I asked, 'a free land market in Russia?' There was no reply. Because they were not translated, some of the American texts were among the most difficult to follow; they were peppered with 'quip-words' and heavy with Germanised English.

Real education, however, came with the sauna baths, the genuine article of Finland. Out of deference for the tender susceptibilities of their foreign guests towards mixed bathing in the nude, an attitude which, incidentally, they found odd and even laughable, the Finns had organised 'Men' and 'Women' sauna evenings. Ignorant and eager to learn, on 'Men's Night', I followed a path through the pinetrees to a low timber structure off the seashore. Inside, tiers of shelving bulging with discarded garments reached to the ceiling.

The babel of a latter day Tower of Babel crackled like static electricity. Over in a corner was a grinning Jimmy McGregor from the Oxford Forestry Institute.

'What do I do, Jimmy?' I felt a lost soul in Hades. Surely with a grin like that, he should know how to help my over-excited nerves. 'First of all', he replied in a pronounced Highland accent, 'Ye tak ewer clothes off'. Once exposed to the world in all my unadorned manhood, the problem of what to do with my hands troubled me – I had lost my pockets! It was solved by a lady in a sandy overall who came up to offer vodka and other local potions. Custom required one to take these as a prelude to the torture chambers. Stronger than Adam in his innocence, I declined the temptation and fled through a nearby doorway, only to slip with a bump on the floor beyond. The floor was a wet, polished stone and the cubicle a lead-clad ice-bath. A cascade of glacier water pummelled with merciless candour my cowering naked body. A door from the ice-bath led to Dante's Inferno. The lofty timbered chamber was packed high with seried ranks of naked bodies whipping each other with birch twigs in an orgy of flagellation and sweating in a steam heat of 120°. A mad Brazillian, ducking the clutches of a furious Finn, was throwing water on to a red-hot stove and driving the heat and humidity beyond danger point. I yammered for the ice douch; but found another exit. Outside was a true Eden of woodland and sea and a long, long causeway running through the Baltic twilight to the water. Loose duckboards rang like a xylophone to my tripping feet running to plunge in the sea. I was too relaxed and relieved to notice the bevy of fair Finnish maidens who had gathered to take up illicit positions at the end of the jetty. What were they doing there? This was Men's Night'! Far out to sea, a head bobbed in the water. It was Ford Sturrock of the Cambridge contingent, a large, silent Scots bachelor whose fear of the female world was driving him further and further from the shore in vain hope that darkness would cover all. To clamber back on to the jetty standing, as it did, six feet out of the water couldn't be done in a 'flash'. Alas, poor Ford had not bargained for the flood-lighting, to be switched on with the passing of the twilight. A decision had to be made between a slow death in the night water or the exposure of his person. Fortunately, he opted for the lesser ordeal. What did it matter? If he had met any of the lassies again, they would not have recognised his face!

In those days of post-war tensions, part of sovereign Finland by Porkkalla, some way to the west of Helsinki, lay under the rapacious paw of the Russian Bear. Leonard Elmhirst with his disciples had set off for the Arctic Circle and left the sauna baths and empty halls. Other plans occupied me. First, was a train journey westwards to Torku, the Finnish seaport on the lip of the Gulf of Bothnia. The main, indeed only, railroute ran through the USSR occupation area of Porkkalla. It was no nominal annexation. Real, effective, fussy custom barriers ringed the place. Travellers without entry visas to the USSR were locked in their compartments, iron blinds fell with a clutter shutting out all vision and daylight. Dim, low wattage naked bulbs threw an eerie light from remote corners of the carriage ceiling. Speech was prohibited. The Finnish engine blowing pathetic 'farewells' in puffs of steam was uncoupled to return to friendly Helsinki. Known as 'the longest tunnel in Europe', the Pokkalla took an hour or so to traverse in the enforced darkness and silence. By the dim light, however, the words of a newly-purchased *Aino Wuolle* were just legible, sufficiently to pick out the Finnish Numbers. By the end of the 'tunnel', it was possible to count in Finnish. There the Russian steam engine, its entrails festooned about its belly and sporting a huge Red Star made way for the more elegant, eager and impatient Finnish counterpart.

At Torku station, friends met me. Apart from the Numbers just mastered, I spoke no Suomalais. The Finns made up for such ill learning and chatted in tolerable English. They were Tora's people. Tora herself, by amazing coincidence, was living with us at Chaucer Road. She was a frank, radiant, buxomly beautiful specimen of Finnish maidenhood; one of a line of nurses, maids, nannies, *au pair* girls, cooks and other aides who, during the traumas of childbirth, had joined the family to help my wife.

Tora's folk were wonderful hosts, delightfully un-English, neither high nor low on the social rung. They set before me an enormous spread of Finnish dishes, hot meats, cold meats, cooked fish, raw fish, fowl and venison, salads and soups and spices, vegetables in countless varieties, cheeses, cakes and candies. All jumbled on the table together with drinks in abundance. There were no starters, no enders. Diners ranged at will, tasting and mixing, eating and chewing, piling disbanded crocks and reaching for clean ones, continuously from midday till early evening. Unfortunately, the Cinderella hour struck and with it a rush for the airport and the homebound plane. Lack of time cheated me of the sauna bath which followed the day's entertainment. The family sauna was constructed in a cellar beneath the house proper. There, every Friday at least, the family, guests and servants would repair *tout nu* for a twig-flicking session. Farewell Finland! I had promised to return; and did so years later when Tora's charges had grown to sturdy youths.

❧ ☙

Finland for Finland's sake had drawn me to that brave and beautiful country which the stubborn patriotism of her people had saved from a process that had grafted other free peoples of the Baltic States on to the ungainly trunk of the USSR. To make this confession in no way belittles what was contributed by friends and colleagues at Otaniemi about the implications of technical change for world agriculture and the feeding of the nations. Inorganic farming, biological engineering, mammoth machines and other unheard of mechanical aids lay on the new horizons of progress, horizons far beyond the reach of age-old customs, entrenched traditions and the pinched pockets of peasant farmers or even of their more affluent yeoman neighbours. The new ideas called for new means and new money and raised questions of provision, investment and the order of command over capital and land. A family farmer who, as tenant, could afford new machinery on a scale unknown to his forebears would, as a freeholder in possession, be strapped for cash to buy his land and buildings. The ways in which land and the buildings upon it were held and provided would be crucial in the new age.

At the centre of this world debate lay the new concept of 'estate capital', gleaming and polished from the research workshops of the Department of Estate Management at Cambridge. Even now this idea was hardly recognised, let alone understood, by the agricultural economists and their more generalised brethren. The property rights which fashioned it, the strength and range of them, and the character of the hands which held those rights, public and private, predetermined the motives which moved the investment and use of estate capital. An article of mine introducing and defining these ideas in the winter edition of the *Times Agricultural Review* had coined the term 'estate economy' to denote the abstract image by which the practical functions of estate capital were determined. It was gratifying that the lawyers understood us better than the economists. Blindness of pride not of intellect was probably hindering the economists. For the time being, our research was to run a kind of three-field rotation – estate economy, estate management and estate capital.

When academics play with new thought forms and devise neologisms to fit, they run the risk of writing a vocabulary of weasel words, words which are either superfluous or spurious. Should, however, the new language be taken up by the professional in practice, not as a pretence of learning but because it helps to marshall thoughts, describe and present ideas and bring improved cogency to the daily task, it will pass the test of validity. So it was with 'Estate Economy'. Sir Henry Wells at that juncture had taken over from his illustrious father, Sir William Wells, the mantle of Hippocrates to the surveyors' profession. He acknowledged the paucity of its academic standing and the difficulty of attracting the best brains of a rising generation. He realised that the profession was still in search of an intellectual *persona* to justify a stall in the halls of the learned professions. Even so, as late as 1960, Henry Wells when presenting the Wells Report on future educational policy to the RICS supposed 'that the most satisfactory normal way of entering the surveyor's profession is as an articled clerk under which the master-employer accepts by legal agreement obligations to his pupil'.

It was that Report, written by Henry and his mates, which first recognised the contribution from Cambridge of the concept of 'estate economy' as a thought form, comprehensive yet *sui generis* in which the bits and pieces of applied knowledge imparted to an articled pupil by his betters could cohere in a deeper intellectual unity, a philosophy and a discipline. Consequently a Paper on *Estate Economy* was introduced by the Wells Report into the Final qualifying examination. Robert Steel, an Under Secretary of the RICS and later its Secretary-General, speaking to the Wells Report at the RICS Annual Conference at Bangor in 1960 generously acknowledged the debt the profession owed to the new thinking and research at Cambridge and said:

> 'the proposed syllabus for the General Section of the Institution Examinations includes, for the first time, the subject of *Estate Economy*. This subject is the direct product of the new approach to the Institution's educational policy; but its significance in the surveyor's academic equipment has largely been brought to light by the fundamental thinking that has inspired the research programme on which the Estate Management Department at Cambridge has been engaged for the last few years.'

The new thinking in the RICS owed nothing to spontaneous revelation. Ideas were conveyed there from the University by a kind of shuttle-cock diplomacy. Leading among the emissaries was J.F.Q. Switzer; 'Jeffrey' to those who like Henry Wells and I had the good fortune to count him among our special friends. Without putting too fine a point upon it, Jeffrey Switzer taught Henry what we were thinking at Cambridge and carried draft paragraphs, phrases, words, commas, colons and stops backwards and forwards to shape the text and embody within it the new ideas. We were proud and pleased by this recognition and acceptance. Back at the University such cosy intimacy between us and the profession was fated to be held to our disadvantage when, later, the battle for the Land Economy Tripos was joined with the ranks of the prejudiced.

As yet, however, that displeasure was in the future but not so distant as to muffle altogether the rumbling of far off war drums. Jeffrey Switzer was a tower of strength in those years of overture before the struggle. He was young, handsome and cast in a mould between Father O'Flynn and Ian Smith of Rhodesia. Jeffrey from the genii among his Irish ancestors had inherited an engaging, penetrating wit, ready courage and natural charm which, now and again, would combine somewhat incongruously with a taut serious-mindedness and display of *de rigueur* to curb, so he would claim, my wilder excesses of thought, manner and action. This prim trait probably derived from his Swiss forbears, the protestant émigrés to Ireland from the Tyrol valleys of 16th century Switzerland. As Secretary of the Board of Estate Management and also to the Trustees of the Estate Management Development Fund, he had an authority and opinion to be respected among those who were to shape the fortunes of the future.

It was a happy day when, persuaded by the Board of Estate Management, the University nominated Dr. Josef Popkiewicz of the Technical University of Wrocław as a visitor under the foreign University Interchange Scheme. In response the University at Wrocław had Jeffrey and myself back to Poland in the summer of 1958.

In July, Jeffrey Switzer and I were under the machine-guns of the Iron Curtain at Razvalov, an inconspicuous border-post between Czechoslovakia and West Germany. War was in the air. Three days earlier, military thugs egged on by Nasser in Egypt had murdered King Feisal and the Crown Prince of Iraq. Next day, American marines waded ashore at Beirut; Syrian troops were massing on the Jordanian border; Russia was offering immediate military aid to Nasser; and, on that very morning, 2000 British paratroopers had flown to the aid of King Hussain – hardly a propitious moment to dive under the wrong side of the Iron Curtain.

Retreat was cut off by the click of the border gates behind us. A hundred yards in front a smothered track, once a metalled road, ran eastwards under a tangle of matted grasses and self-sown thicket. No human had gone that way for many a year. Suddenly, to the left, a window opened in an armoured *celnice* and a military voice boomed 'Denman: Switzer'. They'd got us. Two belted and spurred officials, revolvers bumping from the hips, descended the steps and without a word proceeded to rape our stately middle-aged Armstrong Siddeley, boarded, unloaded and stripped her. A third swaggerer waving and pointing his pistol signalled to have the baggage opened and tipped on the concrete. An hour later, the first lot were still screwing back the panelling and boards in the car. On the concrete, the last suitcase had been repacked, the lid sat upon, fastened, locked and vetted by the *policista* who, dusting his hands together, turned to march off to the *celnice* in a show of impatience.

'Achtung!' yelled Jeffrey. Tightening his belt in fury, the man came back. Meantime, Jeffrey with calculated impertinence had reopened his pack and was deliberately and in slow-motion undoing his tie. Once off, he delved into the lower layers of the rumpled clothing, extracted an enormous red replacement and bowing to the *policista* proceeded with outstretched neck and lifted chin to bedeck himself. Then clicking his heels, he gave the glowering officialdom an elaborate salute, shouted 'Vertig' and made for the car. The road switched abruptly to the left, leaving the green ghost-way to run on among the pine trees. There were no responding smiles as we waved *sbohem* and disappeared under the Curtain.

The ravages of communism had not wholly expunged the beauty and loveliness of the ancient city of King Wenceslas. A pall of sadness mingled with shame hung over the Voltava, the flaking stone buildings, the unkempt public gardens and the rusty tramways, in nothing resembling the brave defiance we were to encounter in the rebuilt mediaeval splendour of the *Starego Miasta*, the old city, of Warsaw. In Prague, however, there were niches of care and concern from which the State had lifted its lifeless hands, notably the treasure chamber in the cathedral of St. Vitus. There, on the return journey and to our irrepressible amusement, the lady curator when presenting a magnificent Byzantine monstrance had, inadvertently, called it a 'monstrosity'. With a repeated 'Vy do you keep on larfinck?' the dear women rebuked our laughter and mounting bewilderment.

War hysteria had somewhat abated by then but on our first arriving in the city, deprived as it was of all communication with the West, rumour ran rife.

That evening among the dining tables of the Atlantic Hotel the stories reached to a pitch of horror. A saucy youth on a business trail from the Argentine had just flown in with the news, the only news, from London, and was shaking his head sadly over a couple of tremulous American tourists whose destination in the morning, he dolefully assured them, would not be homewards but eastwards to the salt mines beyond the Urals. Egress from Czechoslovakia was as troublesome as getting out of Hampton Court Maze. The maps were right, our sense of direction was right, but not our knowledge of the Czech language. Pointing to the map, we sought assurance of the road. Every time the locals would say 'Ah! No'. Eventually, having returned on our tracks four or five times, desperation set in and we gave a lift to two giggly maidens who explained that 'Arno' in Czech means 'Yes, that's right'.

Whereas the Czechs had lost their soul in a sullen sadness, the Poles, we found, were a people of mirth and piety. The devil of communism had, as on the banks of the Voltava, left footprints along the wide valley of the Vistula from Wrocław to Warszawa; roofless houses, loose telegraph poles swinging from twisted wires, gaunt socialist architecture, disrepair and dilapidation. But this people had learnt with Sir Thomas More that 'The Devill, the prowde spirite, cannot endure to be mocked'. We had come among them, to the lecture rooms of the Technical University of Wrocław to tell audiences who hung on our every word, of the land and planning policies of Britain. Our very being there was evidence of the new freedom, a consequence, so they told us, of the 'October Revolution' in 1957. Student riots against the Government had been held in check by a shrewd Mother Church and the hard-headed Deans and Professors of the Universities, among them Professor Stys. The wisdom and voice of the academic establishment had stood behind First Secretary Gomulka who, first as a patriot and secondly as a communist, had been sent to Moscow after the Hungarian uprising. Unlike Imre Nagy, he had returned ostensibly as Moscow's face-saving man but with a free ticket in his pocket to alleviate some of the social pressures bearing so heavily on the Poles. Stys, after we had contrasted Attlee's socialist controls with MacMillan's Tory liberalities, stood up in public to demand a bipartite, democratic Parliament for Poland.

Polish humour often has a practical turn and almost always a political one. Jozef Popkiewicz, our host, lived in a stylish house with a spacious garden. The place was last repaired and repainted by the Germans before the war.

Jozef was not short of white lead and turpentine to paint it but had he lifted a brush to the exposed woodwork, Comrade Popkiewicz would have been denounced as a 'deviationist', a renegade among the faithful. On the first morning under his roof, Jozef, who was a genial intellectual modelled to the curvilinear proportions of the late Richard Dimbleby, woke me. With a beaming 'dzien dobry', he proceeded to explain that the crack in the wash-hand basin was 'political' and the missing plug, now somewhere in the sewers, was a small necessity which had gone out of fashion. After such extenuating observations, he opened my eyes to the empirical wonders of Polish international monetary exchange.

''Ere', he said, 'are 4000 zloty. When you get back to London, give my brother Sabin, who lives in Ealing, £10. He will buy slimming pills for me'. My mouth fell open. The official rate of exchange was 11 *zloty* to the £1; the tourist rate was 70 *zloty* to the £1; he was offering 400 *zloty* to the £1. We could buy *benzena* at the equivalent of 1/6 per gallon!

Deviationists in those days were a species of heretic among the party faithful. While officialdom smelt them out, our friends at the University who, incidentally, invariably used classical Latin to converse among themselves in public, poked fun at the witch-hunting. 'If you go into the country', Stys spoke with feigned gravity, 'and there's nothing in the country but plenty in the town, that's left deviationism. If you go into the country and there's plenty in the country but nothing in the town, that's right deviationism. If you go into the country and there's nothing in the country and nothing in the town, that's toeing the party line. But if you go into the country and there's plenty in the country and plenty in the town, those are the horrors of capitalism'.

'You people from Cambridge', he once said to us with a cocked eyebrow, 'don't know the difference between capitalism and communism? So I will tell you: capitalism is the exploitation of man by man; and communism is the exact opposite'.

Delightful. Regaled with such cracks and endless humour, the working days went by all too quickly, when, towards our last weekend, Popkiewicz suddenly announced a trip to the Tatra Mountains, to Zakapane. Next day, by noon, with Wrocław some 50 kilometres behind us over a narrow bumpy road, our hearts at ease and our spirits high, an ominous knocking possessed the engine. The whine became a screech, the speedometer dropped from 60 kph, through 40, 30, 20 to 5 and on the next slope the Armstrong started to run backwards. Speech was unnecessary, the truth was patent. Nemesis had caught up with us. A comical smile spread over the face of our Polish friend who wagged his head. A week back on the autobahn, coasting with some speed down a hill, I had unthinkingly jammed the preselector gear lever into reverse. There had been a tearing and rending beneath the floorboards and the swish of a dying clutch. Popkiewicz without a word jumped out and wandered away into the distance. An hour later he arrived back in a taxi. There are no repair garages in Poland, but all taximen are trained mechanics. His worthy taximan took us in hand. By dusk a bedraggled caravan, the taxi and the Armstrong with its collapsed clutch, limped into the nearest town. Temporary accommodation was found for the night. The Armstrong was banished, we knew not where. Over the next three days, the wily fingers of the taxi driver and his mates rebuilt the clutch.

This episode had its compensations. Instead of the mountains, we enjoyed the hospitality of new friends who, but for the accident, would have remained unknown. From them came insights of the folklore by which non-communist families live in a communist land. Nothing is written. Parents are vigilant, ever looking back over the shoulder, careful lest they are discovered, even by their own children at times, to be deliberately contradicting the 'truth' imparted to the young at school. Many school teachers, however, were hand in hand with parents. Both accepted and practised the folklore methods. When, for example, the school taught that the Red Cross was founded by Russian nurses working in indescribably inhuman conditions in the Crimea and that stories of an unqualified English busybody called Florence Nightingale were false capitalist propaganda, the children would recognise that the truth lay to the contrary by the inflection of the teacher's speech.

'Florence Nightingale' An English nurse? Founder of the Red Cross? Oh. No! How Ridiculous'. Polish children discovered the truth by looking at their teachers through the wrong end of the telescope. University teachers, especially those who conversed in Latin and went so far as to invite 'capitalist pigs' from Cambridge to lecture to their students, would probably never be school teachers. Even so *urzednik* could deprive them of the joys of parenthood. What they told the children sitting wrapped in cosy blankets by brown-coal fires of evenings after the tub, those children never forgot, especially if their fathers, like Professor Stys, were imprisoned for the sake of liberty and truth.

Once back across the West German border, we pitched in immense relief for a picnic off *Deutsche Wurst*. Lolling in the tall, dry, uncut grass of an Austrian meadow, our thoughts went eastwards to the 'back there', to the imprisonment, to the people locked in fear, to the most inexplicable enigma of Twentieth Century civilisation. How would it all end? This system that wrought so savagely against the human spirit. Relief was weighted with a twinge of guilt. Back there, in the land of peoples' imprisonment, our hearts had met with wonderful new friends pleading 'to keep the window open to the West'. Yet, here we were, free as the dancing mountain stream that sang to the meadow as it passed. Perhaps we would have been better scholars if we were impelled to camouflage speech in the syntax of the Latin tongue.

There was, nevertheless, something narcotic about the imprisoned lands of Eastern Europe. We were drugged by an inescapable craving to return, a 'fix' pulling against prudent revulsion. No sooner back in Cambridge than the obsessive 'pull' was harassing better judgment. The irrational won. By autumn of the following year, Jeffrey and I had together ducked under the Iron Curtain again. This time into a land of repression far more ruthless than either courageous Poland or stoic Czechoslovakia. Karl Marx and Prussianism had combined to forge the fearful realm of the Deutsche Demokratische Republik.

The Karl Marx Universität in Leipsig, kidding itself that Karl Marx had founded the place 550 years ago, was bent on having friend and foe to a centenary celebration. Cambridge University was among others invited, providing its representative brought to the pow-pow some suitable thoughts on *Das Problem der Betriebsgrosse in der Landwirtschaft*. The Registry in Cambridge spotted the 'land' allusion on the invitation and sent the summons to the Board of Estate Management. The Board decided upon whom they could best do without and nominated me. There was a merciful condition that if I could not meet the date because of other oversea's commitments, Jeffrey Switzer should face the Marxian jamboree. Both our names therefore were sent to Leipsig. And lo! two *Einladung* came back. So we both went.

Khrushchev and Mao were shaking hands in Peking. 'You've never had it so good' sentiments had returned MacMillan to power in the UK along with a new woman MP – Margaret Thatcher – on the back benches, when Jeffrey and I were in the British Consulate in West Berlin. The Consulate were impotent to help us. 'You are hoping to go to the DDR, you say?' The junior who received us was a merry cynic.

'Yes', we said. 'Why so incredulous?'

'Oh nothing, really, but so far as we are concerned,' he continued 'there is no such place as the DDR. We can give you no consular protection. But best of luck. Just two pieces of advice: don't change East Deutsche Marks in West Berlin and don't take any photographs once you are lost over the Border. Wiedersehn!'

There was no Berlin Wall. Freedom and fear, nevertheless, met at the *Grenze*, a lonely station on a ghost railway that ran between phantom platforms now dead to the tramp of feet. Leipsig our ultimate destination meant making for the Ostbahnhof, the most desolate and dreary station on earth. Before boarding, we diverted to the *Herrentoilette*, a row of filthy underground cabins beneath a black, leaky vault. Every door was a panel of peeling paintwork and had been robbed of handle and fastening. The doors swung loose to expose broken, cracked, age-begrimmed pedestals, seatless and jagged with evil-smelling dentures of broken porcelain. There was no other facility. To function was an acrobatic feat, one leg twisted sideways to secure balance and the other thrust out horizontally to hold the door. At intervals, an angry banging on a door would signal that one's neighbour was seeking the equivalent of an East German toilet roll. When my time came, after a few piston strokes against the door with the horizontal leg to draw attention, a grimy hand flicked a 3″ square patch of sodden, grey blotting-paper over the exposed lintel. It fluttered down to the bepuddled floor. Could it be that the supply and quality of toilet-paper were sound indices of the condition of a nation's economy? If so the DDR was in poor shape!

The DDR in 1959 was a very dark land. Big Brother's eyes swept the open places, gouged out every corner and cranny and pierced all dissimulating veils; the diktat of the State was palpable; communism there, unlike in Poland, was no joke, no object of ridicule. A suffocation of spirit assailed us as the train from East Berlin trundled into the night and into the heart of the Peoples' Republic. With much clatter and clanging, the carriages would repeatedly stop, shiver and start again, sometimes in a station, sometimes not, as if uncertain of their destination. Between the stops, the *Polizei* did their work, ferreting through the carriages, flashing torches and highlighting the drama by random gunfire. Occasionally they carried from the train a stretcher bearing the ominous outline of a human body under a draped tarpaulin. We were evidently the only English-speaking passengers on the train and yet as it drew into the bombed-out station of Leipsig, every public *Lautsprecher* proclaimed greetings to the Conference delegates in English and gave instructions where to go.

The Celebrations had been deliberately planned in this Moscow-cultivated garden of the truly faithful to bring the Poles to heel. Of first importance were the DDR state farms (*sovkhos*) and the collective farms (*kolkhos*), the icons of the communist agricultural credo. The Poles, however, on the sidelines, were for peasant farming and the peasants were for Poland, not on economic grounds but politically. The mammoth collectives by contrast were dependent for essential heavy equipment, implements and machinery upon the giant steel mills of Czechoslovakia. Poland under peasant hands therefore enjoyed a relative national independence, albeit in conformity with the tenets of Marx-Leninism.

Friends, including Jozef Popkiewicz who had come over from Poland, urged us to speak up in public; to decline to do so was running the risk of our names going on record as being among the 'Fellow travellers' or worse. I was called to speak 'as the only Englishman in sight' (Jeffrey was either behind a pillar or was counted an Irishman) My voice was heard from a platform of communists, upholding the virtues of the landlord and tenant system based on private property as the optimum *Betriebsgrosse in der Landwirtschaft*. Groans and murmurings of dissent followed me off the platform.

A fellow waving a sheet of paper rushed up, brandished it in my face and stuffed it in my pocket. He wanted to remind me, so his message ran, that in England the landlords were driving the people off the land to make room for sheep runs. Like the historians of the Karl Marx University, he was four hundred years out of date. The doggerel he quoted –

> 'Sheepe have eate up our medows and our downes,
> Our corne, our wood, whole villages and townes';

and which he had ascribed to Sir Thomas More (hardly a contemporary of our times) were in fact lines from Thomas Bastard's *Chrestoleros*, written in 1598 in the days of the Tudor enclosures.

My words had irritated the party-faithful. They gathered to pound us through the night-drinking hours in Goethe's Bierkeller with endless polemical dialectic. This was innocuous academic exchange and in contrast to the assault which the East German Broadcasting Network made upon me. The offensive came from behind the curtains of the Conference Hall as the delegates dispersed. Button-holed by an importunate radio-interviewer, I was pressed to broadcast over the DDR Radio. Seemingly an expectant Europe was waiting for Cambridge University's impressions of the DDR. The trap was sprung, but I declined the bait. 'You will cut out all my criticisms and use my voice to broadcast doctored comment', I objected.

'That would be more than my job was worth', my assailant interviewer was trying to be assuring and thrust the microphone at me.

'Before you start recording', I cautioned, 'let me tell you, off the record, what I intend to say. Here in Leipsig, over your city buildings and squares, and in the country, across lanes and farmsteads, are festooned great red and while banners extolling ten years of peace, plenty and the joy of socialism. Underneath them slouch the drawn, grey faces of reality, in contradiction of the brave banners. This is what I will say'.

'Hmm! That won't do'. He looked peeved and tried another tack. 'Would you, perhaps, tell us how you found the Conference?'

'Worthwhile', my reply was genuine and unhesitant, 'scholarly and informative but marred by the political agent from Moscow whose hour-long unannounced interjections on Marx-Leninism bored us to sleep and dislocated the programmes'. The patient interviewer gave me a wry smile, detached the microphone and left us to the misery and darkness of our capitalist unbelief.

The red and white banners over the approaches to the State Farm were, probably, there to impress the delegates to the Conference. Two buses laden to capacity had been used the day previous to take them to a display of the wonders of this enterprise. En route, the buses passed through Eisleben, where Martin Luther was born and where also he was buried. On the outskirts of the town was a great bust of Lenin standing in a sea of freshly cut floral tributes. Everyone, with cameras at the ready, was ordered out of the coaches to make obeisance to this idol. Later, when crossing the city square, Jeffrey and I spotted the Statue of Martin Luther. We got up, banged on the windows of the coach and shouted 'Stoppen bitte'. Amazed, the driver pulled up in alarm. We solemnly disembarked and with deliberate ostentation walked to the statue and photographed it from all angles. Then, smiling a happy *Schönen Dank*, we clambered back on board. That evening, on the way back, the supervising policeman in charge, nicknamed *Der Schwarzmann*, pointed to a huge buzzard hovering in the scarlet sunset.

'A free bird,' he exclaimed in a fair English and for our exclusive benefit, 'flying over a free land'.

'Yah,' I replied, 'looking for his prey'.

Unknown to the *Schwarzmann*, a farmhand had whispered to us in good English behind a cowshed tales of human misery and cruelty. He was the only survivor of a once proud family whose ancestors were laid generations deep in the very soil of the State Farm. The Communists had slaughtered his father, mother and all other relatives, confiscated the land and set him to menial labour; since then they had picked his brains as the only source of farming know-how about the place. Those whispered words of pathos by the cowshed will never be forgotten.

In the years to come no one could recall whether the Celebrations at Liepsig that dreary bleak October ever solved *Das Problem der Betriebsgrosse in der Landwirtschaft*. In my view, the problem in the DDR arose not so much from *Betriebsgrosse*, the size of an enterprise, as from the misery and dejection of those who worked it.

Messages ran loose over the capricious telepathy which informs the academic world of what's going on. The new interest in land problems at Cambridge evoked curiosity and response from far-away, unexpected places. The Leipsig invitation was one of the signals of recognition. Earlier, in the Spring of the same year, a letter had come from a Professor Giangastone Bolla of Florence University calling the First General Assembly of a newly constituted Institute of International and Comparative Agrarian Law, an august body under the patronage of the Italian Government. *Participanti* were being invited from some forty countries across the globe to concentrate on the problems of: (a) landownership and land tenure; and (b) credit security for capital investment in agriculture, nation by nation. The story from *Gran Bretagna*, the latter said, was to be left in my sole care.

Landownership was the new key we were trying to fashion and turn as a help to unravelling the problems of resource distribution. That landownership had been overlooked was understandable in the collectivist ethos of the post-war years. One could study 'land use' in a

146

detached, abstract way, as the self-satisfied Left always does, without running into awkward obstacles, like private property and capital. Land under socialism belongs to the State, it has no market and, according to Marxists, no value. In socialism's demi-world, land-ownership is of little consequence. Applied economists, the neoclassical theorists and rougher relatives among the macroeconomists had fought shy of the institution of property rights in land. At Cambridge, we were seeing them in a new light and had come to realise that landownership was a power-base, an essential factor to be recognised in any analysis of resource distribution.

Definitions were in hand at Cambridge. 'Estate Capital' denoted the resource, a measurable quantum of property power over land and buildings exercisable within specific decision-making units. So far so good. But the whole calculus still wanted a universal term for the unit itself. The 'estate in land' was coined and had been well-ventilated in my *Estate Capital*, published in 1957. This form of the fundamental unit, wholly self-contained and seised of self-potency for action, was given a special airing the following year. Three leading voices – Sir Solly Zuckerman, Enoch Powell and Lord Chief Justice Parker – alongside others less well-known, were invited to give a series of public Lectures at Cambridge on the theme of the national estate, its resources, develop-ment and restrictive power over private liberties. One by one, each generalised national pattern was questioned and shown to be a mosaic composed of the cells of a national honeycomb of 'estates in land'. Debate went further and pointed out that little is known yet of the distribution and character of these fundamental units to justify the assumptions on which the national patterns were presented. The book that contained the Lectures – *Land Ownership and Resources* – and other supporting literature had reached Florence well before Easter 1959, hence my unexpected invitation to join the *Prima Assemblea*.

Florence, a city of sandstone, sienna and deep sepia shadows, the mistress of mediaeval architectural splendours, became known to me over the next six years. The early anonymity of her interstices gave way to familiar doorsteps, courtyards, balustrades and chambers year by year. The standing International Committee of the *Instituto* had made me a member. An essential reconnoitre in 1959 of the Florentine background was a wonderful opportunity for a three-weeks family tour of Europe, from Cambridge to Rome and back by Florence and the Austrian snows.

If previous calamities are portents, we should never have taken that journey. The plan was to use the Armstrong Siddeley, the same sedate motor car which had collapsed on its way to the Tatra Mountains in

Poland. It was mid-March, two weeks before the start of the Florentine venture, when Neil Elliott proposed a tour of his family estate in the valley of the Stinchar, close by the lands of a place called Dinwoodie in Ayrshire. His neighbour, the Laird of Dinwoodie, whom he had not met lulled our consciences. He had persuaded Elliott to bring his Cambridge friend as confidant and adviser to a consultation where would be disclosed the mysteries of a momentous scientific break-through of a portent and magnitude that would transform the global supplies of energy. By the time the Armstrong had made Robin Hood country on the way north, a strange tapping and rattling had developed under the bonnet. Neil, who heard the ominous sound as the car came up his drive, took a bit of firm assuring to get him aboard. Pessimism was alien to him. The fact that such a man was cautious should have warned me. Somewhere by Dotheboys Hall on the Brough road over the Pennines, an almighty bang beneath the steering column brought the stately vehicle to a sudden stop. Under the lifted hood were revealed four piston ends snapped in half and poking through a shattered crankcase.

After two hours of tramping miles and thumbing numerous lifts, we bade the invalid motor car a sad farewell, left it at Temple Sowerby and proceeded north in a ramshackle, hired contraption, half car, half tractor. It was well into the night and into the vortex of a 60 mph. howling gale before the headlights of the hybrid conveyance caught in their beam a gaunt, solitary signpost. Bending in the wind and pointing into impenetrable darkness, the ominous post read 'Dinwoodie'. Beyond its swaying finger, a pot-holed track fringed with sloughs led to the edge of an escarpment. Three hundred feet below, a single light was playing tricks through the tossing fir trees and sending what resembled frenetic morse signals from the dark depths. Descent over the rough, sinuous track took most of an hour. The guiding star of light had long been lost before the solemn, lightless face of a baleful dwelling was made out in the darkness.

'Ah!' moaned the shadowy figure that answered the bell-peal, 'So you've arrived. Come in'.

Through a tunnel lined by gloomy, indeterminate shapes, the silent figure made signs to us to follow him into a dimly-lit lounge. Neither sip nor sup had passed our lips all day and there were no signs that the lack of victuals would be made good. Weak whisky was the only relief over two hours of the most weird and exhausting conversation I had ever encountered. Elliott was alright. He sat facing me from behind the back of our host who had suddenly become wildly excited. Thus it was that I faced his frenzied antics while my friend spent the time sticking his thumbs in his ears and wriggling his fingers in unmistakable gestures. Choked with suppressed laughter, Neil's signals confirming my own fears. I could do nothing. Desperate for a break, I enquired after our host's wife.

'My wife is not here', confessed the laird 'and I have only just returned myself'. Nothing doing there. So I asked where he had come from.'Ah!', he nodded slowly 'that's a good question. If only I knew'.

He was a small man, thin on top, a fresh complexion and blue, gimlet eyes which bored into one with the utmost conviction.

'My proposition', he continued oblivious of our consternation, 'my proposition, the reason for your coming, is to tell you of the proven faith I have in my ability, given the right backing, to make petroleum out of cow dung'.

The exposition of this incredible thesis went on and on into the dreary hours of a never-ending night. Eventually, we tottered to bed starving, bemused and heavily suspicious that the unfamiliar whisky was probably petrol made from cow dung.

We were too exhausted to care. The sane light of morning, a whiff of a substantial breakfast served by a local varlet did much to restore pose to our souls. Both of us, however, were resolved upon a single course: to bid farewell without delay. After the Stinchar valley, a telephone call put through to Harrogate from the Peebles Hydro summoned my generous brother's Bentley to meet us at Temple Sowerby. Promise followed promise from the garage but the Armstrong was never restored. So a hiked-up mortgage and a new California took us to Florence. Despite Dinwoodie, we ran on Shell petroleum and not a drop of cow-dung!

In Florence, the senior professors planning the *Prima Assemblea* welcomed the spontaneous response from Cambridge, gave me a kind of metaphorical toga among the internationals and refused to discuss any 'ifs' and 'maybe's'. Unconditionally, I was part of the new edifice. All they wanted to know was when they could have my address on behalf of *Gran Bretagna*.

The Italians are ready talkers but poor listeners. Habitually they run out of time. Italian time-keeping required a postponement until April 1960. The extra time was well utilised. My Paper on *Agricultural Law and the Ownership of Land in England* could be delayed so as to tip into it the latest statistics from the national farm rents survey we were completing for the Government. Other original material from the body of 'Estate Capital' made up the bulk of the text. The Italians followed the new Gospel. The opening caption in the Italian version read: *Riassunto: 'The Estate in Land'. Elemento unitario di proprieta.*

The essential thought behind the Cambridge philosophy was under-stood and heartily welcomed but 'the estate in land' had no equivalent in Italian ('estate' to them meant 'summer') nor, indeed, in any other of the languages used at the grand Assembly. The lesson from the linguistic confusion was readily heeded. The 'Estate in Land' was dropped as a denomination and, taking a cue from the Italian *elemento unitario di proprieta*, the English 'proprietary land unit' put in its place.

❦

The clear-eyed Bolla, concerned as he was to comprehend the place of landownership and tenure in modern economies, called his new crea-tion 'The International Institute of Comparative Agrarian Law'. This indirectness was a symptom of the prevailing confusion which accepted the proprietary structure of land as a consequence of and incidental to its use and not the reason for it. The World Bank laboured and stumbled for forty years under a similar illusion. At Cambridge, like Luther at Wittenberg, we were nailing our theses to college door-posts: 'Landownership is cause not consequence'.

Like Luther, our daring set us apart. When in 1959 the ten-year old Nature Conservancy backed by the failure of the Zuckerman Commit-tee on Forestry, Agriculture and Marginal Land pointed out to the Government that resource development policies were hampered for want of proper, adequate information on land holding, it was to Cambridge that the Conservancy turned to remedy certain of the deficiencies. A Royal Commission, after some years, had published its Report on Commons whose use patterns were manifestly the outcome of ancient customary tenures and statute law. There were 1.25m. acres of commonland in England and Wales and the management of it in the future, as in the past, would be conditioned by the complicated law of commons.

In recognition of our special understanding of the problems, we at Cambridge were asked to undertake a six-years research study of the Commons and Village Greens south of the Border, backed by Nuffield Foundation finance.

Those who acknowledged and respected our *renascence* in thought and our academic achievement were few in number and were more frequently found abroad than at Cambridge. The Department of Estate Management was still encumbered by the fallacy that its professional services were in some way an asset to academic and teaching progress;

when in fact the gulf that had opened between the new adventurers on the academic side and the professionals on the other had reached chasm proportions. The old guard wanted to close it; and the new model army to widen it further. The duality had always been an absurdity and now was an unwarranted anachronism.

Noel Dean, the Director of Estate Management, was due to retire in 1961. Who would succeed him and to what an inheritance? These were questions mulled over and reported on from time to time by a laid-back Committee of the Board of Estate Management. For reasons then incomprehensible and, even today, obscure, I fell into disfavour with the old-guard mafia in the Department. Maybe my step was too sudden and too insolent. There were jealousies, misunderstandings and generation-gap conflicts aplenty. Whatever the cause, after a year, the 'Future Committee' detailed Noel Dean to advise me not to entertain thoughts of taking over from himself in any direction.

Naturally I was downcast, but thought the Department more important than myself. So I replied 'I have put too much life-effort into building up the academic respectability of the Department to want anyone appointed a successor to yourself who would do less than credit to it. You and your advisers must decide what is best'.

Nothing further was heard of the matter until many months later when circumstances had changed. Meanwhile, an interview for a Fellowship at Nuffield College, Oxford had been arranged with the intention of following it up. Although excluded from the 'Future Committee' myself, Jeffrey Switzer was on it and his knowledge, judgment, outlook and grasp of what the future should aim for, I held in highest regard. Following brilliant Fabian tactics, he bent its thinking in directions which some years earlier would have been held as high treason. Late in 1960, the Committee came to a final decision. Its proposal ran the gamut of the Department, General Board, Senate and University – the Department of Estate Management should as from 1962 be exclusively responsible for teaching and research. A new independent Directorate of Lands and Buildings would parallel the old professionals, be answerable to the Financial Board and become incorporated in the administrative structure of the University. Emancipation! The old fallacy would no longer cripple academic progress.

When Sydney my twin died, an unexpected and inexplicable sense of being in a new order of creation possessed me. Never before had life to be lived without my *alter ego*. People's eyes were upon me in a hitherto unknown exposure. My going was in unaccustomed aloneness. A similar sense of severance, an unfamiliar exposure, affected the Department of Estate Management in 1961. There was no hiding behind an *alter ego*, no one to point to, no one to accuse, no extenuatory professional side. The teachers and researchers had asked to be free, to be themselves. The University had given an answer unreserved, forthright and displaying commendable *Realpolitik*: 'very well, then, so be it; get on with it'.

Gentle surgery, no bleeding or jagged edges, cut and cauterised the new Directorate of Lands and Buildings from the old stump. The Council of the Senate reporting the operation to the University deliberately refrained from mentioning, let alone discussing, the teaching obligations and expectations of the rump Department. For that old corpus, still labelled 'Estate Management', the split meant either death or new birth. Most were happy to have it so; we knew the strength of the pulse of life within, its purpose and potential. Nothing could be the same again; nor would those most deeply committed to change want it otherwise. We wanted 'estate management' humanely disposed of, buried with blessings and no regrets. From its burial would spring a new life, a Tripos in Land Economy leading to an honours examination with all supporting research facilities to ensure its vitality.

Sam Weller retired the same year. By the end, he was not unsympathetic towards the new aims but was never fully convinced of anybody's ability to consummate them. Noel Dean, the other original pillar with Weller in the sustaining arch of Estate Management, retired concurrently. Noel took on a growling brief. He muttered assent to changes he little understood and even less welcomed, yet he knew that they would be linked to a past that would stand to the honour of his memory in the new times to come. That is how those of us who had worked with him wished him to understand the metamorphosis. To frame the case for a honours school was a task he perforce left to me and to younger colleagues with a like perception. October 1 1961 was to be the birthday of the new order. If it were not to wither on the stump, a sustainable programme of rebirth would have to be on the agenda by then.

Cambridge life has always been pregnant with shapes, indefinable and amorphous, yet palpable and potent. One of these shapes is called 'a Tripos'. The word derives from the Greek for a three-legged stool. In modern Cambridge idiom it denotes a course of teaching which leads to an honours degree; some years back it was a patrician distinction above the plebeian Ordinary Degree. No one among the old and the young, among Fellows and non-Fellows, among Heads of Houses, tutors, bursars and dons could tell me what a Tripos was – they all knew what it wasn't! By early spring 1961, the Board of Estate Management had sent a well-reasoned statement to the General Board proposing a Tripos in Land Economy. It was not rejected forthwith. I was instructed to prepare along with colleagues a consultation Paper preparatory to formal discussion with representative Committees to be appointed by the Faculty of Law and the Faculty of Economics and Politics.

Whatever the innards of a Tripos should be, one thing seemed certain to friend and foe alike, its ingredients had to fit together like the pieces of a jig-saw puzzle. Although each piece would be identical *per se*, the several parts had to cohere to make a pattern, a unity, a discipline. Land law, for example, deeply studied and fully comprehended, had, nevertheless, to be a vehicle carrying knowledge of the concept of property – the power base of resource distribution basic to the new thinking. Economics, the study of the behaviour of man towards scarcity, was an essential ingredient. But it was imperative to understand the special form of 'scarcity' determined by property law and by technological restraints on development.

Jeffrey Switzer was absent from Cambridge that year. He had been appointed planning adviser to the Government of Malta to tell them what to do with that honey-stone Island. So April found me aboard an Air Malta flight. For four days in the sun, between yachting and picnics in Comino waters, I solicited his wise, amending counsel before facing the Tripos scrutineers. We concocted a mixture, a firm coagulation and presented them with a choice of six menus differing, not in the make-up of the mixtures, but in the size and shape of the courses offered. Care was also taken to illustrate how each option stood as an academic pursuit and entirely innocent of any pretence of professional training. Back in Cambridge, Sir Ivor Jennings, Vice-Chancellor and Downing Professor of the Laws of England, together with his legal colleagues made up the Law Faculty scrutineers. They listened, nodded, coughed and cogitated through a series of summer meetings. Robert Kahn, later Lord Kahn of Hampstead, with his economists' contingent followed suit. Ultimately and unaminously, the two Committees in joint session, proposed the acceptance of a Land Economy Tripos of One Part. The Honours Examination in Land Economy would follow a Qualifying Examination and an Honours Examination in Part of another Tripos.

The scheme in outline ran on the following lines. Land Law and the theory of the institution of propety would link into the study of the use and distribution of resources from international, national and regional standpoints (Social Land Use) and be related to the calculus of definitive property rights (Proprietary Land Use). This definitive and analytical approach would be supported by the study of the principles of evaluation of the use of resources within the social and proprietary land use functions. The symphony was to be rounded off by comparative studies of historical and spatial change. The Qualifying Examination would test the candidate's knowledge of the fundamental subjects of land law, economics and the primary production technologies in town and country. Here was a course symmetrical, consistent, deep and unified – in short, a Tripos.

Sir Ivor; Robert Kahn; the Master of St. John's College, then Chairman of the Board of Estate Management, and others of equal stature gave the proposal a fair wind. Early in the new year, on 24 January 1962 to be precise, a formal proposal for a Land Economy Tripos was publicly endorsed by the Financial Board and the General Board.

The Regent House had yet to have its say. It met to do so in the coming February; forty four years almost to the day since Sir William Dampier-Whetham asked the same Regent House for a grand School of Rural Economy.

The enemy came out of the skirting, vocal and determined. Two champions from the ranks of the antagonists led the attack. They were vociferous, cogent and logical but each in his own way would argue from premises which were misinformed. One, John Ziman, a man of winning and mischievous wit, had obviously invested some trouble to brief himself on the thought, objects and principles which justified the proposed Tripos, yet spoke with a cynical sadness of a 'worthy subject'. He expressed dismay that the University should contemplate teaching it to no other purpose than making money from property management and development.

Speaking of the syllabus, he said, 'It reads as a narrow professional training in the detailed administration of the financial aspects of Land Economy. It is concerned with buying and selling land, not with using land . . . those who study it will see human life and human society through narrow blinkers, as if men were entirely dominated by the little red and black figures in their bank accounts'.

Walking back from the Debate, I challenged this myopic opponent. How, in the name of honest literate man, had he been able to draw his conclusions. Turning to the Land Law Schedule he pointed to the word *profits*. 'There you are', he said, 'you can't get away from it'. With infinite patience and considerate courtesy, I explained that 'profits' was legal shorthand for certain types of incorporeal hereditaments known to the Law as rights *in alieno solo*. We parted in silence.

The other antagonist, Peter Laslett, spoke in a voice of pained anguish, selected purposely, it seemed, to hold in check the tempest of anger within. He, a scholar of distinction among his own and an

154

authority on the history of population, delivered a speech freighted with resentment.

'It was', he said, 'with considerable perturbation that, on opening the *Reporter*, I found that the subject we had chosen to institute as a new subject in the University, having instituted very few since the War, the subject which we had chosen to be pioneers in apparently, was to be Estate Management, renamed Land Economy'. 'My feeling', he went on, 'is that professionalism, the sense of being here to train people for a career, for almost no purpose connected with the intellectual breath of those we have to teach, is far too close to the surface of the Report'. From there, he felt it right 'to take a glance at the profession which may well be, or perhaps can only be, in the mind of those who think it is important to have a Tripos in Estate Management at Cambridge. The operations or the expansion of the economic activities of those who deal in land is indeed an important feature of the contemporary English scene. It is a fact, not I think mentioned in the history of the study as outlined in the Report in front of us, that a very large sum of money was given to this University with the object, I suppose, of encouraging us to give precisely that status, that professional status, to the training of an auctioneer or estate agent which it was felt the University of Cambridge could give to a study being pursued as part of its academic instruction. I feel strongly that we should not take this additional step; my own doubts go as far as to wonder whether even the present course in Estate Management as part of University studies is justified at all'.

After this outburst, I sought out Peter Laslett to try and discover what at a deeper level, below this veneer of criticism, had set him so vehemently against us; knowing all the while in my own heart that he was saying no more than had been spoken in halls, lecture rooms and corridors for nearly twenty years. The purists were against the professional training element in Estate Management, against the lowliness of the profession it served and based their understanding of the subject on the empirical evidence of the work of the Advisory Service. At my request and through his courtesy, Peter Laslett and I met over coffee some days later. The gist of his defiance was a straight-flung denunciation.

'Even since World War One', he said 'the University has offered Ordinary Degree courses affianced to training in professional competence – in Agriculture, Forestry, Estate Management, Architecture, Engineering and the like. Now a movement is afoot to perpetuate this unfortunate degradation of academic learning by elevating these 'studies' to Tripos standard. I and others in the University are determined to stop this. Your subject happens to be the weakest link in the chain and we intend to snap it there'.

That events over the last fifteen years had faced, accepted, answered and rendered unjustifiable his criticisms was a truth he was reluctant to hear or respect. He had thrown down the gauntlet.

So it came about on Saturday 24 March 1962 when the Vice-Chancellor doffed his cap to the assembled Regent House and submitted 'That the recommendation contained in the Report, dated 30 November 1961, of the Board of Estate Management on the establishment of a Land Economy Tripos be approved', the voice of Peter Laslett proclaimed '*non-placet*'.

That could have been the end of much painstaking work over the long weary weeks following a decade of intensive research, academic outreach, philosophical revision and the rejection of many familiar assumptions. Constitutionally, however, a *non-placet* may be rejected by the majority vote of the Regent House. Votes of *placet*, in our view the votes of the righteous, had to outnumber the voices of the Lasleteers. Such was the form of battle. For six weeks, I sat at the telephone. The Vice-Chancellor let it be known that the fatal day of decision would be 5 May 1962. We on the *placet* side needed names for a Flysheet, many and mighty to show where the weight of just opinion lay. Some names were ready to hand, others required a gentle reminder, many needed stalking with craft and cunning. So successful was the ultimate diplomacy that we feared an onset of apathy. A list of no less than a hundred names, many illustrious in the world of learning, Heads of Houses, Professors and other grandees graced the *placet* Flysheet. It was headed I was delighted to see by the name of Noel Annan. For myself, there was no discharge; to the last minute of 5 May 1962 my telephone was red hot.

The *non-placet* Flysheet was terse: 'We urge you to vote *non-placet* on this Report on the following grounds:

> That the case presented there for an honours course is quite unconvincing.
> That there is no important or obvious reason why this study should be given Tripos status.
> That Land Economy if it has a place here, should be a post-graduate study, and professional training in Estate Management should be the concern of a post-graduate institute'.

There were fourteen names on the Flysheet, a few carried great weight and esteem. To his credit, however, the name of John Ziman was not subscribed.

The *placet* Flysheet gave a short historical run in and reminded the University that 'in the working out of the form and content of the proposed course there has been consultation with the Faculty Boards of Law and of Economics and Politics; and in framing it immediate professional requirements have been deliberately ignored'.

The evening before the Vote, I took advantage of my position as a member of Christ's College to dine at High Table. Diners' heads were together laterally. Conversation on my side was very thin and hard to come by, so, inadvertently, I overheard Jack Plumb opposite me say to his neighbour, 'No, not now, later. We are in the presence of the enemy'.

So I knew there would be at least three votes against us. Concern lest apathy should set in was real because when a verdict is sought, it is the votes that count, not the names on the Flysheet. A genuine supporter, seeing a Flysheet so generously endowed with the names of the worthy, might be tempted to blow us a kiss from a college window and leave others to vote. When the moment came, the vote was public, a count of the noses of those present and ranged according to voting intent either side of the Senate House. Just before the count, I noticed the University Archivist, a lady who had yielded heart and head to our cause, sitting among the *non-placets*. That would never do. With a flourish of my gown, I strode across the floor and, remembering my childhood feats at Nuts-in-May, yanked her over the black and white tiles to her proper place among the righteous. When the tellers had done their job, the enemy was outnumbered somewhere in the region of ten to one. The Land Economy Tripos stood to be part of Cambridge history. Prejudice had not been drowned but it had been truly submerged in a tidal wave of goodwill.

7

POLITICS IN BLACK AND WHITE

But Scot with Scot ne'er met so hot,
Or were more in fury seen, Sir,
Than 'twixt Hal and Bob for the famous job –
Who should be Faculty's Dean, Sir.

The Dean of the Faculty, Robert Burns

On a sunny morning in the August of 1964, I was happy, lolling among the daisies on a tennis court, scribbling notes for a stop-gap speech. The lawyer in charge of City affairs who in these days would be known as the Chief Executive of Cambridge City Council had suddenly fallen ill and, thereby, opened an ominous gap among the speakers billed to address the annual get-together of the Association of Municipal Corporations under the Dome at Brighton in September. Apologising for so short a notice, he hoped I could, nonetheless, give them something which would fill the gap and keep the assembled delegates fully engaged over the first afternoon of the Conference. Judging by what happened subsequently, my ill-wishers, who multiplied rapidly after my speech, must have supposed that some political daemon had deliberately driven my pen.

Far from it. While nodding to the daisies and muttering to myself, my musing had followed lines which over the last ten years had become fairly familiar to readers of the national newspapers and of the academic and professional press. Admittedly, to many my writings were controversial. They ran counter to the accepted wisdom of the day and generated opposition among socialist and Fabian journalists. One of my more glorious moments of infamy had followed a series of articles on the Leader Page of *The Times* in 1960. Endowed with virgin statistics from my research into landownership finance and from the new Farm Rents survey, the series had made the point that among the various social categories of post-War Britain, the landlords had suffered the worst setbacks and their incomes were lagging far behind the invigorated advances of business takings and workers' earnings. So allergic was the socialist stomach to this unyielding truth that it caused the comrades to ring the international alarm bells. By mid-morning on the first day of the series, Prague Radio in the name of the Czech Government were vehemently attacking the articles and denouncing me, the author of them, as 'a prominent international delinquent'.

So it was with my words to the Lord Mayors and their attendant acolytes under the Brighton Dome. The words came from the heart, innocent of all invective, stating another clear objective truth, perhaps ingenuously; but, nonetheless, with conviction. Quite incidentally and through none of my doing, the electricity of politics had charged the atmosphere; a General Election was only days away and nationalisation of steel, water and land divided public debate. Labour wanted to set up a Land Commission virtually to appropriate to the State all building land. Clutching my text, I insisted that the opinions to be expressed were entirely my own and were not meant in any way to reflect the outlook of the Conference. Under the caption '*Private Property and Community Needs*', its purpose was to weigh the case for and against the compulsory acquisition of property by the State in a free society. To that I spoke. Led by Alderman Frank Price, the Lord Mayor of Birmingham, worthy delegates rose one after another to condemn the sentiments and, if they had had the chance, to crucify the speaker. Press response ran from the graphic banner headline of the *Cambridge Evening News* – *Don attacks Land Grabs* – through various balanced assessments in *The Times*, *Telegraph* and *Guardian* to the nearest thing I had ever encountered in objective journalism where the *Municipal Review* explained how the Conference chose their speakers:

> 'when a subject has been selected, a speaker is sought who may be expected to provide an informed contribution to the subject in a way likely to encourage discussion by the representatives present; on the last criterion, at least, Dr. Denman's must be adjudged a notable success!'

The flutter and the fury were ephemeral but for me the reverberations of Brighton were dramatic, long-term and momentous. Arthur Jones, Conservative MP for Bedford North, had presided over the afternoon session under the Dome when my Paper was read. Its words had sounded a dulcet tone in his ears, however others may have heard them. Nothing was said at the time. On Christmas Eve, however, among a shower of Christmas cards a letter sounded an echo from Brighton. It was from Edward Heath, the newly-elected Leader of the Tory Party.

The Tories had lost the General Election in October 1964 but by only a shaving of votes. Harold Wilson with his quip 'nice place you have here' had walked into 10 Downing Street backed by a House of Commons over-all majority of only four to keep him there. Thirteen years in office had blunted Tory wits. Now there was time – but not too much of it – to rethink life after Alec Douglas-Home. An Advisory Committee piloted by Ted Heath was busy setting up Policy Committees, the equivalent of later day Think Tanks and centres of policy studies. Ted's letter was a personal invitation to join the Land Policy Committee under Reggie Maudling.

The brief alarmed me for it carried the ambiguous phrase 'to examine the feasibility of proposals to collect for the public purse some of the increases in land values arising from public decisions'. The words suggested that the Tories might have it in mind to follow a will-o'-the-wisp which since Lloyd George's Finance Act 1910 had brought successive land policies to grief. Loyd George had called this chimera the 'unearned increment' in land values and wanted to tax it. He was challenged at the Bill stage by the then shadow Chancellor, Joyston-Hicks, who had recently married a well-to-do heiress and had hyphenated his name with hers. So muddled-headed was the Chancellor, in the view of Joyston-Hicks, that he couldn't even define 'unearned increment'.

'Indeed I can', Lloyd George was on his feet. 'Unearned increment is the hyphen in your name'.

My pen had been active stabbing at Labour's new ideas for reaping 'unearned increment' as Betterment Levy, so the opportunity for critical direct action was welcomed with open arms. The Rt. Hon. Reginald Maudling MP was a man whose outward, languid posture belied a fervent purpose, especially when he expected others to 'get a move on' under the slow tempo of his baton. He was impatient for the Land Policy Group to meet. Early in February 1965 at the latest, if you please. Mostly unknown to each other, the group gathered in an upper chamber beyond the run of a creaky lift that was ever at risk when the portly frame of the new Leader squeezed into it; the address was Old Queen Street. It housed the Conservative Research Department with its out-dated letter-heads with the name of the Rt. Hon. R.A. Butler, CH, MP, as Chairman.

Like Ted Heath, other Tory grandees of the day had large incommodious bodies – Boyle, Butler and Maudling himself. Jammed between bookshelves and the butt-end of a narrow table, Maudling, the ex-Chancellor of the Exchequer, flanked on his right hand by John Boyd-Carpenter who would cock his head like a robin to peck out the points of a discourse, deliberately opened a spate of arguments, lively, contradictory and as many-sided as the number of speakers round the table.

Parliamentarians from both Houses, leaders in the construction industry and two academics had been called to make up the company. The Chairman gave each one of us two weeks to prepare written proposals, policies and supporting observations. To avoid taxation based on the imprecision of valuations instead of on firm market prices and also to avoid the questionable notion of community-created value, I knocked out a scheme for a Proceeds Tax. The idea was to levy tax on the takings from land sales – a notion which eventually fell foul of Treasury experts. Progress was made and more expeditiously after John Boyd-Carpenter had taken over the Chairmanship, towards a coherent policy timed to be a counter to the Government's Land Commission Bill. Credit for shaping cosmos out of the earlier chaos of the Group's deliberations goes to the patient efficiency of the Secretary to the Group, John MacGregor. Now, a quarter of a century later, his is a more serious role as a Cabinet Minister.

Nearly twelve months of the Labour Government with its slender majority had passed before its wisemen and scribes produced the Government's White Paper on land policy. Hot from the press in September 1965, I snatched a copy from HMSO in Holborn and made for a bench on the Embankment to devour its contents. It was descriptive of a subtly-devised scheme to bring all those redevelopment sites visible between Waterloo Bridge and St. Paul's along with urban development potential everywhere into State ownership. Time was short; hence the haste. There were printing presses laden with my 'copy' written in anticipation. The presses were waiting to 'go ahead' or 'hold', according to the verdict of my feverish study of the Government's now manifest declaration of intent. My 'copy' became a booklet, a green-backed critique published by Aims of Industry and available in the hands of its readers a few days later. Unlike the Brighton speech and my *Land in the Market*, the 30th Hobart Paper for the famous Institute of Economic Affairs' scholarly series, this Aims of Industry pamphlet was a polemic aimed at the proposed Land Commission and the nationalisation of urban land at its hands. All three publications strategically spread over the last year of expectation had established my credentials as a conspicuous writer on the right of the political divide. The Land Commission White Paper was the Dunkirk of the struggle, the first serious offensive on the land front. Others were to follow – leasehold enfranchisement, reform of common lands and public access to private land. All these engagements meant action of some kind. The camaraderie of the Land Policy Group was missed. Soon after the second and bigger Labour victory in April 1965, the Group was demobilised in a confetti of thanks from Edward Boyle, now research chief after Ted Heath, John Boyd-Carpenter and others.

For the next two years the cafes, corridors, and committee rooms of both Houses of Parliament became nightly haunts of mine as a so-called expert adviser on land affairs, sitting in on committees, some permanent, some short-lived and some simply *ad hoc*.

The Committee Stage of the Land Commission Bill had spawned a brood of small *ad hoc* clusters, antibodies among the Opposition MPs to help combat the Government's antigens. Tory Front Bench Spokesmen with the aid of a lower command would summon these henchmen, usually at night when 'drop outs' between Division Bells gather to the basement in conclaves. There we would shape and sharpen weapons of defence and offence for the Tory members of the all-Party formal Committees that would meet next morning. Geoffrey Rippon – who had replaced Fred Corfield as Front Bench Spokesmen on Land and Natural Resources – and two 'major-generals', Graham Page and Hugh Rossi, marshalled a few lesser rankings and advisers like myself into the Land Commission bunkers. In the House of Lords, this shadow militia had its counterpart led by Lord Brooke of Cumnor and his aide, Lord Nugent of Guildford. Besides these cosy coteries, there were, in the Commons and the Lords, Party Standing Committees whose members would gather for business before 'drinks' at 6.oopm. Some of these Committees I was invited to address from time to time. The first occasion, organised by Albert Costain MP of civil engineering fame, found me face to face with the Conservative Party Housing and Local Government Parliamentary Committee. My apprehension was duly intensified by the sight of Enoch Powell, square-jawed and resolute, jammed into the back row.

The work-a-day life of an MP and indeed of a Parliamentary Peer is a contradiction of common sense. Hours spent sitting in one or the other Place are taken from time dearly needed to meet an overburdened schedule of correspondence and external engagements. Many then had no secretarial or administrative help save what could by sheer will-power, or by threats, or by love, be recruited and rewarded from a hopelessly inadequate expenses provision. The Peers appeared to be worse off than the ignoble. Lord Brooke of Cumnor, the Chairman of the Committee of Independent Tory Peers, always wrote to me in long-hand – a most impressive courtesy if, in fact, he had a bevy of stenographers at his elbow which was doubtful. Dick Nugent also corresponded by pen. Hardly ever was this the case with an MP. Sometimes an MP would operate independently. Fred Corfield (later Sir Frederick QC) had the complicated commonland legislation entirely to himself among the Opposition MPs. Statutory Instruments made under the provisions of the Commons Registration Act of 1965 were especially tricky. He appeared to think my personal drafts were worth waving in front of the Government from time to time.

Working thus with a Shadow Minister brought me, inevitably into the presence of his substantial counterpart. The Mr Real who confronted Fred Corfield was the Rt Hon Frederick Willey, MP, Minister of Land and Natural Resources. His attitude differed from the approval of my Opposition friends. Willey pedagogically wanted to know why I was doing and saying things he disapproved of. A serious, long-faced figure, he contrasted with his Parliamentary Secretary, Arthur Skeffington, a man whose affable humour disarmed the prickly ill-ease that would overcome me in the company of socialist politicians. Fred Willey, nonetheless, and to his credit, listened to my criticisms of the way he intended to register common rights and provide public access to common land; and with special intensity to my comments on his first approaches to leasehold reform.

'I thought you were a socialist,' was my benign challenging enquiry.

'Yes, indeed, I am,' he replied with a baffled smile.

'Why then,' my probing went on, 'do you propose to give some of the richest people in the realm, the wealthy tenants of Belgravia, the right to force their landlords to sell freeholds to them in a way which will give these well-to-do folk overnight windfall gains and deprive modest investors and pensioners, the shareholders of the insurance companies who own these places, of the proper value of their holdings?'

'I hadn't seen it in that light,' admitted the Minister. Nevertheless, he soon did so and clamped a stopper on enfranchisement to prevent highly-rated tenancies benefitting under his proposed Act. Sadly, I was to be hoisted on that petard of my own making some years later.

To my surprise, Ted Heath and his followers were, for the most part, pragmatists, men of the moment, inclined to be dismissive of intellectual postulates. Action ruled above theory, a distressing precedence in a country where collectivist doctrines, however covertly, had made the running for twenty years and more. Despite my entreaties in the cause of individual liberties, my colleagues of the Land Policy Group were prepared to go along with Betterment Levy, the tax on the supposedly community-created land values born of Fabian presumptions, and to cut it from 60% to 30% rather than ditch it altogether. They judged it less disturbing politically to do so. My pleadings against this and other leftist notions were often, so it seemed at the time, a voice in the wilderness crying against makeshift compromise.

All ears, however, were not deaf, as I found on one notable occasion. In February 1966, I had been invited to address the 17th Annual Local Government Conference in Church House, a kind of Conservative matching of the awesome assembly under Brighton's Dome in 1964. John Boyd-Carpenter was in the Chair. Between us at his right hand sat the Conservative Member for Finchley, Margaret Thatcher, and to his left was the fair face of Jill Knight a leading light on Northampton County Council and Tory candidate for Edgbaston at the forthcoming General Election. Once on my feet, my aim was to bring home to the faithful the practical importance of radical right thinking and to beware of a democracy which could lead to totalitarianism – too much Government, too much State interference, too much bureaucracy. The MP for Finchley, an exquisite creation in ultramarine, who had come only to make her speech, stayed to hear me out and, as she left the platform, came over to ask for my script. She apparently followed my writings in the national papers. Jill Knight, who accompanied me that evening to Ted Heath's dinner party after the Conference, also wanted the script, as she put it, to 'add to my Denman dossier'. Ten days later, Mrs Thatcher wanting to talk further about the Land Commission, invited me to lunch at the House of Commons. There over roast lamb, cabbage, green figs and Blue Nun, I had the satisfying experience of talking to a Tory MP, quick, alert and deeply concerned over how people – the demos – could vote for an oppressive state in the name of democracy and be led away by plausible ploys of expediency. We came to the conclusion that the people of Britain of the generations since the Second World War had had no alternative to vote for. Her intellect was fire, the fire that shines in diamonds.

Moods in Parliament change with readings on the political baro-meter. When on April Fool's Day, 1966, Harold Wilson was restored to Downing Street with a ninety-six over-all majority and the barom-eter reading for the Tory Opposition plunged below the 'hopeless', spirits on the Tory benches in both Houses rose. They jettisoned with thanksgiving the load of worried expediency and on-the-spot policy making. None of my friends in the Commons had been a casualty and Jill Knight had been elected for Edgbaston. They settled down to the relatively relaxed pastime of finding chinks in the Government's armour. Late one evening at the close of a day when Harold Wilson's lot had been exceptionally hard-hearted, probably over the 'charity amendments' to the Land Commission Bill, the *Evening News* had reported monks in Burma setting light to themselves in political protest. Before we broke up, drinks among Geoffrey Rippon's contingent were flowing freely in a corner of a House of Commons' riverside Bar. They were worried about the Government's intransigence. Geoffrey Rippon lent forward, his genial face set in mock solemnity.

'There's only one course of action we can take,' he pontificated. 'Quintin (the future Lord Hailsham was among the company), you are the senior statesman among us; Margaret (turning to Mrs Thatcher) you are the most beautiful. So tomorrow at Prime Minister's Question Time, Margaret, dressed in "weeds", must stand to face the Speaker and declare our intent while Quintin pours petrol over her. I will strike the match. There's no other way to illuminate their darkness.'

Somewhere about this time, Lord Kennett of the Dean from the Government Benches in the 'other Place' moved: 'That this House do now resolve itself into Committee' to take the Land Commission Bill. Now was the moment when the work which over previous weeks had absorbed my time in cabals and committees would be put to the test. Fifty-four noble Lords in 'Discontented' Opposition, led by Lord Brooke of Cumnor, although yeasty with criticisms of every kind, had determined to conserve their fire-power and concentrate it on three Amendments, each of which would, if necessary, be pushed to Division. If they were carried, small transactions, owner-occupied houses and charities would be released from the grizzly embrace of the Land Commission. A seat had been found for me in the Visitor's Lobby. Lord Kennett, known to a wider practice as Wayland Young, author of *Eros Denied, A History of Sex Attitudes Down the Ages*, marshalled the Government Peers. He presented and defended their case with polished ease and was well within sight from my north-west flank. My friends, for the most part, had their backs to me and every now and again would dispatch the Duke of Buccleuch or some lesser emissary to enquire 'How I thought they were doing'. The case for each Amendment had been rehearsed like an actor's script and my own hand had played a special part in helping arrange the tactics of the 'charities' offensive.

In a hundred and more clauses of intricate Parliamentary draftsmanship, the Land Commission Bill had a clear purpose: to impose a charge, a Betterment Levy, on every land and real property transaction throughout the realm, no matter who the parties were, an injustice which in particular caught up in its sweep dealings in land held by charities. Henry Brooke and his henchmen, without the benefit of computer science, weighed all possibilities and chose as the best advocate, qualified by station, experience and political art, to present their case and to confound the weak compromises of the Government, Rab, Lord Butler of Saffron Walden. He was, however, hibernating in the Masters' Lodge at Trinity College, Cambridge. So they detailed me to undertake a special embassade to him.

The noble Master of Trinity, a large man, well-favoured, ponderous of speech lit by unexpected flashes of puckish wit, was well-versed in the art of giving lesser men their head to the relief of his own. My credentials were sound – I had come from Brooke of Cumnor – my request seemed reasonable and my head and heart central to its purpose, so, providing I would draft a brief for him, he would see to the speech. Fortunately, John Bradfield, the ablest Bursar in Cambridge, right-thinking and profoundly astute, looked after Trinity's fortunes. What the Land Commission intended to do to charities was a wickedness which enraged his soul. So with his goodwill, able mind and pen to assist me, a brief was written for Rab accusing the Government of the patent inconsistency of imposing a land tax on charities while exempting them from capital gains tax and all other imposts.

The Land Commission debate that afternoon of early December went on and on. Lord Butler rose to speak four hours into it, long after I had left my seat among the visitors. Understandably, he was not a little put out after 'trying for some time to intervene in the debate'. Gerald Gardiner, the Lord Chancellor, referred to by Lord Hailsham some years later in an interview with John Mortimer as 'that great ass-scetic', offered a thinly-veiled reproof which was not calculated to soothe the temper of Saffron Walden.

'I cannot agree to the suggestion, as I understand it, that charities are being in some way penalised,' said the great ass-cetic. He continued provocatively, 'This is not so, I do not know how long the noble Lord, Lord Butler of Saffron Walden, has been here this afternoon. This is not a tax on anybody; it is a levy on a particular kind of transaction which arises, and arises only when the value of land has been increased by the work of the community.'

Rab's straight thinking was thus confronted with what Henry Brooke earlier in the debate had called the 'Drugged stupor of Socialism'. Unwittingly, the doctrinaire Chancellor had laid a landmine which blew up in the face of the Labour Party some years later when they tried to ward off the retrospective abolition of the Betterment Levy on the ground that it was a tax. The open-eyed Tory Government, by then returned, simply quoted back at them the immortal words of the ass-cetic Socialist Chancellor – 'This is not a tax on anyone'.

The Land Commission debates as a record of performance do not show the Socialist Lord Chancellor in a favourable light. Once he tried to tie the hands of Lord Brooke of Cumnor's Opposition behind their backs. He maintained that the Amendments moved by them were infringements of the privilege of the Commons. The naughty Lords were being asked to amend Part III of a Bill which dealt with Supply and such Amendments could, therefore, be rejected by the Lower House. Clearly, to debate them therefore was a waste of time. 'Let it be so,' was the immediate response of Brooke and his henchmen. 'The entire Bill is not "Supply" so we will use our superior numbers not to vote in the Amendments one by one but to cut out Part III of the Bill altogether. Shorn of all Supply provisions it will go back to the Commons with a great hole in it.'

In December a keepsake came from Rab, a note thanking me for the brief and rejoicing in the defeat of the Government in the Lords. The Commons, early in the New Year found the Land Commission Bill on their doormat, a callous thing with its heart cut out.

◄§ §►

About the same time as politics at home provided an exciting and welcome involvement, politics in Africa opened for me the way to undreamed of Odysseys through countries then unknown to the geography of my wanderlust, countries which, because newly-born, were politically unfamiliar to themselves. Ghana was master-minded by Nkrumah, Nigeria by Azikiwe. What the leaders had wanted to do, force the pace of national independence, had operated downwards through a long line of chained events and ultimately had affected my personal destiny. Kwame Nkrumah – Ossafago, the Saviour, to the unsophisticated Ghanaian – was cast in the mould of an implacable Third World dictator. Nnamdi Azikiwe, in contrast, was a scholar who achieved political ends by adroit intellectual manoeuvres, including the publication of the *West African Pilot*. He used politics to drive the pistons of the machinery of academic development. Thereby he fashioned for himself a 'kingdom' of many thrones from the Governor-Generalship (later the Presidency) of Nigeria to a lifetime appointment as Chancellor and Chairman of the Council of the University of Nigeria at Nsukka in Iboland. Romance and propaganda saw to it that the founding of the University of Nigeria coincided with the declaration of Nigerian independence in 1960. The repercussions caught up on me two years later.

Freed to do what she would with her own, Nigeria after independence either kept intact or transformed the good things left behind by the British colonial administration. Azikiwi let go what he didn't want, which was not very much. Nigeria listened, also, to new voices of people who at her invitation came from Britain, America and elsewhere to give advice, offer the hand of cooperative fellowship and bring with them an unprecedented eyeball-to-eyeball approach in their dealings especially in the realm of higher education. Among the transformations ranked the Nigerian College of Arts, Science and Technology with its separate branches at Enugu, in the east, Zaria, in the north and Ife in the west. A Commission on higher education with Sir Eric Ashby, Master of Clare College, Cambridge in command had in 1960 recommended the absorption of the branches of this old colonial establishment into the new university system.

Azikiwe, opposed to anything that would compete with or rival his University of Nigeria at Nsukka and rather than suppress the College at nearby Enugu, incorporated it into the University as a separate campus. Politically, the link-up made sense but it created an awkward administrative problem, as the heart of the new University stood on Azikiwe family land forty miles into the bush from Enugu. The Enugu campus was thus held at length from the centre of operations at Nsukka by miles of laterite road and tended to generate a life peculiar to itself. My job was to go there, and advise on conceiving and delivering a new birth. The commission meant replacing the elementary teaching in Estate Management and Surveying founded in the old collegiate days with a school worthy of the new University. The new creation would be exclusively at Enugu and would serve the needs of the whole expanse of federated Nigeria. Neither the Inter-University Council nor Sir Andrew Cohen's Department of Technical Co-operation (later the Ministry of Overseas Development) which jointly had asked me to accept the West African assignment had any idea that my going there would forge a personal link of thirty years standing.

The thirty-three pages of Analysis and Curricula in my final Report would have been nothing more than waste paper without teachers to handle the proposed courses. Sadly, at Enugu none of the old staff was suitable and some had gone back to the UK. My aim from the beginning was to avoid appointing expatriate staff and to go for Nigerians wherever possible – a policy at variance with the accepted staffing mode in the University. Among the newly-qualified practitioners in Nigeria were three graduates, two from London and one from Durham. Of the London men, John Umeh was a new recruit to the Enugu staff. He was able, good looking, exceedingly enthusiastic and a traditional Ibo of the Ibos. To please Azikiwe's impatience and get things moving, my proposals provided for fertilisation of new ventures from established Faculties over an initial two years. John Umeh and others were to come back to Cambridge, there to be better schooled, polished and initiated into the mysteries of academic life. No time was lost. Before the year was out, the new courses were being time-tabled and taught, albeit on American lines.

A somewhat similar metamorphosis of college lava to University imago occurred in Ghana. The Technical College at Kumasi became the Kwame Nkrumah University of Science and Technology. The overseas aid people, having discovered me for Nigeria, proposed a parallel service for Kumasi. Nigeria and Ghana consequently became for me the termini of a West African shuttle that ran twice yearly, sometimes more frequently, until the Biafran conflict shut down temporarily the Nigerian station.

Engineers dominated affairs at Kumasi, especially a Dr Levine acting pro-Vice-Chancellor. He was a civil engineer from Israel who had forgotten how to be civil. Fortunately, leave removed him before I could pack my bags in justifiable pique at his rudeness. My reception was reminiscent of early Cambridge. Although the Vice Chancellor, Registrar and others in authority had asked for my assistance, Levine seemed to resent my visit and was derogatory of Estate Management as a companion subject to Surveying. Twum-Barima, the Deputy Registrar, a splendid Ghanaian well-rounded in soul and body, came to the rescue. He took me off to tour the country from Accra in the south to Tamele in the north, introduced me to qualified Ghanaian surveyors and valuers in Government service and generally smoothed a pathway to success. Apart from the temporary impediment of Levine and unlike the Nsukka scene, native scholars led by R.D. Balfour, the Vice-Chancellor, were in control. Also unlike affairs at Enugu, there had been no earlier courses in Estate Management to be dismantled and rebuilt. Foundations had to be dug and something new and in a measure unheard-of erected upon them. Among the qualified surveyors of the Ghana Land Secretariat was Ebeneza (Ben) Acquaye, a London University graduate, clever, handsome and ambitious and willing to venture out on the Kumasi enterprise.

Another promising introduction was Yaw Asante, also a London graduate, goggle-eyed, querulous, genuine in heart and an Ashanti traditionalist. Professionally, at the time, Yaw was City Valuer and Town Planner of Kumasi. He, like Ben, showed much interest in the prospect of a teaching career in the University. Arrangements were made to do for them both what had been already done for John Umeh and to bring them back to Cambridge. Ben came first and Yaw later.

When I first I set foot on West African soil, despite the obvious influence of British administrators, missionaries and traders, Western ways were still only a thin veneer covering the ages-old cultures of traditional African peoples. What was exotic, especially the technical, soon acquired indigenous forms, features and African mannerisms of its own; where else in the world would Bedford lorries out of Luton become Mammy Waggons, those bulging loads of squirming humanity trundling over the laterite from market to market? Sadly, within a year of my first visit, changes for the worse occurred. Old rest-houses, often of timber and red daub where, in close proximity to Hausa travelling merchants and their wares, one slept cocooned in mosquito-nets in the still heat of a tropical night, had been replaced by hideous concrete hotels. Flying beetles colonising the earth floor and primitive pedestal of a picturesque rest-house loo are an art form; beetles crawling up the whitewashed concrete of the 'hotel' toilets are a Western obscenity.

By coincidence, my brother-in-law, Dick Prior, had some months earlier taken charge, as headmaster, of a native, upper-grade school by Awka in the Ibo bush. One night he put me up in the school living quarters, an abode close to the soil and the African way of life. Unglazed wall-apertures open to the night air served as windows. Candles and hand-swung lanterns were the only illumination. Nights too hot for bedclothes were for me too hot for mosquito nets also. The bane of the night was the darkness which dared not be lit up. Ignorant of this, I had gone to bed swinging a lantern. Its crescent of light unextinguished, the lantern was put on the floor between the wall-apertures open to the insect-infested night and the trestle bed. Five minutes later, the king of moths, huge, fashioned in size and form to resemble a US B-52 Bomber attacked the beacon of light head on. Sleep was out of the question. Squadrons of lesser flights were now buzzing round the ceiling. My brother-in-law had been a Wing Commander, but what could he do? I left him to his slumbers and armed with a large, zinc washing bowl made for the gigantic invader. After an hour of cunning, many oaths and infinite patience the 'bomber' was captured under the up-turned bowl. The gigantic wings beat against this zinc 'hanger' and the floor at frequent intervals throughout the night, kicking up a shindy like a back-firing racing motor-bike. The creature was still at it when a welcome dawn rose above the paw-paw trees. Despite the bombing raid, the living rooms of the Okonguu Memorial Grammar School had a lure and charm denied the new hotels in Lagos.

Colonial administrators had, for the most part, wisely respected, protected even, native ways and customs. It was in the wake of foreign merchants and their markets that Western ideas had infiltrated native society. The missionaries, however, with their strange and foreign practices tried in certain directions to force the pace of change. Mistakenly and with more bigotry than benison, they would identify Western social postures, outlooks and values with the truth of the Gospel of Christ they had come to bring. Tensions, clashes and wounded consciences followed where peace, love and harmony should have reigned.

Shortly after arriving in Enugu, I was invited as a visiting speaker into the thick of these controversies on the Ikpa Mission station northwards towards the Benue River. 'Brothers' in collar and tie and 'sisters' with mock cherries in their Sunday bonnets newly-arrived from some American small-town suburb or Wimbledon side street were insisting that native chiefs and lesser men long-wedded under polygamous customs to bevies of beauties should shed all except one of them and go through the colonial monogamic rite of the Marriage Ordnance of Nigeria – which incidentally only applied to Lagos Federal Territory. Other missionaries, born and bred in the place, were standing against the 'Wimbledon Bigotry'. They knew full well that wives abandoned and divorced in droves – a rejection, by the way, contrary to the strict injunction of Christ himself – would be placeless outcasts in local society; that the Gospel of Christ was not conditional upon monogamy; and that the Scriptures took no sides between polygamy and monogamy. Much more disturbing was the manner in which the missionaries when in congregation sat apart from native worshippers. As I squeezed myself on to the native benches, intent on breaking the mould, an awkward silence descended. The jolt was only momentary; the chatter, laughter and feeding of babies on either side soon started up again.

Prancing 'masks' of the Ju Ju, the embodiments of the spirits of the Ibo *mmaw* one evening burst from the bush on to the Nsukka road. Their veritable *coup de théâtre*, alerted me to the numinous powers that haunt the hollows of Iboland. Yet it was not this display of the animistic which stirred me, so much as the astonishing deep wisdom of the Ibo parables. These revealed how impregnable is the citadel of the traditional Ibo soul to the false doctrines of socialism from the 'developed' countries. 'My father,' said the young chief, and expounder of these parables, 'took me as a boy to what he called *good bush*. "What do you see?" he asked. "Tall trees, lesser and small trees", I answered. Then he took me to what he called *bad bush*. "What do you see," he asked again. "Poor, spindly trees, dwarfed and all of equal height," I told him. "Learn the lesson then: tall trees, strong and well formed are the élite trees, necessary to protect the others which grow under the arms of their protection. Only in bad bush are all the trees equal, equal in weakness. Look at my fingers," he continued, and held up an open-spread hand. "Are they all equal? If they were, my hand would be weak like the trees of the bad bush. My fingers are unequal and so together they may give strength to my hand, my wrist and arm. Beware", he added, "of the white man's socialism."

'And of the white man's technology' could have rounded off his adage. It could be more damaging than any irate, prancing Ju Ju. One evening, Dr Bevan, the pro-Vice Chancellor of the University of Ibadan, invited me to dinner. To keep the rendezvous meant crossing the Ibadan campus from the Guest House in deep darkness before the moon rose. In the distance Bevan's bungalow's lights beckoned. Oblivious of cobras and other horrors, my stride made straight for the lights. Four minutes later, I lay a crumpled heap of blood and agony. The razor-sharp rim of a wide concrete flood ditch, the gift of modern technology, lay hidden in the darkness, a menace and a man-trap to all unwary night-walkers. Cut to the shin-bone and spouting blood, I arrived at the Bevans' dinner party as the guests were making for the tables. Happily the local surgeon, Sandy Stuart, was among the company. In a trice he had me whisked away to hospital. Six stitches were necessary. A badly shaken, pale Englishman and a clever, kindly Scots surgeon returned to cold soup and warm consoling friends two hours later. The whole evening, they claimed, had been livened by the episode. Next morning, with a leg swathed in bandages, I limped down to breakfast at the Ikoya Hotel in Lagos to learn of the dastardly murder of President John F. Kennedy. My tiny misfortune fell into right proportion in this world of unexpected tragedies.

Ghana's fortunes would have fared better if Nkrumah had known and heeded the Ibo parable and had put aside his lecture notes from the dismal teachings of Professor Harold Laski at the London School of Economics. A slave to Leninist socialism, Nkrumah by assuming the mantle of a peoples' saviour had made Ghanaians slaves and stifled the free spirit of Ashanti, Fanta and other peoples of Ghana. In Accra I met the same cramping of the spirit, the same claustrophobia that had assailed me in Poland and in other socialist countries of Eastern Europe. Ghana in 1964 was a police state. One evening after dining with Frances Posner, Visiting Professor of French at Lagos University and wife of Michael, a friend of mine among the Fellows of Pembroke, a table of Ghanaians brooded over by a white presence beckoned me to join them. Twum-Barima, the Deputy Registrar of Kumasi was there, Ben Acquaye and others from the Lands Secretariat. The white face belonged to Pogjucki, Nkrumah's Land Policy Adviser, a renegade Pole yet possessor of a British passport. He wanted to know what arrangements were intended for the undergraduates at Kumasi attending my courses to have sessions on Nkrumah-Leninism at the Winneba Ideological Institute. None I assured him, adding that was far from my aim. He lent forward and with a bent smile said, 'You must come and see me soon. Perhaps we could eat together one evening.'

At that point, Twum-Barima's foot kicked me under the table. 'I am in the hands of the University, you know.'

I took the hint and brushed off Pogjucki; then turning the question to Twum-Barima, asked if there would be time to meet Mr Pogjucki.

'Certainly not.' The Registrar was emphatic and pushing his chair back suggested, almost in the same breath, that it was time for us to go. Pogjucki from under lowered lids addressed his gaze to the Ghanaians and said, 'I see. All from the UK. Well, well, well, I shall not forget.'

Once outside, I pleaded with Twum-Barima not to let me cause any awkward unpleasantness.

'Don't bother,' he said. 'You don't understand. Had you accepted Pogjucki's invitation, he would either have had you put on the next flight out of Accra for the UK; or, more likely, have hiked you off to the Castle to be incarcerated at Nkrumah's pleasure. You belong to us. You must come up country with me. Avoid Pogjucki at all costs.'

Next day we were on the road. Among the sights encountered was an immense, white palace high on the slopes of hill country overlooking the coastal plains. 'That's Nkrumah's palace,' Twum-Barima pointed it out. 'I am a socialist; but he's the other kind of socialist.' Twum-Barima had never been to Lagos. We went there together some days later. Overwhelmed by the bustle of the city, its well-stacked stores and general appearance of ease and plenty, my friend declared, 'I'm a capitalist now. This is best for Africa.'

On returning and to hide me from Pogjucki, he took me at night to Kibi to meet his mother for whom he was building a bungalow. 'Is she a widow?' I asked, showing my profound ignorance.

'Oh, no,' he patiently explained. 'My father lives on the outskirts of Kibi with his sisters. Complying with Ashanti custom, he is a peripatetic husband with sixteen wives of whom my mother is one. He visits them as spirit and the flesh desire. Like all loyal sons, it is for me to look after my mother. Come and see.' He was most proud of his filial beneficence; and I was intrigued to encounter an Ashanti version of African polygamy.

Twum-Barima had read history at Leicester University and was of the third generation in his family to have a University education, starting with grandfather at Fourah Bay. He knew his job but never ceased to express wonderment at my own performance. As we journeyed north-westwards over the savannah, south to Sekondi, along the coast and north to the hills about Akosombo, he was astonished at my questioning of clerks, civil servants and commissioners. 'You put questions as if you had been here many times before. How is it?' he would ask. I gave him the enigmatic smile of one who knew something of the universal mysteries of land economy which were so far beyond the understanding of Peter Laslett and other of my critics at Cambridge who in ignorance limited my knowledge to drains and dustbins.

At Yedi on the road north to Tamele, we crossed the Volta in ten minutes, paddling in a crocodile boat. Crossing the same water years later, it was possible to measure the pace and size of Nkrumah's 'developments'. By virtue of them, at the same spot, the Volta, no longer a river, was 30 miles wide. At the Volta Dam, even on that first tour, it was borne in on me how the narrow and short-sighted outlook of expatriate 'professionalism', given to satisfying a Dictator in a hurry, could mortage the future well-being of a country to the dangerous ostentations of a present-day fanatic. The Volta was destined to become a vast inland sea 250 miles long, inundating hundreds upon hundreds of indigenous villages, disturbing the age-long settlement of one in a hundred of all resident Ghanaians and drowning 4% of the land area of the country. When I saw F.J. Dobson, the Canadian engineer in charge of the Volta project, and enquired about the long term land policy of the Volta River Authority, so as to make staffing estimates of this devastatingly ambitious undertaking, his response was, in essence: 'Say, don't ask me. My job is to get those turbines turning. That's what matters.'

Under his company's contract, huts and cabins had been built for thousands of Volta River Authority employees, virtually a new town of shanty quality whose hutments, when the work on the Dam was finished, would be given as compensation to families displaced by the Volta scheme. An indication of the prevailing chaos was apparent when Dobson revealed that the land on which the shacks were erected by the VRA did not belong to the Company and nobody seemed to know whose it was, or even to care. And yet his estimate of future needs for qualified land managers was nil. His was another case of River Blindness, but of an exceptional kind, a variant of the species which attacks engineers and economists, a blindness that fails to see the crucial importance of land and its proper management to the well-being of a nation's economy.

A series of delays had held up my getaway to Ghana in 1964, not that the country in any way took second place to Nigeria in my affections nor because postponement from spring to November meant dodging the rainy season. A long-planned return to Poland seemed possible that summer. To get to Poland had taken two years of playing a kind of rummy, collecting and exchanging a series of application forms and visas with embassy officials in London and Warsaw, manoeuvres complicated at the Cambridge end by the Land Economy Tripos politicking, by illness and other vicissitudes. First Jeffrey Switzer and I said 'Yes' to a fortnight's lecturing; then he cancelled the second week in favour of Colin Kolbert who cried off at short notice. At the last moment, Neil Elliott grabbed the Kolbert visa and flew into Warsaw in July 1964 to meet Jeffrey, Popkiewicz, his son and myself. We had motored up the Vistula valley from Wrocław in my new Wolseley 2600.

Jeffrey and I had broken the outward journey from Cambridge for a day to sample the now more familiar fascinations of Prague. Over seven years, its mediaeval splendours had suffered more neglect under communism. Aristotle's warning: 'That which is most common has the least care bestowed upon it' was manifest along the highways, in public gardens and buildings but most noticeably in the grounds of the State-run Hubertus Hotel. There, a memorial to free enterprise, rampant weeds were competing for the sunlight with smothered tea roses and other barely visible flora of June. Pastmasters now at the art of crossing Czechoslovkia by car, we left the country like the Pied Piper. In the wake of our gleaming blue and grey Wolseley ran strings of kerbside, goggle-eyed citizens and children. We made for the Polish border at Kwdowa Zdroj and reached it in the early afternoon. At Wrocław behind the swing-gates of the Popkiewicz drive (an élite entrance in a communist world) the warm welcome of Joseph and his family awaited us. Once there the rush was on. With no time to count the bouquets (or brick-bats) which might have rewarded Jeffrey's lecture next morning, the Wolseley was lined up at a benzina station, pointed northwards and, with Joseph Popkeiwicz at the wheel, whisked out of Wrocław on trek to Warsaw.

Having handed over the Wolseley to Popkiewicz, our destinies were his. Much he made of them including a two-days sweep through south-eastern Poland to meet shame, beauty and a dull-witted

175   *Politics in Black and White*

police. Oswiecim on the upper reaches of the Vistula to the west of Kracow marks on the map the site of Poland's Belsen, a nadir of human suffering and shame, a grim, yet not inappropriate epitaph to the thousands who fell victim to Hitler's inhuman Gestapo. Over the entrance to this pit of past despair, the legend '*Arbeit machen frei*' wrought in iron still stood, a perpetual, cynical testimony to the devilment once practiced within. In speechless incredulity, we traversed the maze of structures yet intact and tried to take in the significance of the gruesome relics of human hair and discarded spectacles, the illustrated records of the death-masks that once were human faces, the torture cells and the gas chambers. Eventually, as we left, Popkiewicz said, 'We were, you see, on the wrong side geographically of both Germany and Russia.'

Cracow, in contrast, came as a healing balm. It was the most beautiful city in all Poland to grace our spontaneous tour. A cathedral and lesser buildings, whole chapters of history wrought in stone, wide open squares bedecked with trees and flower-beds give vent to seductive alleyways and hint of romance, fortitude and undying faith, the virtues of a deep-rooted Christian culture. Communist vindictiveness had deliberately built the grim, industrial eyesore of Nova Houta alongside this mediaeval gem with the object of destroying its radiance. Fortunately, the radiance was enhanced by the contrast.

At the approaches to Radom, two-thirds of the way to Warsaw, Popkiewicz, who was as happy with the Wolseley and its automatic gear-box as a youth with his first banger, suddenly stopped, got out and beckoned us to follow. 'They want to check-up on your whereabouts,' he enlightened us as we marched into a grimy hutment, obviously a police checkpoint. Grace and beauty were not the strong-points of its décor. A square-faced, pig-eyed *komisarz* sat at a table on which formidable data-forms served as table linen. Given one each to complete and sign, we joined the commissar and Popkiewicz round the table. To the immediate right of Popkiewicz, my place had the advantage over Jeffrey's. His smiling face was two seats away on my right. With Popkiewicz to help me, my task was soon finished and in the Polish language to boot. Jeffrey, hearing only the questions being translated, filled up the answers in English.

Against the question 'Name of employer?', he had written 'Cambridge University, England'. Immediately underneath was the associated question, 'Nature of employment?' To this he had answered 'Illiterate'. The policeman snatched up the completed forms, detained Popkiewicz and ordered us out of the room. Sometime later Popkiewicz emerged with a grin on his face. 'Why so long? What happened?'

We never supposed the *komisarz* would bother to read the forms. 'He glared at me,' said Popkiewicz and wanted to know 'What the capitalist pigs had written.' It had to be translated for him. Surely not literally? How did you translate Jeffrey's "illiterate"?'

'Oh,' said Joseph with a twinkle, 'I said that meant "I am an economist".'

Confident that no self-declared economist would wantonly waste valuable time hanging around a police-station, we pushed on to the Capital, a merry party of Poles and capitalist pigs. At Warsaw airport one capitalist pig was exchanged for another – Neil Elliott, whose plane returned to London with Jeffrey on board.

Neil Elliott, six foot two inches in his shoes, lent over Popkiewicz like a mature Sequoia Gigantia exchanging pleasantries with a holly bush. Joseph returned the lofty greeting in amused bewilderment. He and his wife, when making the bed up for the new visitor, had been told to expect a kind of diminutive troglodyte. At the airport greeting, my face wore a bleary smile of uncertain purpose. Earlier in the day while alone in a restaurant, my teetotaller's ignorance had mistaken Slimovich for orangeade and a glassful of the stuff was having its inevitable effect.

Back at Wrocław, the beds were remade to fit the new arrival to the tune of clinking glasses of cherry brandy and vodka. Between the accidental Slimovich and the 'medicinal' brandy, our travels had run through hundreds of miles of Polish countryside, lineaments of unbroken fields under tall stalking skies. The open fields kindled a memory of the lyrical lines

> In the dark Middle Ages,
> If you trust to history's pages,
> You may search the landscape round,
> Not a hedge is to be found,
> Instead of my field, your field, their field,
> All is one enormous bare field.

The lines had brought home the full meaning of Professor Stys' words when, in our former visit, he had told me how the Polish peasants in Stalinist days would trick the *Schwarzmann*. As this official went round checking up the collectivisation of peasant holdings, searching for boundary stones which were in their lifeblood and which the communists had ordered to be removed, the peasants hoodwinked him. Behind the back of the *Schwarzmann* the lands of the collective farms were divided as of old by blocking the drills of the sowing machines, and, thereby, sowing demarcation patterns in the wheat and other crops; and by daubing walls with coded cement and paint markers – this is for Rudolf, this for Jozef, this for Wicktor and that for Wenus and the corner for Zuzanna.

The inherent individualism of the Polish peasant has a parallel, not a strict similarity, in the way the farming tenant of the manorial commons of England and Wales has always jealously guarded his proprietary rights in the lands of the manor against the encroachments of the lordship and against his fellow tenants and commoners. Because, at the time, I was half way through the extensive research study of common lands, I was supposed to know something about them and, hence, was paraded as an expert on the bill-heads announcing my lecture on the subject to the learned of the University of Wrocław. Popkiewicz, however, thought empirical studies of the farming communities of the foothills of the Tatra mountains should be first on the agenda – besides, he wanted to ride up the mountain railways and sample the eternal snows. So after only a day's rest, it was away again, this time southeastwards to Zakopane, the renowned mountain resort of Poland.

All went well for the first day and night, until the thunderstorms struck. So abominable did the weather become that a dash for home was preferable to mountaineering. Thor in his anger had other ideas. We might have confounded the intentions of the God of Thunder had not Popkiewicz driven the Wolseley as if it had an outmoded gate gear-change. Subconsciously using the brake as a clutch pedal, he would press the accelerator simultaneously. His artlessness, the thunderstorms and the rain-swept road surface spelt disaster. Crossing a bridge humped over a railway ravine, Popkiewicz's double act sent the car into a 90° skid. It hit the bridge parapet with a murderous crunch, bounced off, hit it again, then swung another 90° to make a full right about turn.

Horror upon horror – 'Where's Marcek?' Joseph's son one moment at his side in the front seat, the next was nowhere to be seen. He was lying on his face two hundred yards down the road but mercifully suffering only severe shock. Within minutes, like vultures on red meat, locals of all descriptions were crowding around the miserable spectacle of the crumpled Wolseley whose near-side panelling and radiator had been bent inwards over the engine exposing the steaming entrails. Panicky, irrational, uncontrollable thoughts piled in, one on another. Elliott, all six feet of him, with phlegmatic, indifferent English *sang-froid* declaimed, 'Well, get yourself out of that one!' and retired to the intact back seat with an Agatha Christie. Popkiewicz had disappeared.

Three quarters of an hour later, the portly Professor was observed in the distance with a military figure striding at his side.

The crippled Wolseley surrenders to the Polish Army

On reaching the wreckage, he introduced the Field-marshal of the Polish army. From then on the military took over. An indefinable contrivance – a gun carriage, may be – hooked the woebegone Wolseley to its backside and dragged the bulk some miles to a collective farm on the outskirts of Nowy Targ, the nearest town of any size. Farm mechanics swarmed over the wreck like ants. We left them to it and were taken by the kindly Field-marshal and his aides back to Zakopane.

Next morning saw the classless society of socialism in action. A day in the mountains had been planned by Popkiewicz and his friend the Field-marshal. From daybreak, the proletariat had been massing round the foot-station of the mountain railway. To Elliott and me, it looked like the makings of a morning wait to get near the booking office. We had forgotten that with Lenin some are more equal than others. Up came the Field Marshal swinging his baton. The 'people' were swept aside to allow the Professor, the Field-marshal and the two capitalist pigs into the empty carriage. As we climbed into ever more rarefied air, splendid views over the endless plains below passed in sequence between the pines to take us to an eating house and local observatory among the timeless snows.

On the collective farm, meanwhile, the bashed-in radiator of the

Wolseley was being straightened out, its bits welded together and the entire lighting system of 'Ursus', the collective farm's main tractor, fitted into the wounded car in place of its own discarded kit. It assumed the image of a gigantic blow-fly. Late the following afternoon this amazing apparition took to the road. Because the bonnet flap had been fastened to the chassis by a spring cable, the flap blew up to block the vision of the driver whenever he accelerated above 20 mph. Nevertheless, now on the road with a prayer of thankfulness in our hearts and praise for the wonderful Polish improvisors, we entered the long night. By 6.00am next morning, weary beyond telling, starving and dirty, *chez Popkiewicz* regaled the party with the best of breakfasts.

That evening there was no ducking my scheduled lecture. Neil Elliott was allowed a night's rest. He had come equipped with a Polish translation of his text and a pack of slides. The slides were chosen with the relish of deliberate mischief. With a subtle art, they depicted the glories of a feudal England and the generosity of the Duke of Portland towards his retired tenants. Next morning his audience were subjected to tables of impressive production statistics of British farming and selected slides of Lady Anne Cavendish-Bentick riding to hounds, Dunrobin Castle and the interiors of the Portland tenants' bungalows. The younger elements responded with interruptions of 'propaganda'. In telling contrast, the same pictures were to evoke wild enthusiasm in a gathering of elderly, Hungarian academics in Budapest the following autumn.

Stress, strain and a vestige of shock, the pleadings of our generous Polish hosts and the glint of magnificent shop-displays of cut glass selected for export conspired to give a day of relaxation as a prelude to the return journey. The ultimate payment for the leisure was enforced idleness throughout the daylight hours of the next day at the pleasure of custom's police at Kudowa, the Polish Border post facing Nachod.

By holidaying for a day in Wrocław our visas had expired. Just one day was long enough for the Polish police to regard 'Ursus', the renamed crippled Wolseley, and ourselves with the gravest suspicion. Cuttings from the local press giving news of our visit and the lectures were only half convincing. Telephone calls to Wrocław which took all of the afternoon were necessary before the police let us go.

In contrast, of all the people concerned with the Wolseley débâcle and its consequences, those who were most entertained by it were the customs officials at Dover. They listened to the tale with mounting merriment and were so enthralled that the precious, exceedingly attractive and rare Polish cut glass piled on the back seat passed through the barriers with a wave and a smile. One vase alone, for which in Poland we had paid the equivalent of 25/-, was valued immediately in Cambridge at £50. Capitalists can do well in communist countries.

◦§ §◦

Notwithstanding the wave of democratic goodwill which, at Cambridge, had swept aside the opposition to the Land Economy Tripos, the countenance of certain of the inner circle wore an attitude of disfavour, with a particular frown reserved for myself. My status was paradoxical to my responsibilities and was perhaps deliberately permitted to be so in the hope that I might falter in the discharge of them. Truth could not blink the fact that it had been my thoughts, actions and policies over many years which had been the creative agents of the new learning, honours degree and Land Economy Department. Yet, after acting as a caretaker on probation for a year, I was appointed Head of Department but at a level no higher than a University Lecturer. Circumstances thereby required me to function as a Professor, fulfil the statutory obligations of a University Lecturer and college supervisor, carry responsibility for departmental administration, staffing and financial oversight, be a fulltime Director of Research and formal international advisor. There was, however, satisfaction in being unique, in overcoming odds calculated to undo me and in having the backing of a growing body of influential opinion at home and abroad. The authors of that opinion were at a loss to understand how the University could justify the continued paradox of my position.

Suddenly, by God's grace and the good intentions and favour of patrons I could neither name nor know, a consoling benediction and honour was bestowed upon me, as unexpected as it was welcome. Among the postman's offerings on the doormat one March morning in 1962 was a letter from Sir William Hodge, Master of Pembroke College. He was anxious to know whether I 'would allow us to propose your name to a College Meeting for election to a Fellowship here'. The courtesy, kindness and recognition afforded me by the Master and Fellows of Pembroke completely altered the meaning of life at Cambridge. It did so then and has done so every year since. New energy, confidence and inspiration enabled me to laugh at the University and its strange, unfathomable ways. Pembroke has been a haven of happiness, a home and an honour; the University was ever a battleground. To be offered a College Fellowship meant crossing the chasm which in the esoteric life at Oxford and Cambridge marks off the higher from the lesser privileged. Many value election to a Cambridge College Fellowship above election to a Chair.

As a University Lecturer, I had no claim to a Professorial Fellowship and as a Director of Studies at Pembroke, before the birth of Land Economy, there was little I could give by way of tutorial help. All was of grace. To this day, I have no notion who my backers were. The happy ignorance counterbalanced the darkness which enshrouded the identity of the University mandarins who in those days were blocking my advance.

Emergencies and the unexpected can weigh more painfully on the shoulders of an acting-Head than they do for one endowed with established authority. The pain of the unexpressed was acute in the summer of the caretaker year. Familiar acquaintances suddenly and unexpectedly took on a surrealistic unfamiliarity. E.F. Mills, known to his friends as John, had been appointed to the newly established Directorship of Estate Management in charge of the old professional service – the restraints of an 'acting' status did not incommode his office! John had slipped off for a sunny break and left me, willingly enough, with the double task of 'caretaking' his new Service and heading my own teaching and research Department.

Appointed to the staff of each was a brilliant architect whose pastime was absorbed in formulating symbols of a colour spectrum to match the tonic symbols of music. Out of deference to his memory, for the purpose of this narrative, he shall be called 'Richard Wrathbore'.

When his feet were on the ground, Richard was a very efficient architect. At the time, he was engaged in redesigning and renovating the flat which the University offers, each year, to the visiting Pitt Professor from America. Well in time for the coming academical year, the Professor had arrived in Cambridge and had called at John Mill's office to be told that Richard Wrathbore would be waiting upon him at the flat in Chaucer Road. Two hours later, the visitor returned expressing satisfaction but when asked if he had been met by the architect Richard Wrathbore, he replied with a touch of startled amazement, 'Why no. We were met by a charming lady who had introduced herself as "Alice Wrathbore!". Richard had in the course of the morning assumed an *en travesti* mode and transposed sex roles.

Two days later, to my relief, John Mills returned. Nothing was bruited abroad of these strange occurrences. 'Richard-cum-Alice' had disappeared. Discreet enquiry eventually traced the fugitive to a village on the outskirts of Cambridge. Through an intermediary, arrangements were made for a meeting. John Mills and I embarked on what was, for me, and I believe for John also, one of the weirdest encounters of our fairly mature lives. On the outskirts of the village, behind a large landscape window, was plainly seen as we approached the house the seated form of a woman playing the piano. She answered the door in response to our knock. The eyes and broad lineaments of the facial features of our colleague looked out on us. Otherwise, the dress, the posture, body conformity, womanly up-lift, mannerisms and tears were of a not unbecoming female whom we had never met before – and yet, yes, there was a dreamlike, sisterly resemblance to the Richard we had known. She sat us down for tea after a round of banal greetings. Our purpose was to seek and find the truth while wishing in no way to be unhelpful. Alice's, we learnt, was a sexual identity change, clinically monitored and assisted by top medical supervision and advice. The truth, nevertheless, opened up a long vista of problems to which we could see no immediate answers; driving licence identity? lecture time-table identity? stipend identity? income tax? status in Trinity Hall, a men's college! to say nothing of family relationships – Richard had daughters and a wife. Time found its own answers. At this juncture, we could only sip the Earl Grey, converse in sympathy and make suggestions and proposals. Condolences were *de trop*. Our attitude chimed in unspoken speech with the ambient sense of convalescence. A malady had been cured; a twist in the psyche unravelled. While Richard had been tense, neurotic and anxious, Alice was calm and tranquil. She had come to herself and exuded a palpable serenity. We left, waved goodbye, and heard as we crossed the road to the parked car, the piano start up again.

The Wrathbore episode was handled with most commendable expedition and kindliness. The Vice Chancellor of the day, Sir Ivor Jennings, took matters into his own hands and did not leave them to the mercy of University politics and bureaucracy which to my personal and painful dismay can be soulless, exasperatingly distant and run to the point of appearing inadvertent and indifferent. The protean character of University boards, committees and councils, changing constitutionally year by year as new members come and old ones go, are wholly dependent on briefings (notes) from University officials and servants, not only for intelligence about matters immediately on hand but, as is often the case, for leads on the make-up of advisory bodies on special long term issues.

The seven-years struggle over the status of the Head of the Department of Land Economy shows up something of this sad propensity. Estate Management pursuits at Cambridge were divided into two and the purely academic hived off from the professional to leave an exclusive teaching and research Department with a daunting innovative programme. The Secretary General of the Faculties, so it appeared to me, had been able to persuade his masters, the General Board of the Faculties, that, in principle, the powers of creative conception, application and leadership demanded of anyone required to design and build railway engines and rolling stock, plan and lay out the railroad and, eventually, control the railway system were of a level no higher than the ability of an engine driver. The Board of Estate Management, at the time, in a communication drafted by Kenneth Berrill (later Sir Kenneth of City fame) a member of the Board, had gone to no small pains to try and enlighten the General Board that:

> 'the tasks of devising a suitable honours, course and of directing the Department in this new venture in its early years will require a Departmental Head with outstanding qualities. This is not a task which can be expected of a University Lecturer temporarily filling the post of Head of the Department with a nominal payment for administration.'

The General Board would have none of it. Rather like Pharoah of old, they rejected Berrill's perfectly reasonable request and increased the burden of slavery by refusing any specific elevation of status and reducing the post to a temporary acting Headship.

Going into battle for the Land Economy Tripos without a properly commissioned high command weighed upon me with a sense of injustice which could have been assuaged a little if the General Board had by right thinking given reasons for their cold indifference. We could only assume there was no reason, only prejudice. From the viewpoint of their seats round the table in the Old Schools, not a few of its members were happy to judge our claims and aspirations as baseless forlorn hopes and our case unworthy of the merits we placed upon it. Again, the horizon would have been brighter if consequent upon our winning the laurels of victory, we had been promised an appropriate higher command. As it turned out, the cunning of some devious mind saw to it that my mean status, improvised and justified on the grounds of an unforeseeable outcome when the struggle was on, could be taken as a guide to the future.

A University Lecturer who in the past had led the troops into conflict and on to victory on the evidence of 'his very success' could lead the Department into the future. By such chicanery, the newly-born Department of Land Economy was denied a senior post to be occupied by whoever was in command. The injustice served to sow the seeds of future contests with the General Board in efforts to rectify the deliberate belittlement of the Department and of my status as Head of it.

Although the level at which the Head of the Department of Land Economy stood in the ranks of academics was a matter of intense personal concern to me, it also profoundly affected my staff, especially those whose activities ran beyond narrow domestic confines. Within the University itself, the Department of Land Economy was the single exception among other Departments carrying responsibility for honours teaching and research to have neither a Reader nor a Professor. For the Board of Land Economy to have lain quiet would have put a permanent smile on the faces of certain mandarins at the Old Schools. We, however, were made of sterner stuff. By the summer of 1964, the prestige of the Department of Land Economy and the lowly academic status of its Head showed definite imbalance. Teaching officers on the Board acted to press their colleagues to do something about it and without delay.

A two-pronged attack was mounted. Each arm pursued its own circumlocutory route. A Readership campaign in April 1965 obliged Board members to write cryptic letters to Sir Joseph Hutchinson, the Professor of Agriculture, and to Trevor Thomas, the Bursar of St John's College, proposing names, if they so wished, for a 'named Readership' in the Department. By May, a unanimity had distilled through the process and my name was put forward to the General Board in June for a Readership. Now, indeed, it was high summer and in the sleepy days of the Long Vacation sloth was acceptable. Although the pace of the General Board was never electric, there is no excusing the one and a half years silence which followed, a silence not even punctuated by an acknowledgement. A familiar, stark three-line rejection arrived in August 1967!

During those months of dilatoriness and in contrast to their inertia, the case for a Professorship of Land Economy, the object of the second of the two campaigns, had made considerable progress. After a series of rebuffs throughout 1965–66, the University, by Grace 6 of 27 July 1966, had acceded to the request and an Advisory Committee was commissioned to recommend to the Council of the Senate the name of a suitable candidate for the Professorship. As if to weave more tangles in my web of fate, the end of my five-years Headship of the Department was only a month away. The apathy of the General Board towards the Department and its dismissive attitude to myself allowed the Department to run headless like a decapitated chicken throughout the entire Michaelmas Term 1967. The Vice-Chancellor exploded in anger. To assuage his wrath, a formal letter arrived on 7 December

1967 from the General Board appointing me Head for another five years from 1 October 1967.

The campaign for the Professorship was a combination of special committees, drafts and re-drafts of reports and memoranda. The manoeuvres involved a change of Chairman of the Board of Land Economy and earnest diplomatic exchanges with the Managers of the Development Fund, the financiers of our fortunes. A formal request, well-laced with pungent argument and supported by facts and figures, was sent to the General Board in March 1965. Over a year later, the General Board braced themselves to give the 'thumbs up' sign. By doing so they got themselves off the hook of having to deal with Denman and the messy business of a named Readership. Let him compete in the market place! Doubtless, this dodge in some way explains the long silence over the named Readership business.

So it came about that in September 1967, I had to knuckle down to submitting my name, along with others contending from the four quarters of the Globe, in a contest for a Chair which stood on ground which my own creative efforts, perseverance and sacrifice over the previous twenty years had prepared. The irony was irksome.

The Advisory Committee had been in session for a year before it fixed a closing date for applications. No one outside the Committee could explain the delay. A smog of mystery descended over the scene. Waiting, watching and wondering not only frayed my nerves but also gave time and cause for rumours and suspicions to raise their ugly cobra heads. My constant criticism of the Government, over radio, TV and especially in the Press; my undisguised links with the Tory hierarchy and support of free enterprise politics in lectures, public and private, ran counter to what was acceptable to the learned ranks of economists, left-wing sociologists, historians and lawyers whose views dominated in the Cambridge parlours. There were but few friends in that direction; and such as there were never found me currying their favours. To Government circles my name was anathema. So much so, that the Chairman of the Land Commission was detailed off to warn me in person to desist from writing against the Government's land policies. Unless I did so, he told me, all research funds would be withdrawn from the Department. Undoubtedly there was pressure in high places to keep Denman, the writer of right-wing polemics out of the Chair at Cambridge. Just after the closing date for applications, the *Guardian* wrote a column to the effect that Barbara Castle, then Minister of Transport, was keen for her 'young eagle', Chris Foster, to oppose me. J.W.M. Thompson of the *Spectator* and 'Yorick' a sympathiser in *Time and Tide* joined issue in the Press to suggest that the reason for the delay in coming to the obviously right decision was the absence of any shade of 'pink', to say nothing of 'red' in my political makeup. To back me, Michael Ivens of Aims of Industry had rallied to action stations a phalanx of Tory might, ready for action if the decision went against me as a consequence of what was seen to be political and government pressure. Rab Butler, whom at that time I was helping with a lecture to celebrate the RICS centenary, asked to be kept informed of any evidence of 'dirty politics' usurping justice.

Apart from Jeffrey Switzer, who by great good fortune had become Chairman of the Board of Land Economy, I doubted whether any of the chosen 'Advisors' to the Council of the Senate fully comprehended the concept, content and philosophy of land economy as a subject or could have explained to a candidate for the Chair what would be required, supposing the University were to seek continuity of the subject on the lines of the past eight years. Many of the Advisors, I had reason to believe, were simply playing home politics. At a late stage, Professor Hutchinson proposed a way out of the dilemma of having to recommend Denman for the Chair. His thoughts went far beyond the Advisors' brief. He proposed the setting up of a Land Economy Institute headed by myself with the status of Reader and which could, so he imagined, become an integral part of the Agriculture Faculty.

Patience was running short in many directions. Time after time false hopes that a decision had been taken invaded the atmosphere, reminiscent of the Cardinals struggling in the Vatican to appoint a Pope. Throughout October it became known that 4 November 1967 was the deadline. On that day, as on others, the Advisors met, talked and dispersed. As a consequence much criticism was unleashed, especially among the officers of the Department where the uncertainty and delay were inflicting distress. News then ran of a short list of candidates to be interviewed on 9 December 1967. In the meantime, the Vice-Chancellor, Sir Eric Ashby, for whom the codes of protocol lay subordinate to human considerations, had found an expedience for the two of us to meet. Common problems touching my work in West Africa constituted the *ad hoc* agenda. By the end of the meeting, I knew who the alternative candidates were. So strengthened, I waited upon the Advisory Committee on 9 December, to be cross-questioned at the bar of their judgement. Again there was no decision. Again it was Professor Hutchinson who lobbed a grenade into the proceedings and asked for an adjournment while one of his prodigies was called back from Australia for an interview. By then it was Christmas.

Once into the New Year, there was deadlock: my friends on one side of the table matched exactly the voting power of the opposition.

The Vice-Chancellor not a little irritated by the patent bias and filibusting cast his Chairman's vote in my favour. No hint of this reached me save for a kindly phrased note from Professor Nathaniel Lichfield who obliquely mentioned his 'gladness about a recent decision which had been taken at Cambridge'. Meanwhile my mother lay dying in Scotland. An overnight rail dash to Edinburgh brought me to her bedside; alas, too late, even with Nat's kindness, for me to give her the news which would have lifted her heart with pride and joy before she entered heaven. Again the cold protocol of Cambridge had wrought its mindless hurt. Confirmation of the decision came in a letter from the Vice-Chancellor on 24 January 1968, the day before my mother's funeral. The Council of the Senate had accepted the 'advice', offered me the Chair and had their action duly Graced. Then, as a final dig in the ribs, they postponed my taking it up until April 1968. Again it was April, the month of good omen. Seven years had passed since I had assumed command of the Department of Land Economy.

8

CLIMACTERIC

The frozen river is as mute,
The flowers have dried down to the root.
And why, since these be changed since May,
Shouldst thou change less than they?

Change on Change, Elizabeth Barrett Browning

The establishment of the Chair of Land Economy at Cambridge in 1968 anteceded by a week my fifty-seventh birthday. For me, the elevation was a nodal point on the stem of my years. What had hitherto been merely a testimony of hope was now realised fact, an engrossment written in indelible ink. Looking back, it is seen together with other changes, circumstantial and psychic, as a distinct climacter. Responsibilities at Cambridge burst all earlier bounds. Travelling multiplied. Visiting high and low, I became a kind of apostle conveying new ideas, laying foundations among professionals and academics at home and abroad and especially throughout the Commonwealth. Innately, also, I was possessed of a strange inner assurance of will and command which encouraged a sense of mission. The changes, furthermore, were accompanied by repentance of a stuffy Pharisaic subsconscious and released me from the wearisome grip of unjustifiable self-esteem.

Eight years earlier the then foreseeable course of events had been spiced with a prospect of exiguous success. The future had pointed to my presiding over the adolescence and eventual adulthood of the Department of Land Economy. That the Department should one day absorb what was then half the extensive Faculty of Agriculture was beyond rational prognostication. So, indeed, were the events which led up to that absorbtion and its accompanying upheavals. Seen in retrospect those events go a long way to explain Sir Joseph Hutchinson's otherwise unwarranted animosity towards me.

The beginnings of that sad state of affairs are to be found in Newcastle University. The Vice-Chancellor, Dr Charles Ion Carr Bosanquet was weighing up current and future national demands for undergraduate teaching in Agriculture with a view to reporting his findings and opinions to the Minister of Agriculture, Fisheries and Food. Later, armed with the Bosanquet figures, Sir John Wolfenden, Chairman of the University Grants Committee, communicated with the Vice Chancellor of Cambridge University. He in turn sent a letter to Professor Sir Joseph Hutchinson, couched in the courtly language used by academics when mutually engaged in mortal combat. All three Vice-Chancellors, for Sir John had been one in his day, virtually summoned Cambridge to consider how its Faculty of Agriculture might shed its general courses. However obsequious its language, the letter to Jo Hutchinson was none other than a merciless *coup de grâce*. Its portent was not lightened by the UGC wanting to keep specialist teaching in the agricultural sciences and in agricultural economics. How was what had to be done, best be done? The General Board intervened and passed the question to Dr James Beament (as he then was), Reader in Insect Physiology, invited him to gather a bunch of the learned about him and waited on their deliberations. Proposals were delivered in April 1967, well before the Board of Land Economy heard anything of what was afoot.

Showing through the proposals were unmistakeable signs of a move to subvert the Department of Land Economy and my future prospects by tactics which *ipso facto* threw a life-line to the struggling Sir Jo. Scientists on the staff of the doomed Faculty of Agriculture would find refuge in the Department of Applied Biology. No problem there.

What to do with the agricultural economists of the Farm Economics Branch was the *point d'appui* upon which the Beament move now turned. Of those agricultural economists, some sparkled, some were lack-lustre, some keen on the new life, some sceptical, all had UGC and Ministry of Agriculture money behind them and could have found a proper, if not open-armed, welcome among the waring tribes of the Economics Faculty. To have chosen that option, however, would have lost Sir Jo his opportunity. Beament's merry men weighed it up, only to toss it aside and to report, prompted by the hardly hushed whispers of advice offered by Sir Jo and others, that the Committee:

> 'have been led by their informal soundings of various individuals and groups concerned to command the second of the main solutions they have considered. This solution entails transferring the Branch to the Department of Land Economy, and merging the work and staff of the Branch in a new Department of much wider scope.'

So that was the scheme: a new Department 'of much wider scope'. Here the script was no longer veiled, its revealed subtlety, patent to those with eyes to see, anticipated an ultimate valediction to Land

Economy as we had conceived, fought for and established it. Now, moreover, was a propitious time to strike at Land Economy. No recommendation had yet been made by the Advisory Committee on the Professorship, a procrastination, it now seems, instigated by Sir Joseph. The Department was, at the same time and due to blocking tactics engineeered by the same hand, without a Head. Those of us in the Department and all straight-minded members of the Board of Land Economy became alarmed and suspicious. They read 'take over' where the Beament proposals used 'merging'. The Beament thinking matched too closely for our ease the behaviour of Sir Jo who, although on the Advisory Committee himself, had suggested a Land Economy Institute within what would have been a salvaged Faculty of Agriculture (or perhaps a new Faculty of Agricultural Economics) wherein my wings would have been clipped.

Perhaps the Board of Land Economy would not have stood so firmly against this intrigue if the Beament Committee at the outset had extended common courtesy towards it. On the contrary, they never once prior to writing their Report formally consulted Land Economy with whose fortune they were playing. They even had the audacity to attach to their formal Report a draft revised and extended curriculum for the Land Economy Tripos. The Board of Land Economy itself were in the circumstances a model of discretion and forbearance. They were neither antagonistic nor destructive: let the transfer proceed but as an influx of numbers and strength to enable the Department of Land Economy as now established to flourish to the benefit of the University. Although support for the transfer was genuine, it was also, emphatically conditional. The terms would have to see to it that the new-comers in no way jeopardized the prospects of promotion of the existing staff or of later essential additions to it. Recognising the short-lived memory of the General Board it was required to give a solemn undertaking not to accuse the Department of having an imbalanced staff-student ratio. While these changes were afoot, and thanks to the resolute determination of the Vice Chancellor, the Department of Land Economy was soon to find that its duly-appointed Head and elected Professor lay in one and the same person. Land Economy welcomed the Farm Economics Branch, *pares inter pares*, to join in building a common future. Eventually, the Branch became the Agricultural Economics Unit and brought with it what were for me unfamiliar responsibilities standing on a formal contract to provide services for the Ministry of Agriculture, Fisheries and Food. Here, then, was another change. From henceforth I was a *quasi* civil servant but not altogether back among the red-tape brigade, yet doffing my cap as all civil civil servants should do to their rightful masters, in my case, the farmers and landowners of East Anglia.

❧ ☙

The designation 'Land Economy', which had preceded the Beament Committee by many years and had provided a peg for them to hang their hats on, had, more significantly, opened up for me international opportunities. 'Land Economy', unlike 'Estate Management', did not confuse the lexicographers of the Continent; they could make sense of it in French, German, Russian and other tongues. The term, the ideas and the realms of applied theory it denoted also fitted well into the development policies and land reform programmes of the embryonic Third World. Scholars, there and elsewhere, wanting to understand it, to teach it and open up the new research opportunities sought us out at Cambridge. In particular, they looked to me without hyberbole as a kind of apostle – some might even allow 'Messiah'. The recognition caused me to sit 'in transit' in the Executive Lounges of many international airports. Associated with this expanding overseas activity and reputation were new home developments. The surveyors' profession was trying to establish home bases for overseas outreaches of different kinds. The action meant a place on the newly-constituted International Committee of the RICS.

With the retreating tide of Empire in the 1950s, many colonial Land Commissioners, members of the RICS, highly qualified by training and experience, were stranded on the beaches of premature retirement. Their talents were wasted, their brains idle and their spirits frustrated, especially by the knowledge that former servants now reigned in their seats. Many once-upon-a-time servants, and now 'excellencies' of various grades, recognised the waste as golden handshakes dismissed and dispersed district officers and commissioners, men of long experience and canny knowledge, from their posts throughout the Empire. Timely intervention was made. Need the farewells be absolute? Was there no place for a valediction *nisi*; a conditional discharge dependent on the men of yesterday becoming teachers and trainers of the new masters? The question had a peculiar relevance to land affairs. In the UK the Government listened to it with sympathy and to some purpose. What about the Department of Land Economy at Cambridge? Would it be prepared to recruit demobilised Land Commissioners to promote and run courses in land administration for selected candidates from officialdom throughout the newly independent Commonwealth and elsewhere?

Penury and prejudice stood in the way. Governments overseas were hard up; and the Government at home was tied up by red tape. Money could be found to send worthies to Cambridge but not to pay for the necessary teachers when they got there. Eventually, after some two years bargaining, an acceptable compromise committed the Ministry of Overseas Development to meeting the shortfall which would result should the yearly intake of candidates drop below a given minimum. The prejudice was a home-brew of jealousy and hurt pride. Back in the inter-war years, the University had provided special courses for graduates intending to go overseas to manage the affairs of the Empire in the Tropics. This *bonne raison* on which its Overseas Studies Committee had stood had become infirm and in the 1960s the Committee was trying to prop up its foundations and hang on to teaching posts financed by Government under the old arrangements. Resentment was hardly camouflaged when the Government addressed its request for new services to the untried Department of Land Economy. The Overseas Committee were, also, judging by the remarks of its Chairman at the time, immoderately jealous, as indeed were other smaller University institutions, of the wealth of the Department of Land Economy. He considered the wealth to be misplaced. But for it, we would not be there in his way. The ensuing friction was another growing-pain for Land Economy but one which ultimately proved to be more a blessing than a bane.

The way in which the scholars of the Department of Land Economy thought about land and property rights over it was, at the time, twenty years in advance of the theories and applied wisdom of economists whether in schools of higher learning, Government Departments or financial centres like the World Bank. At that time unenlightened economists, directly or indirectly, shaped the thinking of the Overseas Studies Committee at Cambridge. But the men in the Ministry of Overseas Development who received the pleas for help from the newly-independent Commonwealth were primarily lawyers recruited from the former ranks of the Registrars and Land Commissioners in the field. They knew from experience what and who were required to administer the 'national estates' in Africa, the Near East, Asia, the Caribbean and the Pacific. Instinctively they realised that the perspectives of Land Economy were appropriate for the teaching and research required in the Developing Third World.

Predominant among the influential men who guided the Government's response to the current need for trained land administrators was S. Rowton Simpson. He had become Land Tenure Adviser to ODA after retirement as Registrar-General and Commissioner of Lands in the Sudan, a man of dogmatic views whose larynx appeared to be activated by a sack of gravel. Early in 1967, he had been invited to return to the Sudan to advise on whether the 'Land Regulations' which years before he had scribbled out on his knee while jogging in a train over the primitive track from Port Sudan to Khartoum should be the basis of the land law reform of the country now free from the oversight of his eagle eye. Rowton Simpson on a visit to the Sudan did me the honour of asking me to join him along with David Lloyd, the recently retired Director of the Department of Lands, Mines and Surveys of Fiji. Across the burning sands of Darfur, we found time to consider at first hand the need for the Land Administration Course at Cambridge just established under the expert guidance of Henry West, another Commissioner of Lands and one who had been retired from the service of the Uganda Government. On that trip we were able to design a blueprint for a long-term research project that was to lead under the authorship of Rowton Simpson to the writing of the modern standard work on *Land Law and Registration*.

A man like Rowton Simpson of the lineage of colonial administrators walking through the alleyways of Khartoum North was bound to raise the dust of memories long dormant. More than once, as I recall, an unkempt serf waving outstretched arms, shouting joyful salutations through toothless gums and regardless of man or beast, would run up to him and clasp his feet shouting out: 'Master! Master! The Master has come back. Master you remember me. I was your gardener. You gave me two years for stealing. O Master, say you remember me.'

There was genuine affection for the old administrator who had treated the fellow well and dealt with him fairly. Even a newcomer to the scene, like myself, could see in episodes like these that something more than sound land administration had gone out of the Sudan and other domains of Empire when the British gave them 'freedom'. Outwardly Rowton was a sturdy remnant of the old 'raj', guilty in the eyes of the left-wing ninnies of today of a 'bovine psychology' of an 'outlook crude and raw'. Inwardly, he carried in his heart between his old trilby hat and his sun-bleached shorts a love for the Sudan, the country he had served for a lifetime, that only true self-giving could engender.

Every inch of the map seemed to have a place for him. The 'plane stopped at El Fasher and we went by truck up winding dry wadis and across trackless deserts to the border town of Zalingie. There, while David Lloyd and I talked with the local uniformed civil servants of land registers and the work of tobacco growers who had been smuggled in from sanctions-bound Rhodesia, Rowton was muttering to himself before a vast Overseas Ordnance Survey Map pinned to the wattle and daub wall.

'What are you doing?' I interrupted the mutterings and handed him an iced indigenous beer. The gravel started to grind in his throat. 'I'm looking for Kaffy Kingy,' he growled, 'to see if it is still on the map. Thirty years ago, I was the District Officer in these parts. There was trouble at Kaffy Kingy. My wife and I made a proclamation one Saturday night, "By Monday morning, the long-legged Dinckars from the north will get out of Kaffy Kingy; and the warring Bantu go home to the southland. If this is not done, we will burn Kaffy Kingy." Nothing happened. So at first light on Monday, my wife and I lit tallow torches, mounted our camels, rode into the place and burnt Kaffy Kingy. I can't find it. So I suppose it remains still one with the walls of Jericho. I'll have to tell her.'

Light was fading over Kebkabiya on the way back. In the hope of finding something to drink, we assailed the doors of the only building marking the spot. In the course of these operations, out of a rolling ball of dust in the far distance, a ball which grew to a globe as it came nearer, emerged an enormous lorry. It was driven by a fellow who jumped clear of the cab to announce his arrival. 'I'm Doctor Watson,' he assured us and thrust out an Australian paw. 'Where then,' our dry throats responded, 'is Sherlock Homes?' As if pre-rehearsed there alighted from the opposite side of the vehicle a lithesome Swedish blonde. 'I'm an athropologist,' she introduced herself. 'Ya, Ya, this is Doctor Watson, the water-borne diseases expert from WHO. We have been three months away in the bush and sand, each teaching the other about man, woman and diseases. Right now, we are chasing a leopard. Hoping to get him before dark. He came this way. Have you seen it? Can't stop. He's getting away. By.'

Bed and the craved-for drinks were still hours away into the night and Khartoum a day's flight beyond. By the time we arrived in Khartoum the thought of yet another Land Registry inspection was hard to bear. Happily, the threat was lifted by an invitation to a grand tea-party in the afternoon somewhere out towards Omdurman. It turned out to be one of those dolorous, all-male gatherings, attended by a hundred draped and befezzed grave-eyed Arabs, chewing sweets and sipping tea. Getting there, we were opposed by massed hoards of animals, sheep and goats milling by the thousand, driven in panic to the slaughter by angry, impatient herdsmen. At tea, my companion was a huge, dark Nubian bristling with knives. He sucked his cup of tea with an off-putting hiss like an ill-tuned burglar alarm. In an endeav-our to turn him off this irritating performance, I asked for an

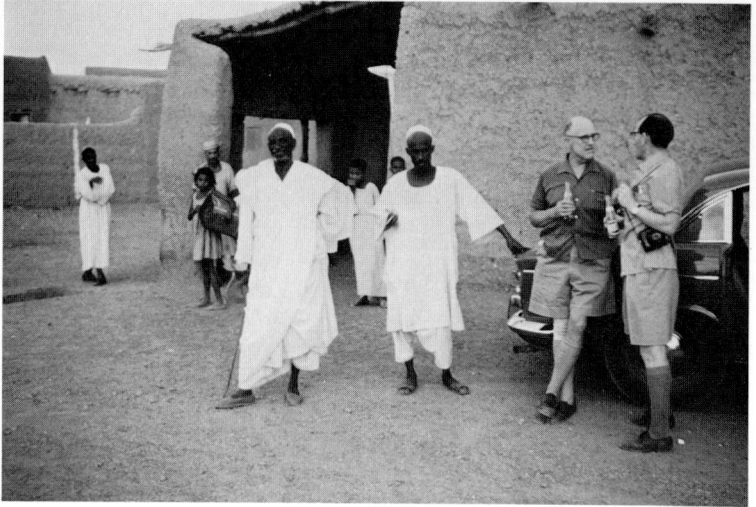

'Kaffy Kingy' at El Fashar, Sudan

explanation of the swirling herds which had held up our coming to the party.

'They are for the Id,' he replied between hissing and coughing.

'Which Id is that?', I pretended to some knowledge.

'We commemorate the sacrifice of Ibriaham's Son on Mount Moriah.' Astonished at his reply and its implications, I pushed the probings further.

'You recognise the sacrifice of Isaac, Abraham's son on Mount Moriah?'

'Isaac. Isaac.' He nearly choked with fury at the blasphemy. 'Isaac. Never.' And he drew a heavy, curved dagger from its sheath. 'It was Ishmael, Ishmael the Son of Promise; not Isaac,' he boomed as he swung the knife towards me with a menacing gesture.

'Oh, I see,' I heard my trembling, infidel voice respond, in a bid to mollify the zealot. I reached for another lump of sugar and prayed silently to the God of Abraham, Isaac and Jacob to remember what happened on Mount Moriah and stay the blade.

My friendship with Henry West, David Lloyd and Rowton Simpson and my custody of the teaching and research in land administration and registration for which they were responsible at Cambridge notched another feature of change on my Totem Pole. Along with it went a change of name – ever after our Sudan adventures, Rowton, for me, acquired the affectionate sobriquet of 'Kaffy Kingy'.

❧ ☙

Working with international clay it was possible to fashion a plausible image of Land Economy as a discipline which could find its place among the humanities. The subject was protean, adaptable and relevant to the varieties of human need and too innovate to be impeded by the bounds of particular cults and customs. Its claim of belonging to the humanities was never the consequence of deliberate purpose.

The claim emerged as a consequence of the numerous get-togethers of nations in a kind of second tier of assemblies under the primacy of the United Nations. The movement was a prominent feature among the international changes which took place in the epoch-making mid-century. Two of these secondary groupings meant much to the surveying profession and notably to the RICS. One was the revival of the *Fédération Internationale des Géomètres* (FIG). The waning energies of this late nineteenth-Century creation were revived by the formation of the EEC. The other grouping was the Commonwealth Foundation. The Prime Ministers of the Commonwealth in 1964 set out to enhance the standing and sharpen the purpose of the professions within the newly independent nations of the Commonwealth. We have noted how professionally qualified civil servants and others were 'retired' early and thereby deprived those nations of their expertise just when development and progress needed it most.

About the time of this international arm-joining and barn dancing, the RICS had its 100th birthday. The birthday was celebrated in the Festival Hall on the banks of the Thames the very year responsibility for the triennial Congress of FIG fell on the RICS. Opportunity, therefore, was taken at that Congress to rally the Commonwealth delegates behind the RICS to meet the request of the Commonwealth Foundation for a body to represent the surveying profession throughout the Commonwealth. The architects of this novel surveyors' Association handsomely acknowledged the Cambridge philosophy by designating the new body '*The Commonwealth Association of Surveying and Land Economy*' (CASLE). Each one of these occasions – the Birthday, the Congress and the Conception – and the events which led up to them increased the pressures that upseated me from being a sedentary philosopher-prophet to become a travelling evangelist.

The run-o'-the-mill chartered surveyor with a hundred years of institutional care behind him was still home-based, instinctively indige-nous in outlook, little inclined to venture abroad and ill-equipped to do so. His colleagues in Europe, tied to the narrow confines of cadastral survey, were even more home-bound. When in FIG Congress if they came face to face with practitioners from the Americas, the Antipodes and the Tropics lack of a common dictionary of terms would beset progress towards a valuable exchange of ideas. So, at the Twelfth Congress in London in 1968, the RICS, confronted with this obstacle, took timely steps to remove it. Explanatory booklets based on two years research were to be prepared and handed to the expected 1000 delegates. Such work was aimed to contribute to knowledge far beyond the immediate know-how for the Congress. When FIG asked me to provide an interpretive *vade mecum* on land systems it was taken as a challenge and seized as an opportunity to illustrate in a practical way important aspects of the universality of land economy.

Little had been written on comparative land tenure, except under the pens of economic historians. On the modern scene there were one or two bulky volumes from America on farm systems but they merely highlighted the propensity of agronomists and economists to overlook the fundamental importance of land. So with a *tabula rasa* in my hand, the summer and early autumn of 1967 were spent crossing and recrossing Europe, from the mountain pastures of Norway with their *felles fjellbeite* to the *mezzadria* of southern Italy. Places and systems already known – the *métayage* of the Loire and the *gospodarstwa indywidualne* of the Vistula – widened the range of comparisons. A spate of letters from Canada, the Middle East and New Zealand furnished materials for an *urbi et orbi* presentation to the FIG London Congress in the September of 1968. On the northern leg of this European research scramble, the family came with me.

Leaning over the gunwale as we left Bergen Harbour, with my eyes to the fells behind 'from whence cometh our help', the clouds that shrouded them reminded me of the future of the Department of Land Economy, shrouded in mist. By the end of the month, the Department would be Headless and the General Board in its folly would be doing nothing about it.

The southern leg of the tour was both useful and therapeutic. The tension of what seemed to me a purposely drawn out torment over the Chair at Cambridge had reduced my nerves to a ganglion of raw ends. So I telephoned Pat Rivet. As my Secretary, Pat Rivet had tramped the commons and village greens of England with me in the closing months of that six-years research. Asked then if she could take leave for a fortnight from her WHO appointment, she responded with ready and

Iván Földes and DRD with what is left of wine and goulash on the Hungarian Collective Farm

delighted alacrity. We motored from Geneva first to see Professor Tanner in Zurich, then to my friend Sorbi in Parma and on to end up with another friend Christodoulu in charge of the Land Reform activities of FAO in Rome. WHO, the United Nations World Health Organisation, could have provided no better treatment to restore my shattered morale. The outcome of that summer's work was a simple classification of land rights – absolute, conditional, private, public, individual and collective. It has found a place in the literature of some permanence and an Italian version was used to map the proprietary structure of Italy's land resources.

The FIG Congress paid lasting social dividends. Among the delegates were many academics who did not naturally belong there. Having helped me with the land systems research they were curious to accept invitations. My colleagues from Nigeria and Ghana helped to swell the Commonwealth contingent and added weight to the discussions that led to the formation of CASLE. And of particular value was the chance to strengthen ties with friends in Hungary and Czechoslovakia.

From Hungary came Iván Földes who held academic posts of distinction in the Universities of Budapest and Pécs. We met first at Florence in 1962 to talk comparative agricultural law and forge a friendship which lasted until his death in 1988. Iván made himself understood but was no master of the English tongue. He shared the natural wit of his countrymen and would refer to Professor Csizmadia, his superior who neither spoke nor readily understood English, as 'Cheese my Dear'. To help Iván squeeze his diminutive figure into the queue of selected delegates to the FIG Congress who were lining up at the Guildhall to meet the dignitaries of State, we draped him in an over-large capitalist tuxedo and persuaded 'Comrade' Sir Desmond Heap, Comptroller of the City of London to part with one of his red tickets, the only form of passport. Two epigrams in Földesian English will always endear his memory. He was, at the time, in the throes of a happy divorce and re-marriage process involving two spouses whom he called 'Elizabeth I' and 'Elizabeth II'. One day he was found making for Market Hill in Cambridge 'looking for "Vooleys"' where he could 'buy voollies for his vimin'. Earlier, he had referred in a lecture that morning to state-owned British Rail as 'Ze stately railway of England'.

On one occasion in Hungary I was with him in circumstances where the trappings of academic office were forgotten and Iván was truly at home among his own. Officially, according to the schedule handed me, we were to visit a collective farm. The prospect looked pretty routine and the weather that February afternoon was cold and sober. On arrival, the party put on the Hungarian equivalent of Wellies and walked along muddy paths to a remote hut on the outer edge of the collective farm. Here Fercuc Nagy, a rotund man of genuine goodwill, the local *bon vivant*, welcomed us under a low doorway into an extensive hut laid out with a few chairs, cooking stove and long trestle tables laden with empty wine glasses. In his hand Nagy carried a *lopo* which he used as a mace to control proceedings. This device properly functioned as a pipette to draw superb Hungarian wine from vats sunken beneath the floor. A guitarist, Benko Andras, eventually took up his position in the near proximity of a massive copper pot of goulash in the charge of Susan, the cook. Nagy, the vinter, intimated his pride in his outpost of free enterprise and swung the *lopo* in rhythm to the rising tempo of the strumming, song and stamping feet. By dark, the true workers with sweat and mud still wet upon them, led by Janos Gulyas and other gnarled rustics, had packed every free cranny. When crammed to bursting and with decibels beating from the players threatening the roof structure, Susan beckoned for help with the copper bath-tub and the goulash. Each song challenged the other in daring and carried the soul of Iván further and further back into the days of his childhood, to those days when 'comrades' and communists had never been heard of. The night was well advanced, bellies truly warmed and feet unsteady from the magical work of the *lopo* when Nagy's guests emerged from the light into a pitch darkness to try and make their way to the highway. Some twenty minutes out into the darkness and being unequipped with gumboots I foundered in a slough. Both shoes came off. The hue and cry of the match-lit hunt that finally found them was the crowning event of the evening. The only line the triumphant crew translated for me was, 'The Englishman's lost his shoes'. I was content. There among the cattle, the cabbages and the vino, beat the true Hungarian heart which the deadly hands of Marxism could never still.

The merry-hearted Nagy defied the Marxian creed in other ways. The high level of production from his small vineyard provided evidence of economic truth that was leading economists and advisers in the corriders of power in Budapest towards dangerous apostasy. Among these enlightened souls worked the renowned Dr Anna Burger. One evening after dinner with Iván Földes, Anna, her husband and Colin Kolbert we expressed horror at the prospect of a Hilton Hotel being built alongside the medieval glory of the church of St Matthias on the Buda bank of the Daunau. Our friends assured us that all was well because the Hilton would be built 'ruin style' in harmony with the ancient splendour. That romance in architecture should be permitted was itself reaction enough. It did not touch the depth of the apostasy which the works of Nagy, the wine bibber, was precipitating in the mind of Anna. Her Government dismayed at the drop in wheat output had tried to stimulate production by raising the state-controlled price for wheat. Production, nevertheless, continued to decline. Output from market gardens and vineyards rose. Slowly the truth dawned: although inherent fertility and the unit costs of production were uniform, there was greater income per hectare from the vineyards and other intensively cultivated land. The margin must represent the relative differences in the value of the land! But this could not be, for Marx had told them that land has no value. Anna confided her fears to me and asked if we would receive a colleague from her staff at the Department of Land Economy to learn *ab initio* how to value land. The last official land values for Hungary had been recorded there over thirty years ago.

On another occasion when visiting Hungary, our family had pushed east into the dolorous misery of Romania. Strangely incongruous with its poverty and squalor was the mythology through which the cheerful, ill-clad street children saw the West. We had spent a night of continuous downpour cramped on the floor of an auto-camp storehouse. It was the only dry accommodation available. At daybreak on a hunt for benzina at Sobes the car was surrounded by screaming children yelling for 'the King's Head'. By signs, patient diplomacy and with their grimy hands fumbling uninvitedly in our pockets, the kids made us understand they were after outdated coins carrying the King's head but would take a 'Queen' as second best – as if she were a lesser deity.

In contrast to Iván and Anna, Professor Fabry of Karlovy University, Prague enjoyed nothing approaching their liberties. In the summer of 1967 he had written a courageous and valiant critique of my draft text for *Rural Land Systems*. With the Russians battering at the ramparts of Prague, however, he was in no position to join the FIG September Congress in London, much as he would have welcomed the opportunity to do so, being a scholar fluent in English and one who had taken 'the King's shilling' as a fighter pilot with the RAF during the Battle of Britain. Some time later, while the Russian place-man Gustav Husak was in power, he visited the Department of Land Economy. He was in a state of constant agitation. However, he was eager to meet British politicians, so I arranged a private lunch for him in the Houses of Parliament with the then little-known MP for Finchley, Margaret Thatcher. Fabry's apprehension approached a state of extreme anxiety. Some time afterwards, when Margaret was Leader of the Conservative Party but prior to her election to the Premiership, she visited Eastern Europe. Soon after getting back she was speaking at a Carlton Club dinner and instanced Fabry's lunch as an example of what was probably then the only current mode by which useful dialogue with the Socialist Bloc could be had and maintained. 'You academics have a contact and rapport which the politicians cannot engineer.' She was insistent and admonitive.

The gestation period of the Commonwealth Association of Surveying and Land Economy (CASLE) was approximately twelve months. By September 1969 ratification of the proposals by any professional society within the Commonwealth eligible to join had to be made. Membership of CASLE was confined to groups, institutions and associations. There was no personal membership. Because of the roles I had played in founding schools of higher learning in Land Economy in Nigeria and Ghana, in bringing many non-practising academic surveyors to the FIG Congress and in the RICS post-Centenary Celebrations promoting degrees in Land Economy throughout Britain, a special standing was afforded me in CASLE as Adviser on Education. From that standpoint, it was my task to help mould in most of the developing countries of the Commonwealth policies of academic initiation and advance.

The voice that launched the Centenary Celebration proceedings of the RICS in the Festival Hall belonged to Lord Butler, Master of Trinity College. We knew each other well from the days of the Land Commission battles, since when he had been my passenger and colleague on lecture trips to Swinton College. So when he was invited to give the Centenary Address, Rab asked me to help him. The high druids of the RICS responsible for proceedings were happy with this development, for their thoughts could be my thoughts and my thoughts Rab's thoughts. Between us, Rab was guided to give a splendid exhortatory speech declaiming that the hour had come when seats of higher learning should provide degree courses for the profession. His speech was a kind of epistle ready for me to carry on a mission to the Universities.

Such a mission was entrusted to my care as an item on the Centenary Year agenda. Reading University sulked at the outset; any money going for higher education they reckoned should be theirs. There was little cause for concern. Liverpool University made a valiant effort, shaped ideas and had them accepted by the profession. Their conception, sadly, was stillborn for want of money. Edinburgh showed little energetic enthusiasm. Aberdeen alone responded with a purposeful intent conditional upon funds. The willing heart was rewarded when some time later at my RICS Gold Medal Address at Stirling University, the opportunity was taken to chide the Scots for having no Land Economy teaching in their prestige Universities. No sooner had I left the platform than the Director of the MacRobert Trust came forward with the money.

No one in that audience knew of the dramatic events which had attended my canvassing Aberdeen University some months previous. In the eventful summer of 1968, Richard, my younger son, had returned home from Italy accompanied by a friend, Philipo Heilpern. Richard Heilpern, Philipo's father, was the son of a Romanian count and unlike the urchins at Sabres had had an English governess as a child. His upbringing fitted him well when after the revolution he became a refugee in Italy. There he had been appointed Secretary to the Institute of Comparative Agrarian Law at Florence and hence was linked indirectly with the FIG Congress. The other Richard, my son, and Philipo were with me on the journey to Scotland to open negotiations with the Principal of Aberdeen University. In a moment of paternal extravagance, the Wolseley was entrusted to Richard and Philipo at Kenmore. My journey continued by train. Waiting with Irene Martin and her daughter Elizabeth at the George Hotel in Edinburgh for Richard to arrive for dinner that evening, Reception suddenly called urgently for a 'Mr Penman'. My name is seldom right, so I responded.

Richard at the other end was speaking from hospital near Loch Lomondside, the Wolseley was a wreck and Philipo had been left stranded in Inverary. The evening shattered, we hired a car from Turnhouse Airport and drove into the night, first to Alexandria Hospital and then deep into the Highlands. Next morning, an hour or so after midnight, Philipo was found with a balmoral perched cockily on his Italian locks sitting under a solitary lamp post on Inverary quayside. Dawn had broken over an awakening Glasgow before we eventually got to bed. Fortunately the Aberdeen talks had gone well and success only awaited the kindness of the MacRobert Trustees.

❧ ❧

During the FIG Congress, an unexpected and welcome invitation to peddle my wares in the beautiful campus of the University of British Columbia (UBC) was pressed on me by Philip White. Philip's exceptional intellect is able to perform in the raucous international property markets and court at the same time the intellectual virtues of academic transcendence. As Dean of the Faculty of Commerce and Business Administration in British Columbia, he arranged my air tickets, accommodation and an exacting lecture programme. At the time, Jeffrey Switzer was on a continent-hopping mission carrying the century-old flag of the RICS around the world and was due in Vancouver to coincide with the dates suggested for my visit. Jeffrey could well have filled the bill in my place. Philip, however, insisted on having my maverick philosophies and refused any substitute.

Geographically, British Columbia is a monument to the majesty of coast-line and mountain and substantiates the legend that God's apprentice hand had made Scotland and the Isles, and later as a master craftsman He had made British Columbia. Politically and architectually, Vancouver took me back to Eastbourne as I remember it between the Wars when its local Tory MP would reign unopposed. Mayor Fullerton of Vancouver when we met was so prolix and forthright in declaring his loyalties that I found myself on the point of apologising for being left-wing. He introduced himself by slamming on a Jacobean table a metallic emblem and bellowing, 'Say, that's me; the six-fingered Red Hand of Ulster. Everything is falling apart,' he continued, bemoaning the fate of Canada. 'We haven't a flag anymore, only a bacon-packer's symbol.' Others told me that folk on Vancouver Island were more British than the British. Violent thunderstorms prevented my testing this hearsay evidence. The tiny seaplane intent on getting to Victoria was tossed off the island headlands like a Maple Leaf in a hurricane. Driven back to the mainland, it deposited me a shamefaced victim of acute air-sickness – a malaise experienced neither before nor since.

UBC taught me to recognise and understand an intellectual blindness brought on by thought, habit and unconscious prejudgement. The political ambience of the University, unlike contemporary Cambridge, had a slight stain of pink here and there but was not dyed deep in the red of socialist prejudice. Even so, the misunderstandings I had met in venomous left-wing reactions elsewhere were there displayed and had to be overcome by patience and prolonged debate. Indicative of the blindness was the assumption of my audience that whenever I spoke of 'property', especially as the springboard of economic decision-making, my reference was taken as being exclusively to private property; no one gave any place to public property – a shortsightedness which even the renowned Proudhon suffered from. Likewise with 'planning'. All references were prejudged to denote State planning, the means by which the State controls and determines the use patterns of land and natural resources. My insistence that planners have no power to do but only to prohibit was met with incredulity until these false assumptions were exposed and the vision cleared. Thus it was that I came away much helped and seeing more clearly how, back home in Cambridge, the self-styled people's thinkers, some of whom had wielded majority votes in my disfavour, had failed to understand my thinking and added obscurity to prejudice. They had been blinded by thought habits. Some even admitted to the fact in later days.

Apparently, according to my friends and informants among the minorities, there was, also, in the majority camp the notion that I was an obscurantist diehard Christian incapable of *a priori* rationalisation. Such predilection to judge me would have been acceptable if its grounds were true. It would even have been tolerable if the ill-disposed had confronted me face to face with their verdicts and not left them to filter through the miasma of Senior Combination Room gossip.

❧

The avowal of faith is all too readily denounced by the agnostic as an act of irrational intransigence. In all truth, the denunciation is itself a refutation of rational espistemology which acknowledges faith as the only means by which man apprehends knowledge of the divine Absolute. Scripture from Creation to the Cross points to faith as the sole corollary of revelation, God's exposure of Himself to man. Faith is the only way man comes to know Christ, living and transcendant; the only way by which a man knows His redemptive love and the power of His resurrection. There was a moment in spacetime, the *Einmaligkeit* of Emil Brunner's *The Mediator*, when the cosmic Christ, the Lamb of God slain from the foundation of the world, walked incarnate on the waters of Gennesaret. This truth from eternity manifest in time can only be apprehended by faith. Witness to the break-in of the Eternal into the time processes of history is not obdurate intransigence but the declaration of truth as faith apprehends it.

Let me admit from experience that faith can get it wrong. As the years receded there evolved within me a proclivity to make for myself a set of petty rules and regulations, a kind of personal scroll of *Halachah*, to use a Jewish idiom. It was a heart-comforting 'law' to live by. Under the scroll, my self-righteousness grew fully and falsely justified. By bowing down rigorously to its scruples, to this totem of taboos and commands, an unhealthy stubbornness stifled the Spirit. On such a pattern lived the Pharisees of old. So did I – a veritable negative practitioner of the Faith.

Man's earnest desire to be justified, to be God's 'blue-eyed boy', generates a deceit of self-righteousness and sears the conscience. Scraping callous off the conscience is as painful as it is liberating and is most certainly a climacteric. St Peter went through it on the house-top in Joppa. God revealed to him the truth that Peter's old conscience had got things wrong. The Lord turned his thinking about and gave him a new conscience to enable life to be a walk with joy and to strengthen him to enter the home of the hitherto 'unclean' Gentile, Cornelius the Roman.

My little *Torah* was obsessive and insistent. It not only turned me into a Pharisee but, as with all other 'law-keeping as a means to salvation' processes, was a denial of faith in the complete efficacy of Christ's redemption. In itself, it was a most miserable affair. The legal 'righteousness' held me apart from spontaneous, open-hearted relationships with colleagues, friends, acquaintances and strangers. I, a veritable matrix of neuroses – inhibitions, prohibitions and denials – held aloof from the world – no alcohol, no dancing, no happy hours with non-Christians (whoever they may be) no gardening on Sunday, no kissing the girls. Verily, it was a negative life of paucity of spirit when Christ had promised life abundant.

My sons, Jonathan and Richard, God bless them, unknowingly but surely came to my rescue. They went for the Pharisee within me and challenged the brute to explain himself. This he tried to do but couldn't. So he metaphorically cut off his phylacteries and gave up being a Pharisee to become 'as other men are'. Henceforth, I did things, said things and thought things well outwith the strictures of my old conscience. In the sequel, I stood up a better integrated fellow thereby. This climacteric was no born-again experience, nor a descent into irresponsible antinomianism. The Lord never left me. He let me wander round and round, back and forth, searching for His love path. There came to me a healthy agnosticism in the matter of daily directions. This caused me to enquire the way at each cross-roads. The life of petty rigidities was gone. In this sense, life had become a half and half affair. But the central pivot around which my tergiversations rotated stood immovable: the Eternal Rock, Jesus Christ, the same yesterday, today, forever.

9

UNDER THE PEACOCK THRONE

And should you think yourselves by me deceived,
Here's one, the Astrologer, may be believed.

Faust Part Two, Goethe

In the summer of 1967, the Shahanshah of Iran published his personal account of the social and economic upheavals which were changing the face of his country and which, dressed up in the robes of peace and progress, he was pleased to call 'The White Revolution of Shah and People'. At the time Iran was a closed book to me. It was, therefore, with a mixture of enthusiasm and caution – another half and half affair – that I responded to a sudden and unexpected telephone call from Jack Hamson, Professor of Comparative Law at Cambridge, asking if I could help Haleh Afshar, the daughter of a Persian friend of his. Haleh wished to come to Cambridge as a research graduate to study Persian land reform, one of the Shah's progressive accomplishments. She was a delight. Her vivacious Persian loveliness was lit by a sparkling intellect shining out of the oriental beauty of a pair of does' eyes. As a gifted linguist she spoke fluent English and French, but was more hesitant in her native Farsi. Among her many attractions was an enigmatic political contrariness. Who else but Haleh Afshar would breathe socialist fire and fury as she waltzed out of a Parisian shoe-shop clutching no less than six pairs of new shoes? Her father, Hassan, was an academic lawyer of international repute, scion of a renowned Persian lineage who adored his pretty daughter. Both lived a life of affluent ease behind the high walls of the family estate in upper-class Tehran.

None was more worthy, more capable and more willing to play the part of *homme d'affaires* in arranging for me the required entrée into academic circles in Iran than Hassan Afshar. Away from Tehran, his habit was to reside in Paris. So on Candlemas Eve 1968, I dined with him and Haleh in the Hotel Maurice. It was in the days when I was chasing Continental sources to help with the Land Systems research. Helping me in Paris was one of the very first graduates in Land

Economy, Charles Spencer Barnard. Hassan was a little hazy about Land Economy. So Charles joined us and proved to be an excellent ambassador for the new subject.

Little time was lost. With a cleared conscience between the end of lectures and marking exam-papers, BOAC flew me to Tehran one midnight in May 1968. Vaguely aware of my purpose, I was, nevertheless, in complete ignorance of how it might be achieved and of where to go. Recognition, beyond the Customs, of Hassan Afshar, the only one among the millions of the Shah's subjects known to me, was reassuring relief. Standing by him and wearing a singularly warm smile of welcome was Mansour Emami, an official from the Ministry of Land Reform and Rural Cooperation. Greatly to my amazement, my programme seemed to be in his hands. This impression was securely confirmed early next morning. The events of the day were guided by the Minister's International Liasion Officer, Madame Zahra Samii. She emphatically informed me by a forceful gesture of her well-manicured hands and sleek head that I was the guest of Dr Valian, His Excellency the Minister of Land Reform, and I should be careful not to mention Haleh Afshar's interests. What had I come for? How should the mission proceed, if I had to seek a *congé d'accorder* from this charming self-appointed companion at every turn? This, however, was the Middle East and within hours of arrival I had learnt one of the secrets of its way of life – to wait and see. *Patience passe science*, as the French have it.

For the following nine days, I 'trod the crimson carpet and breathed the perfumed air' as these luxuries were afforded to VIP guests of the Crown and Government. Nothing of opulence, care and kindness was denied me on a carefully planned tour. It led south to far Shiraz and north to the littoral of the Caspian Sea. Outstanding were Arabian nights in the splendours of the Shah Abbas Hotel in Isfahan where a skilful, romantic architecture had captured, in inlaid marble, traditional decor, alcove and courtyard, the spirit of the ancient caravansarai that had occupied the site centuries ago on the caravan routes from India to Damascus. Governors, Heads of Cooperatives and lesser officials met us at every turn to extol the virtues of the renaissance which land reform had brought to the rural scene. Visits to Universities were not ruled out but were far from being prominent. Only at Shiraz was it possible to discuss with officialdom plans for field study for Haleh. In the process I encountered a serious impediment. As Haleh was single, it would not be permissible, warned Enayatillah my informant, for her to pursue field studies without a chaperone.

The tight-packed itinerary allowed for serious business to alternate with rushed sightseeing; with a glimpse of the blind white fish moving in the waters of the *qanat* under the tomb of Hafiz the poet; and a dash to Persepolis and the ruins of Shushan, the Palace of Darius where aforetime he had kept Daniel in the lion's den. These are fleeting memories. More lasting was the discovery made among the illiterate *nassaq*-holders of the southern villages of the other face of the Persian land reform. Emami's father was an honest *patron* whose lands had been expropriated under the reforms in favour of the villagers (*zarein*). These poor folk came to their *patron* at the instance of the *mullas*, bearing sixty percent of the harvest and thus discharging the will of Allah. Patiently, the good man handed the render back and explained how under land reform he had no right to it. But with their ears open to the *mullas*, the *zarein* refused to listen. At the same time, the Shah was demanding the equivalent of thirty percent of the harvest to be paid into the Agricultural Development Bank as annual instalments of the cost of the newly-won land titles. Between the Shah and the *mullas* there was nothing left!

Within twenty-four hours of leaving Shiraz, the illiteracy of the south had been exchanged for the shrewd acumen of the relatively prosperous rice farmers north of the Elburz Mountains. Paddy fields, tea plantations and orange groves had blossomed under the hands of their emancipated proprietors. The success demonstrated, by contrast with the south, how injudicious any generalised assessment of the Shah's reforms could be.

Business and pleasure were well mixed. Once upon a time, the Caspian seaboard had been a favourite Riviera for Edwardian high society. Tattered traces remained, the Pahlavi Hotel amongst them. Emami in disdain of the old watering hole made for a local eating place, hard against the sands of the Caspian beach. Its reputation was founded on sturgeon, caviar, indigenous wines and vodka. Constellations of stars were high and bright in the darkening indigo of the clear sky before we staggered from the repast, leaving empty bottles on the table. Conversation as well as the food had stolen time. Looking back over recent adventures, I had asked Emami why he had left me to sweat in the Departure Lounge at Shiraz Airport until the very last moment when the doors of the 'plane were closing upon us. He was sorry and apologetic.

'I had gone to worship at the shrine of my Fathers,' he explained, 'and could not get back.' 'The Prophet has taught us that all virtuous men follow the faith of their Fathers,' he continued. 'You are a virtuous man and I am a virtuous man. I hope, then, that you are a Christian. If I had been born where you were born I would be a Christian. And if you had been born where I was born in Shiraz, you would be a Muslim. It is so with all virtuous men.'

Later, the two 'virtuous' men made across the sand to regain their parked truck. Alas, the vodka and caviar had told upon our sense of direction. Within minutes of starting the engine, the truck was driven zig-zagging into treacherous quicksands. Shouts of alarm drew a host of night wanderers to the rescue. With commandeered tractors, ropes and shovels they laboured well past midnight to raise the sinking vehicle.

Further along the coast at Babulsar next morning fortuitously bobbing in the sea was the one thing most needed to help fulfil my obligations to the Minister – a ready-to-hand secretary! Her name was Elizabeth. She was a nurse who had trained at the Cottage Hospital in Eastbourne under my friends Snowball and Harris. Now working in Iran and with a day's leave to spare Elizabeth was swimming in the Caspian as if she were off Eastbourne pier. For me, next morning, there was a deadline to meet in Tehran, a TV and Radio broadcast. Before then the text had to be written and typed for the eyes of the Minister. In exchange for dinner with her friends that evening and a free ride for the bunch of them to Tehran in the morning, Elizabeth found the paper, the time and the willingness to do the job.

What at the end of this romp through a strange, fascinating and friendly country had been achieved? There was the first-hand evidence of an extensive and new land reform policy in action – that was valuable. But the original purpose of going to Iran, to arrange field studies for Haleh Afshar, was far from being consummated. Indeed, there was little prospect of my ever coming back to the country. Consequently, I felt no constraint on offering constructive criticism. No doubt a swift and radical change was enveloping the countryside, so swift and radical in fact that it seemed to be more a political expedient that a genuine, effective redistribution of natural resources. The central theme of my broadcast, therefore, was to ask what the Government intended to put in place of the traditional *khans*, the great landowners who down the ages had been the patrons of the rural social order. To remove or weaken them, without effective and acceptable replacements, could undermine any immediate and long-standing good. This I pointed out in my broadcast. Then, with a farewell to Dr Valian the Minister, I slung his gift of an attractive Persian carpet over my shoulder and made for the airport; there to be waved off to Tel Aviv by Emami and a bevy of maidens in party dresses. My wonder at the Minister's kind courtesies and the generous proportions of his valedictory dinner of the previous evening and at the social significance of its

Swollen waters after rain in northern Persia

company went no further than my enjoyment of them. Likewise, memory of the strange conversation had with Emami on the last evening in Babulsar when he deliberately, almost forcibly, sought my views on the place and virtues of monarchy in a modern democratic world went unpondered.

Beyond Tel Aviv lay the six months of the pressurised summer of 1968; Rab Butler's Paper and the Centenary Celebrations of the RICS; a fleeting visit to West Africa; the last excursion on European land tenure comparisons to Hungary and Romania; and the 12th International Congress of FIG. So apart from a few 'thank you' letters, keeping in touch with Zahra Samii and making a mental note that, with no prospect of seeing Iran again, the programme for Haleh Afshar would have to be rethought, all things Persian – caviar, carpets and Cheloe-Kebab – were of the past. Or so it seemed.

❧ ❧

Some six weeks later, my hand trembled over writing a lunch cheque in the Carlton Club. My guest who had been left sipping his wine had gravely disturbed my equilibrium. The jolt had illuminated the

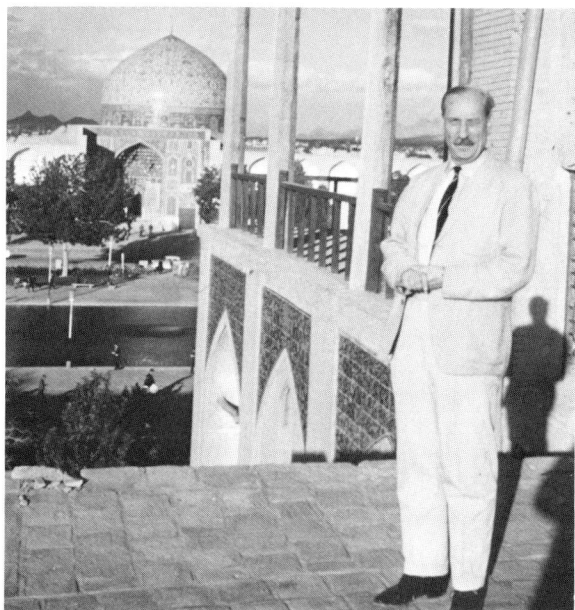

DRD casts evening shadows over the Blue Mosques of Isfahan

contents of a short enigmatic note he had sent the previous week. The note had begun with the abruptness of a thunder-clap. 'You will not know me.' was the startling introduction, followed by 'I am coming to London to meet you on a mission of grave importance from the Middle East Oil Consortium. The nature of the mission will be explained when we meet.'

As an avid reader of John Buchan, whose heroes of international intrigue often met in the Carlton Club, I had suggested to this mysterious correspondent that it would preserve the spirit of his communication if we were to do likewise. His name was Charles Lister. Over the coming years we were to become firm friends. But the message he had brought to the lunch-table had left me shaken and incredulous. 'I have come', he confided, 'on behalf of the Oil Consortium to ask if you would undertake to become a stalking horse between the Ministers of State in Iran and the Shah. We, at the request of the Ministers, especially the Minister of Land Reform, have been looking for a person to fit this job for some time.'

'Why me?' I exclaimed. 'I have only visited Iran for two weeks in my life and was shown what the authorities wanted me to see. I know nothing of the country. If such a person is needed, why not ask the United Nations?'

213 *Under the Peacock Throne*

'They'd only send a left-wing Swede', he laughed.

'Why not an American, then?' I persisted in my defence.

'They don't understand monarchy. Anyway, I'm not here to discuss alternatives. You were out in Iran as a guest of the Minister of Land Reform in May and your name has been given to the Consortium as the man he would like to have. What do you say?'

I left him, went to sign the cheque with a trembling hand and to weigh up this amazing request, knowing instinctively it was too preposterous to contemplate. Over coffee, Charles the plenipotentiary was asked to thank the Minister for his trust in me, for his offer and the amazing prospect it afforded; but to say 'No' – knowing so little of Iran, it would in my judgement have been improper to do anything else than decline. Then a small voice within whispered – 'Make a counter offer'.

'Would it be more courteous of me', I enquired, 'to thank His Excellency and propose that if he could arrange for me to visit Iran for a second time and allow me to explore the country on an itinerary of my own choosing, I would give him an answer one way or the other?' 'Done' said Charles, and so we parted. He was obviously much relieved but left me bemused and incapable of clearly understanding what my impetuosity had committed me to.

Early in September, a massive earthquake, high on the Richter Scale, devastated much of the Province of Khurasan in north east Iran and, by my reckoning, had given their Excellencies every excuse for not having an ignorant alien roaming at will throughout the land. In short, my counter proposal was a dead duck. But not a bit of it. Soon afterwards a 'happy to accept' cable arrived calling me back to Iran at the end of the month. I went.

For some days, short in leisure but long in luxury, relieved of every expense and provided withal with a 'purse' of 20,0000 rials cash 'lest the Oil Consortium had forgotten the razor-blades', my chauffeur-driven car took me each morning from the ugly Hilton Hotel to the portals of the Ministry of Land Reform and Rural Cooperation. There the Minister, Dr Valian, precipitated me into the babel of a high-ranking Committee. My responsibility was to control and advise it. The Committee had been convened to devise policies and plans for an Iranian National Land Reform Institute with a place on the world stage between the Chilean Centre in the west and the Land Reform Centre in Taiwan in the east. Neither Valian nor Mareafat, the Director of Rural Cooperatives; neither Zahra Samii nor Emami my companion of the former visit; nor anyone else showed awareness of my 'compromise offer'. By inviting me to Iran again, the Minister had surely implied acceptance. For a fleeting second, Zahra had hinted at my being a go-between and besides the Ministers and the Shah had added the Universities.

Not until Michaelmas did this pattern of activity change. Then I was asked where I would like to go. With an open map before me, my eyes lit upon Zabol, far from roads and air routes, where the backside of Iran butts into the Afghan Mountains. There also the mighty Hermand River drains to its inland debouchment north of the wastes of Baluchestan. Zabol: the very pronouncement of the name sounds like a final doom. It lies in Sistan, the province in which the earliest attempts at land reform under Reza Shah, the Shah's father, had come to ignominious grief. If present day land reform intended to make amends for the farming follies of past generations, Sistan above all other provinces had first claim upon its ameliorating aims. More sadly than elsewhere, the mistakes of the past had turned pastures into *mavat*, the dead lands of Persia, slopes into bare sun-baked mountain sides and rivers into barren scorched wadis.

A 250 kilometres drive through the desert from Zaredan, the capital, showed proof enough of the past misdeeds and failures. Emami had come with me and was jumpy on the journey. He was a man of built-up places, uncomfortable among the black, low-slung muslin tents of the Afghan nomads and their haughty, far-ranging camels and not a little frenzied by rumours of banditry in the lower hills. Even the absurdity of a lop-sided Pepsi Cola notice blazing in defiance of the broiling heat on the daub of a lone hamlet wall failed to raise his spirits. Zabol when reached was suffering a raging sand-storm. It resembled nothing so much as a cloudburst on the M11, except for the shadowy figure of a diminutive donkey carrying a bearded prophet-like rider swaying against the swirling sand. That evening the sense of the macabre was heightened by a drive to the gaunt workings of the Zehak Dam. The sky was a flaming sienna, shrouded from the dying sun by high clouds of whirling sand. Flung against the crimson and orange reared a massive erection of towering staincheons and girders. This portal to the kingdom of Persia seemed too gaunt and ascendant to fit into any human dimension. In eerie regal splendour it threw into abasement the lowliness of the surrounding desert.

Nothing I had met hitherto quite matched the novel domestic comforts of soup, dates as big as apples, and on marble floors, mats and blankets provided for the night in Zabol. As we lay on the marble in the blankets, the stars took possession of the heavens and Emami related for my edification something of the mysteries of a pilgrimage to Mecca. It was hearsay evidence. He had never made the mission himself for reasons which assuaged his own conscience. First, before any expenditure on a pilgrimage, would come the needs of one's family and close friends. Only when these had been met, either by oneself or others, would a man be spiritually and morally fit for the undertaking. These proper claims upon Emami had to-date stood in the way. He was young, however, and there was time enough. Good intentions were, so he implied, as good as deeds. Describing some of the unavoidable earthly needs of the pilgrims, he also portrayed the place and purpose of the *sigheh* in Persian marital arrangements. The Prophet, it was explained, permitted a pilgrim with the legalising assistance of a *mulla* to enter into a contract of concubinage whereby the pilgrim took to himself a *sigheh* for a pre-stated period – a kind of leasehold marriage. Sometime later on a subsequent visit to Iran and at an official party, his words revived when, with a broad grin on his face, a colleague informed me of his *mulla* status and suggested he should arrange a *sigheh* contract for me with my dancing partner!

❧ ☙

October 1968 was well on its way before BOAC for the second time lifted my feet off Iranian soil. Zahra saw me away. Emami, so he told me later, was scratching his head in desperation in another part of the airport. A long instructive farewell from Valian had dislocated the time-table. Embarassingly thankful for my help so far, he came straight to the point and asked for a formal Report without delay on my proposals for research policy and the establishment of a Rural Research Centre for Iran to cover investigation of land reform and all associated activities. As he accompanied me to the waiting Zahra, he hoped for my speedy return because, as he put it, he 'wished to againge owdiance with "is Impural Majesty"'. Another hint? Be that as it may, a blueprint and outline structure for a Rural Research Centre was prepared with dispatch and sent to Valian within a week of my arrival back in Cambridge.

Only by stages was the vast, intricate pattern of the Shah's land reforms unwrapped for me. Soon after the end of the Second World War, the young Shah had launched the distribution of the lands of the numerous royal villages of Pahlavi and had set up the Bank Omran to provide back-up finance to ensure success. Ten years later, the Government, pushed by the Shah, set about the distribution of public lands, *khaleseh*. The royal liberality and the State largess were not wholly popular with the religious and landed classes; some suspected that the growling of the neighbouring Russian Bear was unduly vocal in the ear of the inexperienced King. However, on the evidence of his autobiographical declaration of thoughts, actions and policies – *Mission for my Country* – it was not so. Opposition to the Shah's reforms hardened when, in the early years of the next decade he turned the attention of the *Majlis* to the private lands of high and low. An Election in 1961 had returned a biased Assembly packed with landlords' deputies. The Shah refused to reconvene it. Assisted by his Cabinet, he ruled for a year by royal *farman* and brought in his radical land reform legislation. In January 1963 (*Bahman 6,1341*) the Shah called for a public referendum seeking the support of peasants and ordinary folk to his *Revolution of Shah and People*. The sweeping land reforms were first among its aims. Thus armed with the shout of the *vox populi*, he called the *Majlis* to bow to it and to prescribe and pass the necessary legislation. There were three phases to the reforms: first, the distribution of the lands of the great landowners, the *khans*; second, the granting of options to sell, divide or lease village land; and third, the final, arbitrary coercion which vested all unsold and undivided land in the occupying tenantry and converted rent payments to annual land cost instalments. My report to the Minister of Land Reform and Rural Cooperation landed on his desk just two months before the start of the Third Phase, in December 1968.

A contract to give formal professional advice to the Minister now tied me to his side. It required my being in Iran two or three times a year. Besides which there was a personal understanding between Valian and myself which meant my looking after his interests in the UK and elsewhere whenever possible and necessary. On a return visit in February 1969, I found the Rural Research Centre well-established and expectantly waiting for a series of teach-ins on what should come next. By June, more specific problems called me back again. These required research studies of farm corporations. In the meantime, Valian had intended to visit the Netherlands, Ireland and the UK and had me running round, making contacts and planning social occasions. Grandees from the University, the Embassies and from political circles were invited to dine with the Minister in Cambridge in April after Valian's extensive tour of Ireland. At the last minute, another Iranian earthquake disrupted everything. The short postponement kept Valian back until July when he attended the Royal Show. From there he was taken to Welbeck at the invitation of the Duke of Portland, a courtesy arranged by Neil Elliott, the Duke's Commissioner. Valian was following preparations made by Zahra Samii who had gone to Welbeck in the Spring. The Minister invited Neil Elliott to Iran to inspect the budding farm corporations with a view to offering his

professional advice on management. Valian's postponed hotel bookings in Ireland were confirmed in the name of myself and Haleh Afshar.
We went to make new arrangements for the summer and to take
advantage of the links forged with the Irish Land Commission. The
land reform pundits of Ireland, who had been lined up to meet the
Iranian Minister, welcomed the Iranian scholar, gaped at her beauty
and were captivated by her ready wit. She had decided to compare the
Irish and Iranian land reforms. Those who knew most and best of the
Irish story were there to hold her hand and guide her feet from Dublin
to Galway, from Clare to Cork.

Valian, not to be outdone, was dined in the September by a posse of
Tory MPs at the Carlton Club. A friendship flourished between us
particularly after the upheavals which were to toss him out of
Government to become a Grand Trustee of the Holy Shrine of Imam
Reza and Governor General of Khorassan. The friendship ran through
his days in exile under Ayatollah Khomeini, to Valian's early death in
America. His personality contrasted in style and method to those of
other notables of Court and State, in especial from men who had held
the office of Prime Minister, like Amir Assadollah Alam, then Minister
of the Imperial Court; Sharif Emami, President of the Senate; and the
reigning Prime Minister, Amir-Abbas Hoveyda. Valian first and foremost was a soldier but unlike his old chief General Esma'il Riahi
preferred the discipline of the barrack square to Riahi's ambassadorial
diplomacy. He had taken over the portfolio of Land Reform and Rural
Cooperation from Riahi when the General became Iranian Ambassador to the Netherlands. Maybe, time would have interposed differences
and misunderstandings between Valian and myself, if each of us had
been able to command the felicity of spontaneous and apt speech. As it
was, his English was halting and my Farsi non-existent. Because his
reputed army vulgarity and hot temper never registered with me, only
his excellent qualities of courage, kindness, generosity and passionate
devotion to his job and the Shah distilled into our converse. Valian
showed up best in mufti with an old cloth cap on his head, notably
when as their familiar Governor he moved among the people of
Mashad.

In December 1969, I followed him back to Iran to join in the
triumph of the Rural Research Centre. Under the efficient Dr Goodarzi
the Centre's organisation had become most effective. An inaugural
National Conference on land reform, rural cooperation and farm
corporations was in process. It was expected of me to cheer them on, to
suggest ways and means of moving towards an International Conference in two years time and, as an ambassador at large, to help bring it

off. A hand wrought silver casket, inscribed with the commemorative date '1.12.1969' was presented to me. It was an outward token of something accomplished; but it set me wondering what next was in the cook's mind to keep me in the kitchen waiting for a royal summons to attend the drawing room.

On the side, I had become a kind of emissary with a personal mission between Ghana and Iran. In her childhood, Zahra Samii had lived in Moscow with her father the Iranian Ambassador to the USSR. The Samiis' diplomatic service neighbours were the Stafford-Cripps'. So it was that Zahra became a firm friend of Peggy, the daughter of Sir Stafford Cripps. Peggy made world news by marrying Joe Apieh, a prominant, practising barrister of the Kumasi Bar in Ghana. The two young Apieh daughers, redheaded, Aryan-featured and ebony-skinned were among the most beautiful girls I have ever seen. My mutual friendship with Peggy and Zahra commissioned a courier role between them. One evening, as part of my commission, I found myself eavesdropping on the repartee of these two raconteurs over dinner at St George's Hotel, Langham Place. Peggy was by then a hardened 'Ghana mammie'. In response to Zahra's probing, she related tales of her matrimonial experiences under Ashanti custom. As I listened, I grew more and more attentive. The tales were colourful and fascinating to say the least. Any 'mammie' *par exemple* concerned for the fortune of her family may request a leader of society, *à la* Joe Apieh, to sire her daughter *en passant*. Such nodding attention would ensure that at least a modicum of exalted blood was mixed with the family genes. Unlike the *sigheh* contract in Iran, this facility to meet a perceived social need would be initiated by importunate pleading from the female side.

❧ ☙

Unknown to me, I was under a form of critical arrest by *Savak*, the watchdog of the Shah. By one ostensible means or another, my presence was assured in Iran long enough for friendly and other agents of the secret police to compile a 'Denman dossier' of habits, speech and thinking. How else could the police risk leaving me alone, *tête-à-tête*, with the Shahanshah? Their methods, procedures and intentions were quite unknown to me. As with the ways of God over the affairs of men, time would eventually disclose the pattern of the dispositions and the reason for them. Illumination came in the closing stages of the National RRC Conference. Goodarzi took me aside and in *sotto voce* told me of the hope that the Shah shared with Valian that I would be willing to write an official history of the Persian Land Reform to be published in English. The purpose of the request was recognised as something beyond its superficiality; along with the RRC commission, it was an intermediate step on the winding path to him who sat on the Peacock Throne; immediate steps were *ultra licitum*.

In his book *Mission for my Country*, the Shah had written 'My heart is with our village people and as I think of the future I believe I see a magnificent vista lying ahead'. His words suggested the title for my book – *The King's Vista; a Land Reform which has changed the Face of Persia*. It took over three years to prepare and was published in 1973. Apart from a tightly packed programme of travel elsewhere and of home affairs in 1970 and 1971, those years also saw the final writing, proof-reading and publication of my introductory work *Land Use: An Introduction to Proprietary Land Use Analysis*. The publishers were getting impatient so the Persian writing had to take second place.

The suggested synopsis and chapter headings for *The King's Vista* were handed to Valian in September. By that time, he also was getting impatient, not with my hold-ups but with delays that had so far prevented my seeing the Shah. He was edgy about this for reasons which were only manifest a year later. Professor Ann Lambton's seminal work, *The Persian Land Reform – 1962–1966*, had been published the previous year. She had travelled widely in Persia and bent her scholarly ear close to the ground as she tramped the villages. She was in consequence critical of the ways in which the Government had gone about the business. While not in any sense opposed to the land reform in principle, Ann Lambton had not in the eyes of his Government given the Shah full marks. Valian was not pleased. He thought the book was bent one-sidedly towards the peasant, was unsympathetic of the Government's problems and showed scant understanding of the mind of the Shah. The book hit him with a force which would have made less of an impact but for the undoubted eminence of Professor Lambton as a Persian authority and her genuine concern and compassion for the country and its people. Maybe she had pressed the finger of truth too firmly on sore spots. By early autumn the upset had passed. Valian's impatience for me to see the Shah remained. It was not the Lambton book that jaded his nerves. For two days I was waiting expectantly while the Shah coped with an unexpected delegation from the British Foreign Office over what remained of British naval presence in the Gulf.

Such daliance fitted ill with plans which included an imminent return journey *via* Shell International's Agricultural Extension Centre at

Borgo a Mozzano in Tuscany. The Centre was arranging with me details of a Shell Seminar on Land Reform in Persia to be held at Shiraz in the coming November. There was no ill temper, only shortage of time. The best the impatient Valian could do was to beg me to go on a kind of 'red alert', ready to come to Iran to meet the Shah at forty-eight hours notice.

Christmas and New Year intervened. Somewhere about Candlemas 1971 a cable arrived to say a certain Personage would be wintering at St Moritz and Valian would come to London to accompany me to meet him. Meanwhile, to make good a promise made to the Oil Consortium, I had met James Prior, then Minister of Agriculture, to try and coax him to Iran to boost UK exports of livestock. Speaking for himself, he was eager to go but the bureaucrats' idea of 'fitness of season' threatened to stand in the way. True to form, as soon as Valian arrived in the UK all arrangements were cancelled. The Shah was delayed in Iran haggling with the oil companies over prices. Three days later, they were shaking hands and the previously arranged programme for St Mortiz revived. An excited Minister waited upon me at the Hilton Hotel in Park Lane only to blow my expectations to the wind. He was off that very hour; back to Iran to grab his slice of the new oil cake. The Audience, nonetheless, was on. The Shah *en route* for St Moritz would meet me at Zurich. A strict briefing of the 'real' questions to be put to the Shah was hardly completed before James Prior, the Minister, arrived to join Valian for lunch.

Two days later on closing my front door vaguely intent upon going 'blind' to Zurich, as I had had no instructions, the telephone rang behind me. I left the door ajar to answer it.

'Ullo, is sat Professor Demaan? Zees is see Minister, Iranian Embassy. tomorrow you have Oudience with His Imperial Majesty in Zurich. Fly from Heathrow at 10.30 in the morning. You will be met by Mademoselle Davodi. She will be at your every disposal, Professor. Will that be alright?' That evening, at the Excelsior Hotel, a negro cast in the mould of a heavyweight boxer, handed over an embarassingly large envelope, marked 'highly secret', to be given personally to His Majesty. Apparently, it was X-ray photographs of the Shah's kidneys. Mademoiselle Davodi, a red-headed Persian beauty came up to every expectation, despite the ridiculing banter of my Shell friends in London who were convinced she would turn out to be a fraud.

High in the hills above Zurich, the Grand Hotel Dolder housed for the nonce the royal personification of the Peacock Throne. At Zurich airport a mammoth limousine with the diminutive Mademoiselle Davodi lost in its ample folds whisked us away to meet the Iranian Ambassador to Berne. He stood anxious and stammering at the portals of the Hotel.

With haste bordering on the indecorous, a flunky rushed me to a first floor bedroom which needed seven-leagued boots to measure the distance from the bed to the *en suite* bathroom. Halfway across it, the telephone rang. It was the uneasy Ambassador.

'His Imperial Majesty can see you now'

'Sorry' I replied, 'I have no clothes'.

Five minutes away down the corridor, a panting porter was trundling my cases. Moments later, half way through changing trousers came another telephone summons, urgent and almost plaintive, 'His Imperial Majesty will see you now'. There was nothing for it. I explained that His Imperial Majesty could see me with or without my trousers. He had a choice! Then strapping my belt and fiddling with the last fly-button, I leapt down the grand staircase six at a time into a stately array of courtiers.

A curtain was drawn across a corner of the hall and gave the impression of a clandestine séance in progress. Ambassador Esphandary parted the folds for me to walk through. Against the far wall sat a man with a newspaper close to his face, motionless. As if playing Peep-O, he slowly lowered the shield. The Ambassador departed and the Royal hand proffered a chair. Now it so happened, that the threatened Third Phase of the Land Reform, the arbitrary eviction of landowners with its guillotine procedure for ending all disputes, was in the air. Valian had urged me to direct talk towards the question of coercion. The Shah, however, seemed disinterested. Because I had come from Nkrumah's Ghana, he wanted to discuss the difference, as he saw it, between a Dictator like Nkrumah and the occupant of the Peacock Throne of Persia. Oil was another topic. 'Your people get five times the amount of money for my oil than I do, and I intend to put it right. Do you know', he continued to chat as if to sieze any deviation away from land reform, 'what country I fear most? Not Russia. Oh No! It's India. India is so very weak against the hordes of China'. Eventually, by an oblique inference I hinted at the disquiet in his Minister's mind over the wisdom of pushing coercion. His ears were not open. So I left with mixed emotions marinaded in failure.

The Royal head had nodded approval of the book. So that was alright. Valian, however, some months later was still a haunted man. A telltale sign when next we met were his open fly-buttons. In kindness I suggested his dress needed adjusting. Moreover, his Ministry had changed its name, another omen for good or ill? With me, his main concern was to know how I found the Shah. What was the imperial attitude to an arbitrary end of land reform. At that stage, I was of little help. So I left him preoccupied but better dressed.

On Farmer's Day, 23 September 1971 the truth was out. Over the previous three days, events had moved swiftly. Within a day of arrival, I was in the Sa'abad Palace. Waiting for me was Valian. He was distraught and decidedly rumpled. As I entered the Royal ante-chamber, he button-holed me and seemed on the verge of collapse, from anger or despair or both. 'I am', he exploded, 'Minister of Rural Affairs for Iran. I know where there are four lights and there should be five; where there are three lights and there should be four; I know where there is darkness in Iran and there should be light! When you came through that door', and he waved his arm in the direction of the sentry, 'there was a man with a gun. He asked your name. If you had not given it, he would have shot you – that is his job. I am Minister of Rural Affairs, but I have no gun for my job. Everything in rural Iran depends on me. Standing in my way, making me powerless, are the Minister of Power and the Minister of Agriculture.'

At that precise moment, as if on a Hollywood set, massive ornate swing-doors opened upon a palatial reception chamber, more fitting by far than the becurtained cubby-hole of the Hotel Dolder in Zurich. Once again the Shah welcomed me to his side. We spoke of the book and of the pioneering work of his father Shah Reza. Then swinging his chair more purposely towards me he said, 'Tomorrow, I shall make an important announcement'.

What will that be, Majesty?', I deferred to his interjection.

'I shall announce the completion of land reform in Iran. By the morning of Farmers' Day, the day after tomorrow, all will have been accomplished.' His English was fluent. Then in a flash, I realised Valian's dilemma. The exercise of the new coercive law to declare all tenanted farms compulsorily enfranchised had guillotined the land reform process. No land reform: no Minister of Land Reform. So let him be an impotent Minister of Rural Affairs.

'Congratulations, Majesty'. I followed the royal enthusiasm, bowed my head one and a half inches in profound *gravitas* and sought permission to make a comment. He nodded. 'Might I submit that you are with your land reform no further than what Churchill would have called "The End of the Beginning". Your rural folk will need twenty-five years of clear guidance to enable them to gain the full benefit of what you have given them. You will need a Supremo over the intricate affairs of rural Iran.' He was listening. So I pressed my advantage 'What power has your Minister of Rural Affairs over the Minister of Agriculture and the Minister of Power? Control should be with the new Minister, should it not?' The Shah nodded a slow, ponderous Royal assent. 'We shall give close attention to that'. Much encouraged, I retreated backwards from the Royal presence.

Next morning the *Kayan International* carried banner headlines: 'Shah announces end of Land Reform. But warns of twenty five years of adjustment.' Valian whom I saw twice again on that trip seemed in lighter spirits. Cyrus Khatibi, a splendid fellow, whom Valian had assigned to me as an aide, lest I should flounder in error or misjudgement in my writings, assured me that his Minister had been tired out with lack of sleep and overwork. Maybe he was right. My thoughts ran in another direction.

The grand Seminar held at Shiraz under the auspices of Shell International was in a measure part of the massive postponed Cloth of Gold celebrations of the Peacock Throne. The Shah had refused to be formally acknowledged as the successor of Cyrus the Persian until his reformed country was worth governing. Now Persia gathered all kings and satraps of the world to Persepolis under the Royal hospitality of a specially-built tented city. For me, however, the memorable event of that year was a lunch given in Tehran by Valian in my honour. As his right-hand neighbour at table I inclined my head to ask whether he ever got his gun.

'Oh, Yes, indeed' he smiled. 'Thank you very much'.

⋖ई ईॐ

Later audiences with the Shah never measured up to the drama of Farmers' Day Eve, 1971, although for colour and glittering high ceremonial, the Royal Birthday celebrations in 1973 were in a class of their own. Whether out of kindness to the Shah or to myself, or for some other reason, the birthday proceedings listed me first among the after-breakfast well-wishers to exchange my congratulations for a handshake. There was, perhaps, a simple explanation for the precedence; I had brought with me as a birthday present a handsome, pre-publication copy of *The King's Vista*, specially bound in gilt-embossed leather. The Shah had been asking for the book and the manner in which he now received it conveyed the impression that it pleased greatly. Whatever his pleasure may have been, it was as nothing to the degree of my personal relief.

Publication of *The King's Vista* lifted a nagging load from my mind. Everyone in Iran, from the Shah downwards, had been unstinting in help and support, in particular Valian who showed personal delight in having my secretaries, Carol Bradshaw and Alison Hill, from

Cambridge to oversee the scrivening. The contract bound the Shah and his Government to meet the cost of providing a secretariat from Cambridge with facilities to travel throughout Persia. Without Carol and Alison to help me, the Shah would not have had his birthday present on time. Even Cambridge was supportive; 'Gunner' Hazelrigg of the Old Schools when told of my intention to take my secretaries, responded as became his army instinct with the quip, 'Nihil obstat'; but who are you taking to do the typing?'.

There were idle moments, especially for Carol when the mesmerism of sea and sand, sunshine and birdlife would transport her to realms where bosses, books and the daily bustle were of no consequence. One such afternoon, with Carol having nothing better to do than dangle her pretty toes in the tepid water of the King's Hotel swimming pool, was disturbed when David Llewellyn burdened with lamentation came to join us at the poolside. David was a scientist, a guru of high reputation in the Middle East employed by BP to advise the Company and client Governments on development projects. Oil companies and their doings meant politics in the Middle East. The upstart Gadaffi in Libya was threatening to nationalise all BP assets. It was a blackmail attempt to force Britain onto the Arabian side over the occupation by Persia of three islands in the Persian Gulf – the very issue that had stood in the way of my seeing the Shah at an earlier date. David was very sensitive over the standing of BP in the Middle East. Valian had sought his advice on how to handle what promised to be a massive underwater engineering project. The contours of the seabed in the Abadan Harbour approaches were impeding the oil barges. Although professional surveys were the first step in the process, it was understood that whichever country was appointed to make the surveys and give advice would have top priority nationally in the race for the engineering contract. The Dutch were good but David had strongly advised the Persians to give first thought to our people at the Hydraulics Research Station at Wallingford.

'Can you believe it?', he groaned as he lamented the tale, incredulous all the while of what he was trying to tell us. 'Can you believe that Wallingford has turned the invitation down! There must be an explanation. Can you get at the UK Government and tell them what is happening here? I shall use my contacts, of course.'

Three weeks into the New Year, in the middle of the first serious miners' strike and on the eve of his personal triumph in signing the Treaty to take Britain into the Common Market, the Carlton Club decided to give Ted Heath, the Prime Minister, a support dinner. Milling around with the pre-prandial drinks mob, Ted Heath suddenly spotted me. He came up wearing his famous grin and with the question, 'What's all this I hear about you?'. He was referring to a *Sunday Express* column on my Ibo Chieftainship. Brushing aside his curiosity, I told him of the David Llewellyn embroilment.

'I can't believe it'. Ted had in his voice the same incredulity as David. 'Write to me, at once', he urged and walked away without the grin.

So I did. 'Phones rang all day between No. 10 Downing Street, the Ministry of the Environment, the Department of Trade and Industry, the Overseas Development Administration and my Department of Land Economy. When the dust settled, it was Graham Page MP, a Secretary of State in the Ministry of the Environment who unravelled the web. The Minister of State for the Environment had understandably decreed that no Government Centre, such as the Hydraulics Research Station at Wallingford, should accept professional tenders to give advice. All advice tendering should be the province of private practitioners. The Minister had, apparently, not been told that Governments like the Iranian would never seek private practitioner advice on a Government contract. The Ministerial edict meant that the UK in following it would forfeit all chance of a British firm getting the Abadan seabed improvement contract. Once understood, the formidable decree was cancelled and David's lamentations by the pool in Tehran were justified.

All debts of obligation were more than discharged by David Llewllyn some months later and in circumstances which only the most mischievous of the Fates could have ordained. Carol, my right hand support, had again flown to Tehran to crib information from handsome Iranian officials, any one of whom was eager to spend a morning with her translating statistics from his records into her notebook. On the morning of our fate, a breakfast-time call had summoned me to the Ministry of Rural Affairs. Zahra Samii was waiting upon an unresponsive Minister in a way which signalled a false alarm. The day was being wasted. So, without more ado, I whistled up my charioteer, the grizzled driver of a groggy foreign motorcar, sent for Carol to share the back seat and set off on the 400-kilometres track to the blue-mosque city of Isfahan.

For miles on end, the road ran stark across near featureless desert under a broiling sun. As the sun lost height in the west, we lost time and tempers in the ramshackle vehicle. The car had boiled dry six times before the journey was half done. Finally, some 200 miles out of Isfahan in a barren wilderness, a slipping fan-belt severed its remaining cords. The car slithered off the road crest to sink under a cloud of steam in the ditch. Communication with the driver was limited. By signs he indicated his intention of trying to cadge a fan-belt replacement from passing motorists. Already the afternoon was far gone. The poor man's sanguine gesticulations gave way after two hours of fruitless effort to antics of disoriented despair. Some cars would pass in haughty disdain; some would stop and offer advice but no replacement; and others offered spare fan-belts of the wrong size. Traffic dropped to about two cars in the hour as the twilight rapidly faded. Suddenly Carol whose feet, calves and thighs had been armoured against the myriads of mosquitoes by 'St Michael' underwear, and moved by chivalry offered to weave a homemade fan-belt out of the fabric of her tights. The Farsi-speaking driver, to whom the description of such a feat conveyed in sign language would have been more suggestive than understood, was shooed away to make weak signals in the night. The resourceful maiden took what would have been next week's laundry from her legs and twisted and knotted it into an interim halter. With due regard for his Muslim susceptibilities, the improvisation was handed to the chauffeur on his return. Dressed in the Marks and Spencer's lingerie, the fan-blade gave two whirls of inspired hope followed by a long, dying whine. What was left of Carol's tights had strangled the axle-shaft of the fan. Only the mosquitoes rejoiced as they made for Carol Bradshaw's unprotected flesh. Sunlight, by now, was a fading flash of fire along the far horizon. Visibility was reduced to a murky hundred yards and all movement on the highway had ceased. The driver was obviously apprehensive about spending a night in the desert with no protection from rapine and assault. There was no alternative that we could see. In desperation, I straddled the road like a mini Colossus of Rhodes. With legs and arms outstretched I waited and watched on the crest in the hope that on-coming headlights would catch me. After what seemed a lifetime, the barely perceivable horizon began to move. Was it a mirage or advancing lights? The light gathered speed. And Lo! from the dark and the dust, like a veritable gift from Olympus, there appeared the body of a large, solid saloon car. The brakes screeched to a stop as I stood my ground. The divine visitation jumped out, half in anger, half in curiosity, to confront me. There was spontaneous laughter. Dumbfounded, I grasped the hand of David Llewellyn – the most improbable of all improbables. Goddess Athene couldn't have done it better. Carol and I were bundled into the ample rear of this heaven-sent carriage and with a wake of mosquitoes at our tail and a promise to send aid shouted from the new driver to the old, we sped into the night and the delights of Isfahan.

In the twelve months from 1973–1974 Iran changed. What was once a well-governed, orderly, progressive country, a reliable bulwark between a yeasty East and opulent West, a communist North and an unsophisticated South, had become a modern 'Klondike', a nation stampeded by the avarice of the world. Tehran in those days resembled a gigantic anthill into whose composed, efficient activity some ill-working God had thrust a spade and turned the orderly structure open to the helter-skelter of social madness. Anyone without access to influential strings would have to wait from six to twelve months to get a booking in a hotel in Tehran or in the larger cities. All the world's merchants, financiers and speculators were at the doors. The Shah and his OPEC cronies had increased the price of oil fourfold on the world market and had done so at a stroke. The country was flooded with ungovernable wealth. Iran never recovered. Vast chasms opened between the new infinitely rich and the still abject poor, a gaping void into which the inflamed hordes of Khomeini revolutionaries were to pour. When the facts of history are eventually sifted, however, it will, I believe, be seen that the blame should not be laid wholly at the feet of the Shah. He wanted the new wealth kept out of the country until such time as it could be more fairly distributed. Some of his Ministers, those least loyal to him, and certain of the Pahlavis thought otherwise and prevailed over the Shah's better sense.

My position was not changed. If anything it was strengthened. Valian wanted me running in and out between himself and the Shah. A new contract was negotiated, a Decoration was to be awarded me at the next audience and 'land consolidation', what the French call *remembrement* was mooted, ostensibly to justify my seeing him. As before, off-the-record chats were the purpose of these personal, private interviews. The more of them we had, the more confidential the Shah became. On his birthday morning when the book was handed over, there was beneath the royal reserve, plainly a churning anger about something.

'Will you' he eventually confided 'do something for me? Your BBC have made a film misrepresenting the purpose and conduct of our Literacy Corps and its work among the villages. They present me as a ruthless dictator using military might to enforce my ideas on my illiterate and less-privileged people. When, in truth, I am asking the more intelligent of our young men doing national service to volunteer

228

to help the hard-pressed teachers to remove illiteracy from the back-ward. The BBC have been ordered out of Tehran. Will you broadcast over our networks and put the record straight?' This was done, followed by my lodging a formal complaint to the BBC on returning to England.

Owing probably to tensions and misunderstandings about the distri-bution of the new oil wealth, Valian fell out with his colleagues. Rumour had it that he threw an inkpot at the Minister of Agriculture over the Cabinet table. Whatever happened, he left Government to become the Deputy Trustee of the Holy Shrine of Imam Reza in Mashed. There in 1975, I was given the facility of his private airplane and asked to draft a policy for the use of the vast wealth of the Shrine to establish an international Royal Institute of Science.

For reasons probably associated with Valian's shift sideways, the Shah never pinned his award on my humble breast. His Ambassador to the Court of St James, Amir Khosvow Afshar, deputised for him at the Iranian Embassy in London. The investiture was at a small luncheon party which the Ambassador had taken some time and care to prepare and had spared no expense in the process. In particular, he took special pains over selecting and inviting the thirty or so guests. When we all sat down to the six-course spread just after mid-day on 22 April, 1974, amusement and disappointment had for me mixed an emotional cock-tail. The lady for whom at lunch I was holding out the chair on my right-hand was Judith Hart, the recently installed Minister of Overseas Development in the newly-elected Labour Administration, and under-standably was one of the formal guests of the Iranian Government. Among counterparts on my side of the Guests List were Margaret Thatcher and Peter Walker. Peter fell out at the last minute. Margaret Thatcher who to my delight had spared time to join and honour the occasion was much to my chagrin at the other end of the table between the Iranian Ambassador, our host, and the Nigerian Ambassador to France, Leslie Harriman, who had flown in from Paris with his brother, Chief Hope Harriman.

Conversation between Judith Hart and myself resembled a perform-ance on a broken xylophone hammered out at great speed, discordant music which for me, at all events, was highly entertaining. At one point, somewhere about the apple dumplings stage, she piped in a tone of pointedly repressed incredulity, 'You think as people thought in 1848'.

'Does that mean I am wrong?' a dessert spoon half full of apple dumpling was waved in her direction as I commented, 'The Hon. Member for Buckinghamshire, at the time, Benjamin Disraeli was probably right, you know'; and continued, 'You give me the impression that, for you, truth marches forward with time. Is that why you call yourself a Progressive? If you are right, Karl Marx will eventually be outdated. So let's hope you are'. There is no doubt (Baroness) Judith did me a great service. She gave me no time to worry about what to say before being called to my feet to thank the Shah for the Distinguished Order of Homayoun which the Ambassador had draped round my neck. She had so charged my mind with electrified indignation, that I was ready to extol with genuine praise the Peacock Throne of Persia and the Monarch who occupied it.

The speech took up five animated minutes. It was far better than it would have been, thanks to the exchange with Judith Hart. She went back from the memorable occasion to pull out from every Department of her new Ministry all files that had anything to do with the doings of Professor Denman. My name was systematically removed from all commissions, committees and consultancies for which the Ministry was responsible. It didn't hurt very much. Despite her excoriation, I retain memories of a most enjoyable luncheon, thanks to the Peacock Throne.

10

TITLE WIVES AND OTHER HONOURS

There are nine and sixty ways of constructing tribal lays,
    And-every-single-one-of-them-is-right!

In the Neolithic Age, Rudyard Kipling

As a prelude to Spring in 1970 and to bring hope to war-torn Nigeria, General Yakubu Gowon, then Head of State, embarked with fervour on a programme of national reconciliation, rehabilitation and reconstruction. Lagos Airport was still embattled when on a fleeting mission, wholly in step with the General's purposes, my plane out of Brussels touched down on the tarmac one February morning in time for breakfast.

With friends on either side of the battle lines, there was nothing of partisan preference behind my concern for the fate of John Umeh from the Enugu Campus, at the hands of the University of London. Defending his homestead in the Biafran bush, John had no way of sitting the University's written examination for the MSc Degree, although he had submitted his thesis. Unknown to him and suffering from my importunate pleadings, London University eventually agreed to waive the written papers, subject only to the opinion of the West African Examination Board. It was this attitude of clemency that brought me to Lagos that morning to have breakfast with the Secretary of the Board in transit to Accra. Gowon was adamant. To honour his word he had appointed Ukpabi Asika, an Ibo academic, to coordinate refugee relief during the conflict and to oversee the restoration of efficient public administration in the East Central State following the end of hostilities. Even so, leading Ibos were hesitant and fearful to trust Gowon and travel to Lagos and Yoruba country. The airport breakfast was an early and most promising step in a chain of exchanges with John Umeh. Over the summer, his correspondence brought mounting evidence of confidence and trust. Patient persuasion paid off and to my delight John came over to Lagos in the first week of September to meet me and his Yoruba professional friends of pre-conflict days.

231

All air services to Enugu from Lagos and elsewhere in Nigeria were still grounded. The main road access had been blown up early in the war in an act of sabotage which but for a timely warning could have carried me together with other bits and pieces down the Niger to the sea. Nigeria's troubles had sprung from civil unrest in the Northern Region where Ibo domination over the Ahmadu Bello University and the civil service had outraged the native Hausas. They were already disaffected over the murder of Ahmadu Bello in 1966. Now they rose in uncontrolled fury to massacre the Ibos in their midst. The Ibos led by Ojukwu gathered to their homeland for war.

While the civil war in Nigeria hampered progress in that country, in Ghana the birth of an organised indigenous surveying profession was taking place along with rapid advances in related higher education, to put that country in the lead in Africa. Ghana, just then, was heaving with its own troubles which were playtime compared with Nigeria's bloody strife. The country was in the throes of peaceful change from military rule to a peoples' elected Government. For many the switch meant an exchange for the worse. General Afrifa, then in command of the military, had a cool head on his shoulders and had rallied wise men among the civilians to sit with him in Government. Included among them sat Isaac Ofori, a friend of mine and graduate in land economy from Cambridge. Isaac had been appointed by Afrifa to be his Commissioner of Rural Development. The hustings before the Election were jamborees of wild excitement. When in Accra I joined in and was given a T-shirt to support Harry Sawyerr. Harry was a chartered surveyor and the only man standing as an Independent for the city. Pulled over the head, the T-shirt displayed across the back a blatant imprint of Harry's splayed hand and the legend – 'the five fingers of integrity'. Both Harry and the shirt were to have dramatic parts to play later on.

From political excitements in Ghana my homeward run diverted to Kaduna in war-striken Nigeria to pay a fleeting visit to Ahmado Bello University. Duties in Ghana had introduced me to J.Y. Abu, a new recruit to the teaching staff of the Department of Land Economy at Kumasi. Kaduna airport on arrival was an oven, metal-hot under a blazing sun. There were no signs of welcome, recognition or transport. Every face was a stranger's. One by one the arrivals were met by smiling hosts to leave me standing in isolation and loss. An hour had elapsed when in the distance through the doorless aperture of the Arrivals Hall came a plaintive cry, 'Dr Abu, Dr Abu'. A diminutive, birdlike figure, wide-eyed and waving a dirty chit of paper stumbled towards me. 'Dr Abu', it appealed to my face. It meant nothing to me. I had left John Abu in Kumasi. Why should he now be at Kaduna Airport? Round and round the room the seeker after missing Abu uttered his plaint. After a while, he went out and left me in mounting anger. Suddenly enlightenment brighter than the sun struck. Abu? Abu? May be he meant Ahmadu Bello University. 'Are you looking for someone?' I called and ran to catch the small creature. He was nigh to tears but had neither name nor title to help. So I chanced it and posed as Dr Abu. After all, there was a war on and risk-taking was the name of the game. My hosts waiting at the University wondered what on earth had become of me.

Flying the Sahara never failed to hold me in suspense. The vastness of the sand is as mysterious as time and space. Sometimes the desert seemed endless. At others the jumbo jet would fly into the dusk above the tropical homesteads and the evening fires winking below before the teacups were cleared from the cabin. But on Wednesday 26 August 1970, the BOAC flight from Heathrow to Accra arrived there before the Sahara and the tropical rain forest had registered at all. Time was *non est*; and the only space that mattered was the aisle between myself and the girl sitting opposite, a stranger at Heathrow, a firm new friend as we tripped off the plane at Accra. Maybe the occasion for my visit had something to do with the eclipse of time. Everything about the trip was honorary, voluntary, *en fête*, taken only at the last minute and well beyond all contractual obligation.

Only Ghana among all the newly-emancipated countries of the Commonwealth could at the time have provided the limelight appropriate to the ceremonies that awaited me and which, in anticipation, had so surely animated my heliotropic pride. The surveyors of Ghana, with public accord, were about to inaugurate the first national professional Institution and break away from the Royal Institution of Chartered Surveyors. History was being made, both for Ghana and for the surveyors' profession. The campaign was encouraged by CASLE, fortified by the indigenous Department of Land Economy at Kumasi University with its degrees linked to the profession, backed by Government and led by the redoubtable Harry Sawyerr. A crèche of 'midwives' gathered from the great and the grand was being assembled in Accra to assist in the birth of this inimitable child. To be a Fellow of the Ghana Institution of Surveyors would, henceforth, for local folk be more prestigous than being a Fellow of the Royal Institution. When I left Cambridge for Ghana it had been intimated that something exceptional was brewing. That the pomp and ceremony should include me in a *tria iuncta in uno* of Honorary Fellows along with Mr Justice Ollennu, acting President of the Republic, and

Professor Sam Sey, Pro-Vice Chancellor of the Kumasi University, was unexpected honour.

In Harry Sawyerr, a pair of straight-browed eyes defended the prominent malar-bone ramparts of his angular countenance. Harry was among the newly-elected MPs, a man too intelligent to make the same mistake twice. The support he passionately wanted from the Royal Institution of Chartered Surveyors for the new Ghanaian Institution he nearly forfeited by pushing for the ephithet 'chartered' as an adornment to its title. Stepping down, he won support from the President and Secretary General of the RICS and from Sir Oliver Chesterton, President of CASLE. The first two of these lined up with other worthy 'midwives' – General Afrifa, Prime Minister Busia, sundry of the new Ministers and Vice Chancellor Evans-Anfom, of the University of Kumasi – to make the birth ceremonies of the new Institution a memorable, outstanding success. The ceremonial extended from the Wednesday to the Sunday, an unbroken series of jaunts, dinners and speeches to be crowned by a spectacular Ball. The inauguration ceremony itself lasted five hours on the Friday. Harry Sawyerr won justified praise and acclaim. For Edward Battersby, the President of the RICS, the proceedings were an eye-opener. They educated him in the worthiness and potentialities of the work and responsibilities of the RICS overseas.

Marion, my companion of the outward flight, was from Canada and the Canadian High Commission had settled her in a bungalow at Winneba. She had partnered me at the Ball. As we tapped out the High Life rhythm under the rising moon, she let me in on the secret of her double-sided job. Only the transparent genuineness of her recital of it gave credence to what otherwise appeared a most unlikely caper. Much of what she said, moreover, justified the alacrity and unreserve of her acceptance of the invitation to the Ball. Ostensibly, Marion was a teacher provided by Canada to help with educational development. Covertly, so her whispered confidences informed me, she had been hand-picked to be a 'good guy' snooper. Underground racketeers, master-minded from Canada, were known to be running an illicit trade in ivories, gold and fashionable African artefacts, making vast fortunes and depriving Ghana of her legitimate revenues. Marion was commissioned to get alongside the Ghanaians, the influential and the commonplace, and discover what she could of the goings-on. Dancing with Ministers of State, leading professionals and academics within thirty-six hours of arrival in the country made her 'wild in the head space'; so her fine-chiselled lips assured me in a profusion of thanks. Whatever else the Canadians were up to, this perfect model of their womanhood

'Hands up if you know me.' Marion, the Canadian 'teacher' on Cape Coast sands, Ghana

dented the local hotel trade by offering me accommodation at Winneba whenever I was not in Kumasi.

No sooner had the junketing finished than a grand valedictory Committee saw President Battersby and his cohort off at dawn to meet Nigerian colleagues in Lagos. The focus of business was to make plans for a CASLE land tenure Conference but at the Federal Palace Hotel we welcomed John Umeh from Biafra. To us he was as a brother from the dead. For him the meeting must have been truly traumatic. Foes of yesterday were once again old friends. Had they ever been 'the enemy'? Certainly not in the mind of General Gowon. He would have harboured no such thought. Years later in the presence of the General, I, a friend of the Ibos, inadvertently used the word 'enemy'. At once, Gowon took me to task. 'My opening command', he said, 'when Field Marshall of the Federal Forces was to insist on respect for our Ibo brothers. The Biafrans were, in our eyes, gravely mistaken men and women. No one, either in mufti or under arms, was an "enemy"'. The General's noble sentiment had made the happy meeting in Lagos possible and had sealed it with the earnest spirit of the Head of State.

The touch-down visit to Nigeria was all over in a day. Our feet had hardly rested on Nigerian soil before Edward Battersby and I were jammed in an overcrowed airplane *en route* for Rome. Squashed between the window and my seat, the only remaining one on the plane, sat a young

Polish lass, Malgorzata Niczyperowicz. She was travelling in hope of eventually reaching her native Poznan after seeing the Pope in Rome. The only tongue she spoke was Polish. Of any other she was almost totally ignorant. Nevertheless, by the use of sketches, magazine illustrations and signs, she informed me that she had no visa for Italy, nor currency other than Polish *slotys*. So I undertook to shepherd her. Two hours in Leonardo de Vinci Airport were spent wrangling with customs for a temporary passport and visa. By then night had descended. My friend from Cambridge, Colin Kolbert, I knew should be waiting at the Hotel Boston. Another two hours saw us there in a hired car driven through dimly lit streets. Colin, to whom a telephone call had disclosed our plight, had spent the last hour or two persuading the hotel management to let go their only available bedroom for the unexpected Polish guest. Colin, himself, on his outward journey had 'rescued' a British Council teacher on her way to the Eternal City to set up an English school. At midnight all four of us were eating spaghetti on the Via Veneto. Next morning, after paying respects to FAO to justify our existence, all efforts were concentrated on Malgorzata. Alas, the last 'plane that summer of the only Polish airline serving Rome had left an hour ago. There was another day to lose. So Malgorzata was shown St Peters but not the Pope. Daybreak next morning saw her dispatched in trepidation to Vienna with no certain prospect of an onward flight. Two weeks later, however, a postcard to my home in Cambridge carried 'a thousand kisses' and the glad news that she had arrived safe and sound. Thus the Denman Rescue Service, sole proprietors Colin Kolbert and D.R. Denman, was founded to serve distraught maidens flying anywhere on the networks of the world!

⁖

Twelve months almost to the day from the memorable touch-down breakfast with the West African Examination Council in February 1970, Iboland was still without airways. A circuitous air route from Lagos via Ibadan stopped at Benin. From there in a Volkswagen handed me *mit Fahrer* by the British High Commission, I was trundled across the Niger at Abusa into war-torn Onitsha and on to Enugu. The scars and ashes of war were inescapable and were especially noticeable in the University. Buildings, naked and empty, stripped of the decor of civilisation, some cracked, some half-burned, some in total ruin and now the haunts of pillage and plunder, stood where four years earlier the brand new architecture of Azikiwe's Academy had crowned the

Nsukka hillside. On the lower campus at Enugu, the older seat of learning, the hideous skulls of murdered buildings marked sites which sun-filled lecture halls, neat cultivated gardens and bougainvillaea-decked villas had once graced. In contrast, nonetheless, there was a new-born spirit in the place. Ibo scholars, who in the civil conflict had come home to their native land from other Universities throughout Nigeria and elsewhere, were in command. No longer a foundling, the University had come home to the arms of the motherland, there to find unity that put to shame the tensions of pre-war days when aliens had been in charge. My purpose was to take up the reins of my old job with the new men, notably with Dr Njoku, the erstwhile Vice Chancellor of the University of Lagos and now returned to his native Iboland. Within a week, links with Cambridge had been restored, secondments scheduled, inventories of equipment agreed and a promising future mapped out and costed. But before I had girded up my loins for departure, John Umeh had sent a runner to the Rest House to ask me to stand by for a special mission early the next day.

'There are', he said, when we met in the red dawn of the Harmattan, 'as you rightly know, slaves, freemen and nobles in Iboland. Tradition constrains us from making men other than true Ibos nobles among our people. The red-cap chiefs of Nnobi have, however, decided in solemn conclave to break with custom and make you, if you will accept the honour, one of our nobility. Will you accept?'

Before I could reply, John Umeh continued, 'It would mean you coming to Ifite near Nnobi deep in the bush after first visiting the witch doctor. The rituals and ceremonies are protracted and cannot possibly be done in a day. But if you are willing, we can make a start today.' Clearly he was offering me an exceptionally gracious honour. I had no idea what it involved but knew instinctively that to refuse would be near to inflicting an insult. Refusal was to be shunned at all costs. So I thanked him and with the hint of a bow accepted the offer. It was necessary to remind John that to postpone my departure longer than a day or two was not possible.

Among the Ibo nobility of *Ndi Nze* are red-cap *Ozo* chiefs each adorned with kingly titles reflecting an ancient heritage from the days of the Ibo kings. Long symbolic ceremonies were the protocol of enoblement. Each ceremonial, performed before an excited crowd of subjects, was devised to reveal the regality of the beneficiary being 'crowned' with the Ozo titles.

John Umeh explained all this as he drove me over the narrow laterite tracks to find the witch doctor. As if it were a small matter of incidental consequence, he informed me that an Ibo title wife, *(nwunye Ozo)*, was

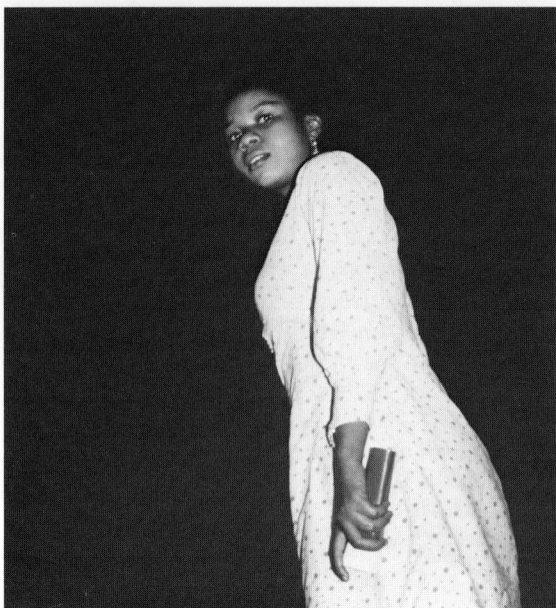

Ngozi, 'The Blessing', by torchlight, dressed for the evening ceremonies

an essential to help with the rituals. More and more intrigued, I enquired if the acquisiton of such a helpmate was a package deal or was there a choice! Comforted by the assurance that it was not only up to me to choose but incumbent upon me to find the lady, we pressed on to the witch doctor's abode. There, resemembling an old-time night-watchman perched in a sentry box, this worthy dignitary beckoned me inside to sit at his feet. After a mumbled incantation and a passage of hands he pronounced me acceptable for the high fate that beckoned some seventy miles away.

Ifite in Nnobi is part of a village complex within the environs of Nnewi in the Awka-heartland. Great excitement reigned that evening. *Itikummanuw* signals had been blown on the *igwe* to summon the Herald of the Grand Mask of the region and to proclaim a high festival.

From an aperture under the palm thatch of the main compound a maiden waved a smiling welcome. 'What is your name', I shouted back in gay abandon.

'I am Ngozi, the Blessing'. Her signals were a summons.

'Wonderful! Will you be my title wife'? I left no time to chance. Thus began the strangest friendship of my life and a future beyond my imagining. For the nonce, Ngozi wild with excitement fled into Onitsha

238

clutching my tangible sterling to buy a rigout suitable for the morrow and its ceremonies. As night fell, Kola nuts of welcome passed from hand to hand in the uncertain light of hurricane lanterns flickering in the huts of local chiefs and *dibias*. There, also, I was instructed how to chalk the sign of a red-capped Ozo title-holder on the swept and beaten earth. In the course of the evening's itinerary, my torch caught in a shaft of light the sheen of Ngozi in her new robe. Resplendent in a mantle of pure white she appeared a veritable Aphrodite against the dark groves of the forest veiled in night.

The compound had until recently been a fortified defence post. The outer wall still carried the defiant legend 'Freedom Fighters' and memories of the civil conflict were still green. The burden of the citation in my honour and of the oration of the reigning Chief presiding over the ceremony next morning was an expression of gratitude addressed to me 'who throughout the troubled years though far away in the land of the White Queen had been a brother, a father and a patron to John Umeh.' For my love, steadfastness and prayerful concern they wanted to give the greatest gift in their power to bestow: the ivory horn of office, the red cap and white eagle feather and four Ozo titles. The ivory horn was a magnificent carved tusk of the rare West African elephant and was prepared to stand on a silver mounting to be engraved with the following dedication:

> Professor D.R.Denman:
> in appreciation of your wisdom
> and knowledge in affairs of men and land,
> your true humanity which
> transcends national boundaries,
> religious creed and race,
> you are hereby admitted to
> the ancient Ozo nobility
> of Ndi Nze; ennobled to
> hold the carved ivory
> staff of office and to use
> the following titles,

> Eze Okwu Oma.   Eze Ugbo Na Ofia.
> Eze Chukwu.   Eze Ifunanya.

> All of Ifite
> Nnobi, this 4th Day of March 1971.

Eze Ifunanya (DRD) as Ozo title holder, Ifite, Nigeria

The morning ended in a phantasmagoria of bands, dancing, the invasion of masks and JuJu prancers and a long gauntlet of men and maidens violently waving goodbye. The farewells were conditional upon my returning. Less than half of the rituals had been performed and it would be something akin to high treason not to come back. Besides which there was Ngozi, the Blessing. 'All women have their dreams', I observed in a rash unthinking moment of departure, 'What is yours?'

'I live' she replied with a bewitching smile, 'in West Africa, half Anglophone, half Francophone. So I want to go to Paris and learn French'. Thunderstruck but undaunted, I assured her the dream would come true. My departure was a baptism in a thunderous tropical downpour. Soaked to the skin, I was bundled into an abandoned military truck for a bone-rattling ride to Benin. Next day, with the military in pursuit of smugglers, a fully armed trigger-happy sergeant pounced on me in Ibadan Airport. Inside my suitcase lay the splendid ivory tusk. In mounting fury he threatened me with arrest. I was just the kind of white-faced pig-rogue they were looking for. In vain, I explained that the ivory horn was my *Ofo*, the token of my new Ozo title. 'We don't make whitemen chiefs in Iboland' he shouted in unbelief and drew a loaded revolver from his hip. 'I am from Rivers', he boomed to justify his denunciation.

'If you are from Rivers', I rose rattled to defend myself 'you will know that I must wear the red cap and white eagle feather in the markets of Iboland.'

'You haven't got a white eagle feather' he parried with a snarl of disdain.

'You haven't looked under the dirty socks'. I tried a touch of humour and scrabbled deep among the dirty linen for the cap and feather. He was wavering but still not fully convinced. Fortunately the searching uncovered the English translation of the citation given me at the initiation by the presiding Chief. The crumpled paper consoled my captor and saved my skin. To my astonishment the bellicose sergeant grovelled on the ground and proffered me his revolver. I lifted him up, handed the gun back, together with a handsome *pourboire*. We parted the best of friends.

The thought of Ngozi and her dreams followed me for the rest of the journey; first to the University of Ife and then on another memorable visit to Ghana. Fortunately, Leslie Harriman, the brother of Chief Hope Harriman, a friend of long-standing and former pupil of Christ's College, Cambridge was the current Nigerian Ambassador to France. Back in Cambridge, I 'phoned him in Paris. He assured me of diplomatic help and advised a short language course at Tours University for my Nwunye Ozo.

Events moved apace. The renewed link with Cambridge had inspired the University of Nigeria with sanguine hopes of setting up a Faculty of Environmental Studies. Although a creation of little practicality, it was sufficient of purpose to get me back to Iboland in December. In the meantime, real dreams akin to nightmares had been the subject of a series of exceptionally strange letters from Ngozi. At morning light, she would wake with her body scratched and sore. She put it down to a night visitation of some kind, probably an animal. At all events, it was an experience beyond rational explanation. The *dibia* (native priest) came up with the explanation that Ngozi was a daughter of the goddess of the Niger, Mammy Water. The goddess required propitiation, the pouring of libations and rendering of sacrifical gifts to the River to appease her anger.

On arrival at Ifite for the continuation of the chieftainship ceremonies, I learnt more of Ngozi's background. Local folklore had it that her ancestry, on both the father's and mother's side, ran back to the Leopard Kings and Eze royals of ancient times. Paramount among the latter was the Eze who, as mythology relates, was the last to have been vouchsafed the privilege of seeing, in the depth of the dry season, the Water Goddess festooned in snakes and sitting on a gusher of pure water from the sunbaked ground. The Leopard Kings last displayed their presence and prowess at the traditional funeral of Grandfather Umeh. On that occasion, a *dibia* (one of the family) in a trance called forth a leopard to ravage and kill the poultry of the neighbourhood. Somewhere in the universal subconscious of the Ibo peoples these forces roam at large and suggest Ngozi's troubles have special links with the past. Whatever their numinous meaning, the troubles were not over by any means.

My return to the seat of chieftainship was to perform the night vigil. Darkness had fallen and a great excitement gripped the village. Surrounding an arena floodlit by naked torches some hundreds of tribesfolk gathered for the Mma dancing. Women and children trembled in fear and anticipation. Ngozi who was sitting next to me on a bench shook like jelly. They came, the weird birdlike creatures advancing to the crescendo of a loud tin-can drumming, withdrawing and coming again.

After two hours or so, I was called to the hut of a local chieftain to meet a kind of vice-regent who stood over me with a live chick in his hand. 'You hear the cry of the chick?' he asked, 'it represents the carping and quarrelling of lesser folk. You must symbolically demon- strate your kingly powers to quell this common disturbance by sitting on the chick when I place it on the floor.' I nodded, incredulous that he could be serious. He was so, nonetheless. At the first attempt I missed the bird which took off with a great squawk fluttering up to the dark, murky ceiling. Eventually, success crowned my efforts in what airmen called 'a pancake landing'. Once seated on the now mute bird, the officiating chief placed the heavy ivory tusk, the *ofo*, in my hands and bid me sit in silence till permitted to rise. Near midnight, Ngozi popped in with a dish of red soup. At one o'clock in the morning a delegation from the paramount chief bid me rise. I sensed it was a mitigation of some kind. 'Why?' I asked, and added 'What is the custom?' The initiate apparently was expected to sit till dawn lit the tree tops. 'Very well, then,' I insisted, 'that I shall do.' The worst of the ordeal was the invasion of the hut by myriads of winged insects, slowly crawling camillion and fast chasing geckos. The drums at dawn sent the message through the forest 'the white man is still sitting'. It was not for long. The oldest chief now in his nineties, totally blind and led by the hand came with an entourage to stand over me. He spoke long in regular cadence. He had come to tell me himself that the elders had decided again to break with custom. Never in Ibo history had five Ozo titles been given to any man. But this morning was the exception. I was to receive the additional title of *Eze di Igbo Mma* (King beloved of the Ibo people). After that I was permitted to rise. Ngozi led me outside to an enormous pot of water heated over an open fire. Her job was to assist with the ceremonial bathing. That over, it was breakfast and dancing which ran on from dawn to the glint of a hot sun shone in the tops of the paw-paw trees.

The next meeting with Ngozi was to be in Paris. The Nigerian flag fluttered proudly from the bonnet of the Embassy limousine as it sped smoothly down Rue Victor Hugo to Le Bourget Airport. The Ambassador had lent his car and much Corps Diplomatique polish besides. Ngozi was due that evening. There was no mistaking the flight, nor the day, nor the hour; what, then, could have happened? As customs cleared, it was patent that the much-heralded Ngozi had not come. Over the hot line to Lagos from the French Embassy there was no trace of her.

When a fortnight later she eventually arrived, there was far less fuss, only an indifferent relief. *En route* to Tehran, I had come to Paris to meet her and to learn from an agitated Nwunye Ozo the story of her false start. Only the strength of my contact with Ambassador Harriman in Paris wangled her visa and passport. They were treasured as divine gifts. She had clutched them to her bosom for the 700 miles of waggon journey back to Enugu from Lagos. On arrival the treasures had vanished. It was black magic. She fled to the *dibia*. He remonstrated with the broken-hearted girl for not propitiating Mammy Water! With a shrug of angry disdain, she returned at once to Lagos and camped for three days on the doorstep of the French Embassy till her doleful tale was believed. Helped by calls to the Embassy in Paris, the missing documents were replaced. The *dibia* back in Iboland was unmoved and gave a solemn warning that if she did not pour libations to the Goddess on the 700 miles of laterite road from Enugu to Lagos, she would either never return to her native home or would come back with great difficulty, sorrow and hardship. This too was laughed at, to be forgotten over the winter months in France. By which time, the bright girl had fallen in love with David Okwudilli. They were to marry according to the rites of the Roman Catholic Church. Her Ozo relationship with myself was unaffected by such mere incidentals.

On Lady Day 1972, I set off from Heathrow on the first leg of the homecoming journey arranging to meet Ngozi in Paris. Under my arm was a typewriter destined for Marion at Winneba in Ghana. With an uncharacteristic show of forethought, I had taken the machine from my Jaguar parked that evening outside the Excelsior Hotel at Heathrow. Clutching the typewriter in one arm and a mammoth valise in the other, I lumbered towards the hotel exit next morning. A kindly porter waylaid me. I was grateful for his help. 'My car', I said 'is over, is over, is over . . .' I stretched my eyes to find it in the car park. The Jaguar had vanished, never to be seen again. The words of the *dibia* came back to trouble me. Time was short. In fifteen minutes, five to inform the police of my plight and ten for a scrambled tax ride, I was outside the entrance doors of Terminal One. As the taxi drew up, the doors closed fast in my face. The entire staff had gone on strike! The dispute lasted three hours. It was a much distraught Ngozi I found languishing to be comforted in the vestibule of the Hotel California when I arrived in Paris. She had with her an enormous chest full of books weighing a hundredweight plus a stack of boxes carrying bridal kit for the forthcoming marriage to David.

'You will never get all this on the aircraft', I exploded in dismay. She retired to her bedroom with a sniff, a whimper and a disconsolate shrug. The evening was miserable and uncongenial – again I remembered the *dibia*. Next morning, Ngozi aired her French as we drove to the Airport. The taxi dumped the baggage on the *trottoir*. *Mon Dieu!* The big valise with my *Ofo* was not there. The taximan valiantly offered to go back to the hotel *pour chercher les baggages*. The crane lifted Ngozi's boxes on to the weighing-in machine and the dial swung with a jerk through 180°! The load was cargo. The cargo depot was three miles away. Another taxi was shouted up. When we got there, the cargo fares took all my spare cash. Time had sped. As we returned to the Departure Lounge, the Tannoy was announcing the positively last call. Just then the taxi driver rushed in, flopped on to a bench and announced my baggage *est perdu*. Ngozi was screeching her head off. There was nothing for it. Drastic measures were called for. The distraught maiden was pushed through the barrier with the typewriter shoved into her reluctant arms. I had no money to give her, only my promise to follow as soon as possible. 'Whatever happens', I called after her, 'don't lose the typewriter'. I remembered the *dibia*.

With Ngozi's wailing still ringing in my ears, I made off in the taxi. Back at the Hotel California, there was my baggage, innocent and intact, standing in the middle of the vestibule. It was a Wednesday, the next Nigerian Airways flight out of Paris to Lagos departed the following Friday. But Air Afrique had a flight *au matin*. Throwing all caution to the winds, I boarded it next day. Never go to Africa without first making cast-iron arrangements to be met at journey's end; this I knew to be a cardinal rule of life. Even so, there was neither time nor opportunity for such precautions. My arrival at Lagos in the gloaming was unwelcomed and unknown. I was alone. The customs official held my passport to his face and scowled – 'You can't come in here'.

'Why on earth not', I swayed with tiredness and mounting anger.

'Your cholera immunity is out of date'.

'That's not necessary for Nigeria'.

'Oh. Ya. It is. You must go back to London'.

'But that's nonsense. There's no flight and I have no money'.

'Your problem. Sit over there'. I remembered the *dibia*!

By midnight, after three hours, a senior Inspector, who had hurried from Lagos city, took charge. He made me swear to have an inoculation immediately on arrival at Enugu. Then he let me go. All transport to Ikeja and the airport hotel had left. Only a broken-down truck was on offer. Alas, as happened elsewhere many centuries ago, there was no room at the Inn. They had me booked in for the previous night. Again, I remembered the *dibia*!

Three days later, Ngozi and I met on the Enugu campus. Her bridal kit had been confiscated at Lagos Airport as she had had no money for the import duty. She had, however, hugged the typewriter for five hours on a painful bench waiting for officials to come from the city to weigh up her story of being an impecunious student returning home with the gift of a typewriter from her Professor. That sad 'student' and bride-to-be looked back with me over her journey and acknowledged that the *dibia's* warnings had been only too true.

John Umeh, indifferent to the dooms of the *dibia* and more inclined to blame Ngozi than conciliate her, had prepared for me a closely-packed agenda of urgent business. High among his priorities was the ceremonies of the last stage in the chieftainship procedures. At the pinnacle of traditional Ibo society, eminent above all others, stand the *Ogbuefi*, the Cow Killers. My fate was to join their exclusive number. Elaborate arrangements awaited us at Ifite. The event was a dismal undertaking and took up the whole of the day. Apportioning the cow carcase was the most messy part of the operation, a protocol that assumed the colouring of the priestly activities of Aaron the High Priest under Levitical law. Jonathan, my elder son, for example, was expected to have the prize of the cow's lights. As he lived in Reigate at the time, there was little hope of them reaching him. The chiefs had to extemporise and find substitute recipients. If by so doing sacred custom had again been breached it was no doubt all in the name of progress. Whatever the view, there was no denying my fifth Ozo title and that I was now *Ogbuefi*.

❧ ❧

If West Africa had little regard for the airtraveller who fails to plan a reliable rendezvous at journey's end, central Africa has no mercy whatever on the unfortunate traveller whose hopes of a pre-arranged meeting are dashed by misadventure on arrival. Experience taught me both lessons. Ngozi's home-coming jaunt was bad enough; but an unexpected dash to TChad in 1972 proved to be far worse. Huntings Surveys, in a welcome deviation from customary practice, had resorted to taking note of African land tenures. An irrigation project loud with political overtones had sprung from the waters and riparian marshlands of Lake TChad. Three nations – Nigeria, TChad and Niger – laid claim to the waters. Whichever of the three could first establish a sound irrigation scheme to use the waters would have the strongest claim to use them. Land tenure was critical as the semi-migratory Falani cattlemen were an obstacle on the pathway to progress. An immediate on-the-spot investigation was to be my tempting brief. Although my tropical kit had only just come back from the laundry, I packed it again, signed up with Huntings and committed my person to the mercies of French international airways in Paris.

The air tickets handed me when translated did not warn of the coming financial collapse of Air Afrique, the company that had issued them. Nor was there a forecast of the dense, early spring fog destined to envelop Orly Airport and ground all aircraft on the morning of departure. I was off to Fort Lamy in TChad. Darkness falls quickly in that remote spot on the banks of the Chari. The UTA flight that had taken on board the abandoned passengers, victims of the financial crash, was itself flying five hours late out of the fog at Le Bourget. It landed at Fort Lamy in the pitch black of a tropical night. The bleak walls and oily floors of the arrival shed of a third rate African airfield can be the nadir of desolation when abandoned of all human life, as was the *dépôt* where the belated flight stranded me. It was some hours before *un simple ouvrier* appeared to whom in schoolboy French I was able to tell of my marooned plight and ask for help. He pointed out a workman's bus. This rickety jalopy was coaxed to take me to the only *auberge* available, a modest hostelry whose decor closely resembled that of the airport shed. One bed remained unoccupied. *Au matin* on taking stock of the situation, I realised there were only four ways out. Only one of them was a valid option. To fly to Kaduna, the largest town some 500 miles from Maidugri in Nigeria where the local Hunting's man was to be found, would mean chartering an aeroplane from Fort Lamy and flying south to Douala in the Cameroun, westwards to Lagos and north to Kano and Kaduna – a distance equal to the journey back to Paris. No cables ran across the TChad,

Cameroun, Nigerian borders, so telephoning and cabling were out. Alternatively, I could dig in at Fort Lamy and wait. Since no message had reached me at the hotel or the Post Office, there was no telling whether anyone knew of my abandonment. The road option meant taking a bus across *la Chari*. Once over that fateful river, I would be in Cameroun and doubly marooned, with no visa to see me back to TChad. As the only sign of life to be seen across the river was the meaningless dawdle of a troupe of elephants, that option was far too risky.

My desperation was deduced from my antics and overwrought speech by a kind-hearted Italian. He had a Land Rover which he was willing to lend if I could find a driver. He was inhibited himself for want of a visa to cross into Cameroun. One of Ngozi's Francophone natives, by the grace of God, found his way into our halting conversation and volunteered to undertake a 'Dr Livingstone' venture with me across four *douanes* from TChad through Cameroun to Nigeria. Money was exchanged between us and with a wan smile, profuse thanks and in desperate hope, I set out across central Africa with my new-found companion. *Un parcours impracticable pendant la saison des pluies* led westwards. Fortunately, the rains were some weeks off. Track and wadis were dust dry. The landscape lay lifeless in a sultry heat. Numerous chattering monkeys swung in the sun baked skeletons of the leafless trees and a few high crested cranes strutted like self-appointed traffic wardens before our on-coming *camion*. At Dikwa, the halfway point *essence pour autos* and *eau potable* strengthened our flagging resolve and allayed fears of disaster which had been creeping upon us. Maidugri, when we got there, was somnolent. The few scattered stalls and ramshackle shops were shut against weekend shoppers. The Post Office was on the point of closure and knew nothing of the whereabouts of Hunting's resident man; there was no numbered Post Box for his mail. We backed off the town and encountered in the adjoining bush a military entourage conveying the Governor of the Province in pomp and regality. This austere personage was not a little irritated when I stopped the cavalcade to ask for directions. Nevertheless, he bent to the task and directed us to the local Rest House. The best *de luxe* bedrooms were equipped with lead handbasins and carbolic soap. There were no sanitary facilities beside a communal latrine. The poverty of the place served me ill that night. In the early hours of the morning, I awoke suffering acute sickness and diarrhoea of a severity never known before. Next day was for me a *dies non*. Too enervated to attend properly to my needs, I had literally crawled to the desk to try and pay the account. There a kindly Irish couple, Dr Johnstone and his wife, took charge. He was a civil engineer on a local project. They picked me up like a sack of rotting potatoes and bundled the 'sack' into the back of their car which became an improvised ambulance. In sweltering heat it made for Kaduna over five hundred miles of dirt track beset with hordes of screeching monkeys. Diagnosis had to wait till experts in the College of Hygiene and Tropical Medicine in London had examined the damage. According to them, I had contracted a killing variety of Amoebic Dysentery. Back in Cambridge and lucky to be there, Huntings with profuse apologies were ready to account for my sore abandonment. Knowing of the financial collapse of Air Afrique, the elusive man on the spot had assumed I would not be coming to Fort Lamy. The ultimate irony was the refusal of the Food and Agriculture Organisation to pay Huntings' consultancy fees on the excuse that I had never turned up at Lake TChad to discharge my contractual obligations.

꧁ ꧂

Ghana and Nigeria honoured me unexpectedly and in modes which pointed to a difference in national character. In Ghana, primarily through the surveyors' profession, I was to have an honoured position among the groupings of a modern restructured society. In Nigeria it was social custom still vibrant with ancient folklore that had designated the titles and status the chiefs and people of Iboland had given me. During the quarter of a century and more that I had spent in those two countries, which had become in their differing ways very dear to me, the peoples of one, Ghana, had tended to accentuate the distinction between the old ways, tribes and kingdoms and the requirements of the technological age to which they aspired. In Nigeria, the rapidly expanding professional and academic classes still held closely to the heritage of the traditional past.

To be *Ogbuefi* among the Ibos was beyond any distinction a foreigner could expect. In Ghana the prestige-dispensers thought differently. Some years after the profession had installed me an Honorary Fellow of the new Ghanaian Institute, the University at Kumasi let it be known through my friend and colleague Professor Ben Acquaye that the local seamstresses were busy sewing a doctor's gown and cap for me. Just then life at Kumasi University and elsewhere in other seats of higher learning was boiling with malcontent under the new Dictator, Ignatius Acheampong. He had usurped Prime Minister Busia's elected authority. So postponement was in the air. Formalities were further delayed by Colin Kolbert's thoughtful and generous undertaking. Colin had gone to immense trouble organising sittings for a formal portrait of me at the hands of Richard Stone, against my retirement. A replica of this worthy picture of 'A Gentleman in a tweed suit' was offered to Kumasi University. Hence the further delay while the logistics of getting the replica over to Ghana were sorted out.

Colin and I went to Accra in May 1979 in time for the Convocation of Kumasi. There were three honorary Degrees to be awarded, including an LL.D for the Archbishop of Kumasi and the D.Sc. for myself. What the bestowers of this grace had not revealed was a ploy to get me to give an impromptu address to Convocation on behalf of all three recipients. A full Doctor's Robe and an ill-fitting Doctor's Cap, far too large for a skull only half the size of the Archbishop's, had me more than uncomfortable in the tropical heat of the expansive hall. Stage fright would certainly have set in had I known of the intention to broadcast my voice throughout West Africa. In another direction there were real distress. Ben Acquaye's mother had that morning died in a far land beyond Takoradi. However, there was no escape. So there I was, facing an audience of 1500 scholars and graduates, to say nothing of the radio listeners in Ghana and beyond, with no time to tie my tie or untie my tongue. At the back of my mind there had always lingered an urge to debunk some of the absurd illogicalities of our ultra biased secular age – such as, for example, the craze to call women 'chairs' because they want to disavow God's creative grace that gives them the beauty and privilege of being women. So I launched sardonically against another solecism with the opening gambit:

'Why, O Ghana, do you allow yourself to be labelled a "developing country"? The developing countries of the world are Australia, New Zealand, Canada and the USA?'

Ben Acquaye who had picked up the broadcast over his car radio, later told me that the listening populace in West Africa were highly delighted and amused. My theme had pin-pointed the Kingdom of Ashante where Asentehenes had reigned for over a thousand years. So it was that in giving thanks to Ghana and the University for bestowing honours on myself and my two colleagues I was able to remind its peoples of a rich past to which they were in danger of giving less honour and respect than its greatness deserved.

ᔑ ᔒ

When the chiefs of Iboland ennobled me among their number, beside granting a prestigious status, they had bestowed upon me the authority of a strange jurisdiction over the marital status of the widows of Iboland. Knowledge of this and its nature have come not by exercise but by instruction from traditionalists and lawyers. The funeral of a man of social stature among the Ibos is a two-phased burial. After the Second Burial, the widow is invited before the red-capped Ozo title holders to answer four questions designed to establish her future place in society. Death does not sever marriage with the Ibos, unless the widow by freedom of choice within her *umunna* (patrilineal line of relatives) names a man to replace the deceased husband. Procedure follows fairly closely the laws of the Book of Leviticus as with so much else in Ibo society. Should the widow name the senior among her brothers-in-law, he, to use the Scriptural phrase, must 'do the duty of a kinsman' and marry her, irrespective of his other wives, if any. The widow may prefer some other man in the *umunna*; if so the obligation to do 'the duty of a kinsman' falls on him. Should she decline to nominate anyone within the *umunna*, she has two options left: either to go back to her own people and leave her sons to be brought up on their father's land by the immediate kindred; or she herself can remain there to possess and till the land and bear and bring up other sons in her husband's name, by paramours. An unmarried widow, so it seems, is traditionally an abomination in Iboland. A devoted unmarried daughter can formally and faithfully become an *idegbe* and opt to live with and care for her unmarried father until he dies and in the meantime to bear and bring up children in the family name.

Marriage under State Ordinance or by rite of the Christian church lies outside these customary choices, commitments and obligations. Sometimes a clash occurs between customary polygamous marriage and monogamous marriage. There was a time in Cambridge when, because I was the only Ozo chief within 3000 miles of St Marks Church, Barton Road, my status involved me in a tangle between the two. Joe, an Ibo graduate pupil of mine, had asked me to preside over a marriage he was undertaking with Obi, his *nwunye* under Ibo custom. Joe was reading for a Ph.D and so was his partner. They were, therefore, hardly illiterate; on the contrary they were a highly intelligent couple. Obi for reasons best known to herself wanted to be publicly married under the rites of the Church. Joe assented out of the kindness of his heart but had no idea what he was doing. The wedding day, the wedding guests, the time and the place were all appointed. It was to be a Saturday. On Thursday evening the Rev. Bill Loveless, the vicar, wisely decided to catechise Joe. Did he know the vows he would be required to make on Saturday? 'No' said the honest man. When told of the vow to love his wife till death do them part and to forego all other, he flatly refused to accede to it and accused the Reverend Minister of purposing to make a dishonest man of him. 'No honest man,' protested Joe, 'could enter into such vows with impunity. How can a man promise to love his wife or any other person all his days? Love is an emotion not an act of will. Come next year, I might hate her. I will stand by my marriage. That I can do. But I cannot, nor can any man, undertake to love his wife for ever. He may well do so. But he should not promise to. And as for rejecting my duties under Ibo custom towards the widows of my brothers, that I cannot do.' Joe helped the Minister and myself out of the impasse in the matter of wife-loving by promising 'to try'. The question of what he would do about his brothers' widows was left to be answered when the time came. Should he predecease Obi, she would be in the clear, if she opted to be.

Under the customary Ibo marriage obligations, a brother-in-law, in no legal sense, 'inherits' his brother's wives. To him belongs the initiative, but no obligation arises unless the widow says 'Yes'. Even so, her consent carries with it no right of conjugal consummation until a public *igbugha okuku*, the 'killing of the change-over fowl' ceremony, has taken place. These marital laws of Iboland are strict and insistent to be obeyed. The BBC TV and Jack de Manio of the old Eight O'clock radio programme were totally ignorant of these niceties when interviewing me after the widespread news of the Ozo title ceremonies. They were much excited by the thought that they had a full-blown bigamist before the camera. Nothing approaching *igbugha okuku* had linked Ngozi with myself. Our friendship and relationship were ceremonial and Platonic.

There was nothing wanton about it. Nevertheless, the introvert within set me musing and meditating on the fundamentals of religious and secular marriage. My incurious Christian assumption had never analysed Christ's injunction: 'What God hath joined together, let no man put asunder'. In His eyes the critical fulcrum on which the unity turns is supposedly God's conjoining, not man's manipulation. Once together, there shall be no cleavage between a man and wife, unless they foresake one another in disloyalty. The linkage is a form of divine creation. Otherwise, even within Christendom, an anthropocentric interpretation of Christ's injunction must lead to one section of the church acting as if it alone could do God's joining together for Him and so invalidating the acts of all other denominations. This of course has happened; within the doctrine of the Roman Catholic Church, *extra ecclesia* union is invalid. Logically then, all children, even of a Protestant Throne, are born out of holy wedlock. A wedding, whether in Cana of Galilee or in some other spot, is an act *in oculis civium* that the bride and bridegroom intend to be conjoined as 'one flesh' within the creative ordinance of God. Not so very long ago in Merrie England, relatives and close friends would conduct newly-weds to the bridal bed and see them into it. The modern honeymoon and fast closed door are questionable escapes from such healthy practicality.

Africa taught me much and caused me furiously to think. Friends over there, sophisticated, intelligent and cultured folk, would point to the state of Western society as the consequence of the tenuous, unnatural unions of monogamy which they would maintain are no sure buttress against promiscuity, as polygamy can be. Life has given me no experience to challenge those views one way or another. Within a long marriage, knit to a wonderfully loving wife, love has blossomed as the radiance and aureate beauty of a tropical flamboyant tree. Other loves, the spontaneous, the self-sown, the cultivated, have, I confess, adorned for me a cherished Grove of Eros, but none so as to dim the paramount flame.

11

DEVOTION AND PROMOTION

And it's time to turn on the old trail, our own trail,
    the out trail,
Pull out, put out, on the Long Trail –
    the trail that is always new.

L'Envoi, Rudyard Kipling

The academic realm has practices which never move in a logical straight line. Certainly one can say this of Cambridge University. If one didn't know the inner secrets, there would be every excuse for thinking that a Professor would be obliged to hack at a rigorous lecturing schedule as a first duty. Whatever happens elsewhere, it is not so at Cambridge. What is required of a Professor by the Statutes of the University is *devotion* 'to the advancement of knowledge in his subject, to give instruction therein and to *promote* the interests of the University as a place of education, religion, learning and research'.

There is no contract with the University to lecture. Within the Colleges runs a tacitly accepted prohibition against professorial supervision. Before taking the Chair of Land Economy, my pen had trailed articles far and wide and publishers had not been reluctant to accept a book or two. Once in the Chair, the pressures of 'devotion' and 'promotion' were very real. Although lecture timetables carried my name, there was no statutory number of lectures to fulfil, as with University Lecturers. Outside the University, appointments came thick and fast to crowd my diary. The future was packed with all manner of deadlines: public speaking dates; submission days for articles popular and professional; opening of conferences and meetings and numerous miscellaneous entries. This pattern of 'devotion' and 'promotion' pertained not only in the UK. It was repeated increasingly throughout the world. Wherever opportunity occurred to put Land Economy on the map, to expound my understanding of it, to employ its analyses to help solve problems and promote knowledge and above all to use its postulates to underpin liberties, there I found time and space to be.

Use of land touches life everywhere. Land meets primary needs –

food, shelter and clothing. Land meets secondary needs – warmth, defence and communications. Flying, even, needs aerodromes and the seabed is land under water. Land can become desert; desert can become sown land. Land can be wasted or developed. Whatever happens to it depends in the first instance upon who owns it and upon the powers and meaning of ownership. Following the Second World War, in Britain for a number of reasons and throughout the world, it became accepted wisdom to plan the use of land by regions or on a national scale. Planning on these principles was not invariably and deliberately political. Nevertheless, it blurred the understanding of the incidence of power. Over the post-War half-century, it had been supposed that planners had the primary say on land use. What they prescribed stood. The notion is untenable. Primary power lies with the owner of the land. Planners can object but if an owner disregards them, they are impotent in the long run *qua* planners. Only by substituting an owner who agrees with them for the recalcitrant who doesn't can the planners' wishes be met. At its heart, this was the philosophical message of Land Economy. Some took more notice of it than others. The first reaction was frequently hostile; on reflection, the unprejudiced would accept its logic. Land management is the practical expression of Land Economy; and in places throughout the world wherever British influence prevails is a respected profession. Ownership of land has been a *au fond*, the difference between East and West. What happens to land is a question no politician can lightly turn aside. The new outlook of Land Economy, therefore, meant much sweeping away of entrenched ideas, new thinking to be mastered, new practices to be followed and planners to be reconciled.

The 1970s were very busy. Just before they opened, an invitation to attend a Conference in Malta turned up. The Commonwealth Human Ecology Council (CHEC) wanted the Department of Land Economy to be represented at its First Conference on Development. Christmas was near, and there were other doubts, besides which no one in Cambridge had heard of the Commonwealth Human Ecology Council. Nevertheless, there were interesting names on the prospectus. Lord Kennett of the Dene was one of them. We had done battle together when he was Government Spokesman on land in the Upper House. Lord Kennett was billed to Chair the proceedings. The invitation was therefore allowed to join the pile of papers on the desk and spared the ignominy of the wastepaper basket.

Ecology, a term defining the department of knowledge concerned with the living conditions of animate creation, other than the human, had been accepted for over a hundred years. That Land Economy could do without *human ecology* no one doubted, but whether *human ecology* in its new guise could do without Land Economy had better be enquired into. So bags were packed and my wife and I set off for Malta. We espied in the Departure Lounge at Heathrow a chattering platoon of newly acquainted folk basking in the presence of Lord Kennett. With them was a tall whiteheaded matronly figure, Zena Daysh, who in carriage and posture took after Queen Mary of Teck. Zena Daysh to those who know her is a prehensile force, grounded in kindness and a kind of wistful wilfulness. 'Zena must be obeyed, you know!' is implicit in her nod. She had been the General Secretary to a group of doctors who during the War had pooled knowledge to find the best ways of feeding Britain and the countries of the Commonwealth. When the War was over, the medics were concerned to make the world a more wholesome place to live in. They realised that health depended upon housing and food; that food depended upon farming; that farming depended upon land; that housing, again, depended upon builders, builders upon economic investment and so on *ad infinitum*. In short we all lived in a kind of House that Jack Built. They designated their thinking – Human Ecology – and Zena became 'the continuity girl' who, unlike the doctors themselves, lived and worked to see the past linked to a future.

If there is such a phenomenon as a human eco-system, one of the best places to observe and study it would surely be a well populated, selfcontained island with a long history. Fortunately for CHEC, the Government and the Royal University of Malta had together under-taken a lengthy case study of the human and physical resources of Malta and were ready and eager to sponsor a Conference. My role was a peripheral one, a trouble shooter standing in for a speaker who at the last moment had defaulted.

October is a pleasant month to visit Malta. Those who had the best of it at the Conference were the consorts. They left their spouses to lecture and argue and went off to tour the island. For my wife, Malta took her back to her salad summers in pre-War Eastbourne, summers she spent in the company of young Maurice Dorman. Maurice was now Sir Maurice, the last Governor of Malta pending independence. These two met again after thirty years under twinkling fairy lights ringing the Governor's Garden Party. On the morrow, the Captain of HMS Ark Royal (then in Valetta Harbour), Sir Hugh Springer (Chairman of the Commonwealth Foundation) John Chadwick (its first Secretary General) myself, my wife, Zena and others rejoiced in the warmth of the renewed friendship and of the sunlight reflected off the highly-polished dining table in the Palace where the Governor entertained us to lunch. For me, the Malta Conference began an association with CHEC which over many years has varied in intensity; and a friendship with Zena that has never been broken. Although she is a dear friend, I cannot admit to always agreeing with her. Her ideas were often difficult to accept and were infrequently incomprehensible.

CHEC, for any who remained loyal to Zena after Malta, meant other Conferences throughout the Commonwealth. Malta was followed by Hong Kong where my spectacles fell over the barbed wire between China and the Colony. CHEC was in those years and ever has been in pursuit of a Holy Grail, elusive, something that was not, is not and yet ever more shall be. Even CHEC's Headquarters were calculated to deepen its riddle and heighten a sense of the incongruous. Plumb on the Cromwell Road, a not unpretentious thoroughfare in London's West End, the seeker after a better quality of life found CHEC. He would climb four flights of a chipped, winding, stone stair in a building, which owing to the landlord's neglect, smelt of decay and neglect, ugly with paintwork worn to the undercoat and capped by leaky roofs supported on rotting woodwork; quarters verging on the uninhabitable. Human Ecology was then, and for me still is, too amorphous a concept to be a scientific discipline, too ambiguous – with its implied assumption of an holistic Absolute – to be a tenable philosophy; and patently not a theosophy. Indeed, it was its claim to a 'wholeness' which logically embraces the spiritual dimension of truth that made me most uncomfortable. At times it appeared a false conceit, a Tower of Babel, a confusion, a medicine for cleaning up the environment but only if taken in very small doses. Certainly, Human Ecology was woolly, vague, indeterminate yet kindly enough for me to stand up at the CHEC conferences in Malta, in London and in Hong Kong to talk about proprietary land use analysis. Just after the Malta Conference, following a misunderstanding with Lord Kennett, Sir Hugh Springer came to CHEC's rescue as its Chairman. Years later, Sir Hugh was honoured with the Governor-Generalship of Barbados. In the interim, he would effectively play a number of parts on the world stage; notably as Director of the Commonwealth Universities Association and President of the Commonwealth Foundation. It was this body which 'carried the bag' that helped to finance the work of CHEC, CASLE and the Professional Associations. By wearing these two hats, Sir Hugh made possible what was CHEC's most congruous function, then and for many years after. Academically and professionally, CHEC faced problems which are inherent in all forms of derivative knowledge; namely, the need to weave together a consistent whole, drawn from purer more circumscribed sources. Only CHEC was in a position to bring together representatives from other Associations to sit round a table and determine how each of their several departments of knowledge contributed to the condition of the human environment. From this lead, there emerged a standing committee engaged in cooperative effort and action, not only regarding the environment but in other directions also.

More concrete in substance than Human Ecology was the Human Environment. The concept was debatable with less opaqueness but could rouse more rancour. So it transpired at the UN Conference on the Human Environment in Stockholm in 1972. Led there by Zena, I was also being pushed from behind by Raine, Lady Dartmouth, as she then was. Lady Dartmouth headed the British Government's contingent alongside Peter Walker, the Secretary of State for the Environment. The Countess had put into my hand her guidebook *How do you want to live?* written for the occasion. The Conference was divided by the fundamental cleft between Government delegates and non-governmental representatives. It was not without confusion and a lack of direction especially among the NGOs. Rancour was rife. The discord split opinions and friendships, profoundly so when France and New Zealand clashed over the former's intent to test nuclear bombs in the Pacific Ocean. Both sides of the cleft shared the incipient pain of a newly felt disease which none could aptly define, diagnose or cure. Contrasts were everywhere. Stockholm's splendid Nordic skyline reaching to a blue June heaven harboured crude pornography in the dirty windows and dusty boutiques at its foot. Raine Dartmouth's carefully produced thoughtful *vade-mecum* competed for publicity with Pow Wow, the 'other' human environment get-together. That outrage was promoted by the 'sandals and nuts' brigades who in their ultimate tableaux paraded naked in the streets, *sans* clothes, *sans* respect for accepted conventions, *sans* evidence of the use of soap and water, contemptuous of the Conference which they regarded as the outscourings of a bourgeoisie-polluted earth. Only when CHEC called the envoys of the Commonwealth together was there an atmosphere redolent with a spirit of quiet reason. The Conference, despite its resolution to set up a United Nations Environment Programme under the Canadian, Maurice Strong, sent me home musing on a world moving towards self-destruction. Its most constructive feature was the impromptu display of the affinities and cohesion of the Commonwealth wherein heartfelt, genuine handshakes were still currency between its peoples.

Countess, Lady Dartmouth changed her name and social environment soon after the Conference. Our paths no longer crossed.

It was different with Fred Cleary who had introduced us to each other the previous year. We both were contributors to his City of London environment conference along with Duncan Sandys and others at the Guildhall. Fred, as Chairman of the Haslemere Estates, one of the few property companies to come through the financial crisis of the early Seventies with comparatively few bruises, was reshaping Covent Garden. Raine Dartmouth was Chairman of a special Development Committee set up to keep an eye on him. Fred's ample person, an outward and visible sign of the compassionate and generous heart within, competed in width with Raine Dartmouth's parasol hats under which she hid features of exceptional grace and beauty. Raine's hats remained to possess my memory. Fred became a 'knight of the shining helmet' in the Trojan wars yet to be fought to right the wrongs at Cambridge University.

It happened on this wise. When the Agricultural Economics Branch joined Land Economy, there was an inherent flaw in the marriage. The bride brought no dowry of undergradutates with her. The General Board of the Faculties were cautioned, pleaded with and warned. They nodded and said nothing. A criterion of good management which was applied to judge the worthiness of teaching Departments was the staff-student ratio. In Land Economy, the marriage gravely upset this ratio. A plea to the General Board asked them to recognise this, to make an exception and to excuse the imbalance; after all the General Board had created the imbalance in the first place. Justice was required lest the imbalance jeopardise the just promotion of existing teachers and research officers in Land Economy proper when the time came to consider their right deserts or to stand in the way of recruiting additional Land Economy staff. When, back in 1969, the Department of Land Economy with a shotgun at its head had pronounced the irrevocable 'I will', the flaw was duly acknowledged and understood. There was little fear at the time because trust was placed in the supposed sense of justice and right thinking of the current members of the General Board. Rule by committee, however, has a poor memory; a truth overlooked at the time. When therefore some three years later requests were made for a number of perfectly proper upgradings and appointments, the General Board had a neurotic fit of selfrighteousness and convened a Needs Committee to consider these 'outrageous' requests. The Needs Committee refused the lot. Face to face confrontation ensued.

The adamantine inquisition would have nothing to do with the idea of the past justifying the present. The bare fact of the imbalance stood as their totem pole. None would exonerate their nefarious import.

Injustice was compounded with injustice since the well-endowed Development Fund and its Managers were more than able and certainly willing to meet the costs of what was so justly asked for. The establishment were tainted at the time with the false probity of socialism. The General Board's disapproval, apart from the ratio evidence, was justified to Land Economy on the hollow grounds of denying a wealthy Department its just due because other Departments were not similarly endowed. Faced by such inexcusable injustice, the old anxiety revived. Those of us who had lived through the last twenty-five years at Cambridge, however, were older and wiser now. We knew that such wrong thinking was the work of little-minded men with small hearts. We knew, also, that Colleges could act where the General Board was impotent. Then it was that Fred Cleary came in.

In quiet, tasty opulence, Haslemere Estates occupied one of the redbrick mansions beneath which at lunchtime Rolls Royces would swish round Carlos Place to the Connaught Hotel opposite. Fred's desk matched in massiveness his person. Colin Kolbert, representing Magdalene College, came with me to see him. Fred opened the door, signalled us to take the two chairs facing his own and returned with a smile to wait upon us.

'Fred,' I opened up, 'we've come to ask if you can help us.'

''Pends what you want,' he replied.

'Would you endow Magdalene College with a seven years covenant to establish a Fellowship in Land Economy at the College?'

'What happens after seven years,' he replied and looked worried.

'I would have to come and talk to you again.' It was the only obvious response. He pushed back his chair, rose with the studied, slowly-moving care of a wealthy elderly gentleman, took his silver-knobbed walking stick and bowler hat from the door and walked out. Shaken, Colin and I were at a loss to know what to do. Fred stood in the doorway and pronounced what sounded to us like final doom. 'I'm not interested in anything that stops' he said and turned on his heel to go.

'But, Fred,' I called after him, 'you wouldn't expect us to ask for £100,000 to set up a Fellowship in perpetuity, would you?'

'If you don't ask, you don't get,' he called back.

'Right!' We were in unison. 'A hundred thousand pounds,' I bawled down the passage.

Fred returned to his chair. '£50,000 in cash'. His grin widened as he spoke and he added, 'And fifty thousand pounds worth of Haslemere shares which will be four times their value by the year end.'

Hardly believing our ears, we settled back to chat idly of the College and its wants. Suddenly the new benefactor became greatly excited. He realised that Magdalene College housed the Pepys Library. Now Fred was the Secretary of the City of London Pepysian Society. Here he was indirectly an unwitting benefactor of the Holy Shrine at Cambridge. The gift to Land Economy at Magdalene proved over the years to be only the beginning of a sequence of generous donations to the College and the Pepys Library. Not only so, Fred had set a precedent and given a magnificent push to the start of what became known as the Fellowship Scheme.

Some two years or so before the debt deluge of the later 1970s swept into oblivion men whose names had become beacon lights in the property world, Fred Cleary had roped me into the affairs of the Association of Land and Property Owners (ALPO) to be author, adviser and general trouble-shooter. The father figures of ALPO were, on paper at least, immensely wealthy, sported large or very fast cars, were well-dressed after their fashion, trusted too simply in weighty mortgages and were eager to be seen as caring, generous and tender.

Conspicuous among them was Barry East, the Chairman of Town and City Properties. Barry, to his credit, had risen from small beginnings, had a natural nose for what was listing in the winds of good fortune and had won renown as a property man of substance. When we first met each other, Barry had a one-eyed impression of me. One of his eyes was black, puckered and bruised from a brawl. Fred Cleary, Barry and myself made up an ALPO threesome to campaign, under the slogan *Down to Earth*, for the abolition of the Labour Government's Land Commission.

Barry East cultivated an histrionic posture for business interviews with either friends or foe. Outwards, from the parapet of a monumental desk towards the entrance of his office, swept fifty feet of deep, springy carpet. Anyone who entered his doorway had to cross the intimidating expanse. An intended, fearful apprehension gripped the solar plexus long before the visitor was within earshot of the Chairman. Barry was virtually enthroned. Although not physically on bended knees before him, the spirit of all but those courtiers who were vouchsafed the highest regard was cowed. Personally, I suffered abasement before him in the august throne-room-cum-office while Barry slowly nodded his head — his peculiar way of waving the sceptre. These daunting theatricals were Barry's way of saying 'Yes'. He agreed to finance a Fellowship for Land Economy at St Catharine's College and probably at Pembroke also. Once off his throne, Barry's satrap posturing was put aside. He became his natural jovial self, a delightful companion. We lunched with Professor Rich, the Master of St Catherine's, Dr Caesar also of St Cath's and others to clinch his offer.

Like Barry East, Gabriel Harrison, the Chairman of Amalgamated Properties, was but a few years away from financial disaster when he too gave his generous hand to fund a Fellowship at University College (as the Wolfson College of today was then called). Pembroke College, unhappy at the prospect of having tied-fellowships, nevertheless accepted a renewable donation to fund graduate research in land studies from the Duke of Westminster and the Grosvenor Estates. St Edmund's, the Catholic Foundation, also played a slightly anomalous part in the Fellowship Scheme. Patrick Drudy holding an assistant post on the staff of the Department of Land Economy and whose just claim to promotion the General Board had turned down, was given a Fellowship at St Edmund's financed directly by the Managers of the Development Fund. Hence it came about that by New Year 1974, sundry College Fellowships or something akin were in place to complement teaching and research in Land Economy. Resistance to the General Board's injustice had made Land Economy stronger than ever. A red sunset of promise heralded the future.

The Managers of the Development Fund were 'perplexed' by the rejection by the General Board of the perfectly justifiable claims of staff promotions and appointments. The General Board's attitude virtually stopped the Managers fulfilling the intentions of Lord Samuel. Litigation against the University would probably have been successful. The Managers dismissed that temptation, not because of the weakness of the case but because, for some time, they were at loggerheads with the University Treasurer and Financial Board over investment policy. Out of a deferential respect for the Treasurer and Financial Board, the Managers had agreed to the bulk of the Development Fund being invested in the University's Amalgamated Fund. Capital was judiciously husbanded, yet grew very slowly and gave only a relatively modest yield. Down in the Antipodes, the owners of a presentable building on Toorak Road, Melbourne, Australia were toying with the idea of selling it as a 'lease-back'. The proposed leasehold covenant would be a sound one; indeed, from the viewpoint of the Managers a twofold blessing as the leaseholders would be Chestertons, the leading London-based surveyors of which the Chairman of the Managers was the Senior Partner. By the wizardry of modern communication, these intentions had reached the ears of the Bursar of Pembroke College, Cambridge. He was fascinated, drawn to the purchase but unable to move for want of ready cash – the owners wanted at least £250,000 jingling in their pockets. At a dramatic meeting in January 1971 the Managers decided to join hands with Pembroke on the deal. The College were to put £50,000 into the joint-venture. In a dash for freedom, the Managers asked the Treasurer to sell the necessary Units and thereby wean some of the Development Fund away from the Treasurer's Nursery and the tucked-in blankets of his cosy cot. Australia's property boom lasted longer than its counterpart in Britain. Toorak Road was sold in the nick of time. An absolute gain was made on the floor of the Australian land market, another harvest was reaped by converting Australian strong dollars to weak sterling and further triumphs recorded when the swollen proceeds were tipped into 'bombed out' Gilts on the London Stock Exchange. The total reward of the speculation was the conversion of an income of £32,000 to one of £100,000 in the space of three years. My success outside the suzerainty of the Treasurer received scant praise. I recalled the days when I cut the red tape in the Air Ministry. The Managers were jubilant. The gains had more than secured the future.

❦ ❧

Prime Minister Ted Heath's hold on No 10 Downing Street in 1970 was too tenuous by far to loosen the grip of the Welfare State in many directions. Looking for private monies to help cultivate University growth, even in the short run, was not *à la mode*, hardly praiseworthy, not to be encouraged and by some regarded as downright unethical. The generous hands of the Samuels, the Clearys and the Barry Easts were not the hands of the sovereign people; they were hands suspected in socialist circles of some kind of *lèse-majesté*. Who knows, if strong enough, such wickedness might break the mould of higher educational welfare and deprive the taxpayers of the pleasure, through their handmaiden the University Grants Committee, of sustaining the people's universities. The happenings at Cambridge, however (the Benefaction and the few newly-endowed College Fellowships), were not going to break any mould. These profitable generosities were mere abrasions scratched on the surface of the welfare chalice. When, however, Dr Pauley, a medical practitioner, wrote to *The Times* proposing the establishment of a University outside public accountability and free of dependence upon State money, 'his foolhardy idea', if ever it materialised, would certainly crack the mould.

An Independent University? How absurd! But why not? Splendid! The notion was most readily welcomed by free enterprise promoters, by free market economists, by votaries of individual freedom. Mrs Thatcher and her aphorisms were long off in the future, but here was someone wanting a University to 'stand on its own two feet'. The letter from first-reading inspired and moved with excitement Ralph Harris and Arthur Seldon, the founder-thinkers of the Institute of Economic Affairs, then no older than late childhood. Ralph and Arthur were friends of mine. They commissioned Professor Harry Ferns to write an Occasional Paper *Towards an Independent University* and sought help in other ways. With customary enthusiasm, they gathered Pauley, myself and a number of other 'pilgrims' and set out on what was to prove a long and arduous pilgrimage. Universities need money and they need land. Where was the money to come from? Where could a site be found? The first question I left others to answer. The second nagged at me. With one or two like burdened folk attempts were made to answer it. Foolhardy visions, mirages, and impalpable imageries floated before our over excited minds. When the initial fever abated, better sense prevailed. The Academic Committee – Pauley, Harris, Seldon and the rest – went in search of sites. They had a day out at Richard Wellesley's estate at Buckland in the Vale of the White Horse. Richard and his wife set the lunch while the academics rummaged round in wellington boots. Planning permission and money to develop were needed but Buckland's wonderful setting and near proximity to Oxford and London were very attractive.

Four hundred miles north, the regal towers of Taymouth Castle built to entice Queen Victoria to live there reared skywards over the fair lands of the Campbells of Bredalbane. Castle and towers belonged to the Mactaggart family and all were happy to have a free University there. Planning permission would only involve changing user as the accommodation was intact and ample. The problem was access; graduates, students, visitors and staff would find it a faraway place. At the opposite extreme of choice was a bare site offered by the London School of Economics somewhere in south London. Disappointment followed disappointment and frustration, frustration. Without planning permission, appeals for money were hopes dispersed on the wind; without money seeking planning permission had little hope of success. It was not a chicken and egg problem – there was neither chicken nor egg.

Since the War, notably so under Labour Governments, it had become customary for leading firms of surveyors to socialise at lunch with carefully selected guests either in the firm's own penthouse or elsewhere. The level of taxation and the tax laws may have had something to do with it. Whatever the explanation, many opportunities came my way to accept with thanks a proffered seat at these gatherings. Acceptance was simply an example of 'devotion and promotion' under University Statutes. On what was to be a red-letter day for the Independent University project, 24 November 1970, Donaldsons, a firm of surveyors of justifiable renown, invited a jovial company to share their hospitality round Disraeli's famous table in the Junior Carlton Club. By good fortune, I had to retrieve from time to time the fallen napkin of my right-hand neighbour, Fred Pooley, the County Planning Officer for Buckinghamshire. Dominant in my mind were two worries: one, the philosophy of the proprietory land unit, the foundation upon which Land Economy was erected, and its relation to practical town and country planning; two, how after lunch was I going to persuade Cluttons to be well-disposed towards the idea of a free University at Buckland on the land of their client Richard Wellesley. Fred Pooley listened patiently to my outpourings until we reached the crème brulé. Then he turned to me with a wink and a dig in the ribs and said, 'You're out of date. Theory's all very well but what about practice?'.

'*Touché!*' I exclaimed. 'So what?'

'How are you going to get County Planning Officers to recognise the primary power of property?' questioned Pooley. He was trying to pinion me. 'You tell me,' I said with a deliberate acknowledgement of his sagacity.

'Come to Buckingham' quoth he, and started drawing illustrations on the table-cloth with his fork. 'There the ratepayers, the landowners, the local authorities, the Planning Officer and sundry burgesses have formed a landowning Company. The burgesses of Buckingham can be shareholders but more importantly the landowners within the environs of Buckingham have brought their land into the company in exchange for shares and the goodwill of the Planning Officer. We are all in it together.'

Admittedly, I didn't follow him all the way but knew he was on the right track. 'So one has only to persuade the Company to agree to a project and, on the nod, planning permission, the land and goodwill are packaged together in a magical world at Buckingham. Is that how it works?'

'Yes,' he said, 'something like that.'

'Is all Buckingham bespoke?' I had suddenly got an idea. The Planning Officer shook his head.

'What about a University?' It was a moment when time stood still.

'Gee what an idea! Wonderful! Something to counteract the ghastly Open University that Harold Wilson has planted on us at Milton Keynes. Why do you ask?'.

My post-lunch appointment with Cluttons became more and more remote as I brought Fred Pooley up-to-date. After a bit, I got to the nub of the matter and said, 'So you see we are stuck; no money because no planning permission, no planning permission because no money.'

'What do you mean, no planning permission?' grinned Fred. 'You've got planning permission. I'm giving it to you now!'

After Christmas and when the New Year was out of the way, a mid-January post brought a letter from Fred Pooley with a request to call on him at his home in Princes Risborough. I wondered what was up and went over.

'We are going out. And you are taking me,' he said with a knowing smile when we met on his doorstep. The famous Bell Inn at Aston Clinton was the destination. In the warmth of its flagstone parlour Sir Ralph Verney, the Chairman of the Buckinghamshire County Council's Planning Committee, was waiting. Anyone who has dined at The Bell will know what comes next. But the wonder of the dinner and its spell were totally eclipsed by astonishment as I picked up a blue and white brochure lying on the dining table and read the bold caption,

BUCKINGHAM UNIVERSITY.

In laminated pristine perfection were drawings of façades, sections, interior plans, exterior plans, blurbs extolling Buckingham, the countryside, the schooling and the future success awaiting any who would be so lucky as to graduate from 'Buckingham University'. Fred Pooley had enthused his staff over Christmas to put together this mock prospectus. The dinner and Sir Ralph's presence at table had been arranged to impose reality upon Fred Pooley's determination and hopes. Planning permission was welcomed with joy, a site chosen, money found, an architect appointed, building plans agreed. All was set fair and running well when in 1974 Buckingham Town 'went Labour' and elected Robert Maxwell as its MP. Troubles born of prejudice, envy and left-wing bigotry came to oppose us. The lefties wanted to trample the enterprise in the dust. Recovery was long and hard but eventually achieved.

Not only Buckingham University came out of the Donaldson lunch. What encouraged me also was the approval which a brilliant and perspicacious Planning Officer had given to much of my thinking. His approbation and the Buckingham experiment with corporate cooperation between planners and landowners cast rays of light on my way which illuminated the path forward for many years to come. It bore upon future research programmes in two fundamental directions and provided a compass bearing by which, had it thought fit, a right-thinking Government could have steered its land policy.

To introduce the landowning factor into the land planning equation was peculiarly difficult in Britain because we have no means of telling how many estates in land there are, where they lie and who owns them. On the continent of Europe it is different. National cadastres are operative. They record land holdings, land values and ownerships. So throughout the summer of 1973 I conducted a research tour to discover how cadastres affected, if at all, the planning process. The study was positive and revealing. Cadastres were critical to the continental planning process either side of the Iron Curtain.They had been a factor in shaping development round the Ringstrasse of Vienna, the new autoroutes and autobahns round Lake Geneva and in replanning the city of Dortmund. In Budapest there was no question who owned the land. Even there, however, no man could live free of a charge of trespass unless the law gave him property rights over his home and in a number of other ways. Householders had to have something like property rights derived from the supreme State titles. Boundaries between 'ours' and 'theirs' were also decisive for Government corporations. Rights so apportioned possessed the character of property rights. The State had to do the apportionment and had neither land values nor markets to help it. There were, however, other bargaining counters. Those who manipulated them did so to their own advantage and became 'quasi' landowners, public and private. Their activities had an *a priori* affect on planning patterns. West of the Iron Curtain, much of my research time was spent in Dortmund. The city needed opening up, wanted space around its ancient buildings, had to prevent smoke once out of chimneys blowing back through the windows, traffic had to be kept moving and much else besides. The objects were achieved by devising an *Umlegungskarte*. The chart and documentation, based on cadastres, brought landowners and planners together.

Returning to the UK with files bursting with German, Swiss, Austrian and Hungarian texts, the material was used to bolster theories and to pursue three empirical and practical ends. First among these was the Warburton Lecture. Manchester University had included it in their autumn programme of public lectures and had honoured me with the job. The bulk of the Dortmund evidence went into it. My second concern was to compare the Continental evidence with the British scene. Comparison revealed how woefully this country needed a national cadastre. The lack coloured much of my current advice to the Conservative Party whose land policies were for a time in the hands of Graham Page, MP for Crosby. After his early death, they were masterminded by Hugh Rossi MP. The Tories were disinclined to follow my advice. National cadastres might, in their view, become the backdoor entrance to land nationalisation. A compromise in 1979 was the Land Act which requires local authorities to set up a register of their land holdings. Nevertheless, land registration and cadastre were germane to the place of land in the future economy of this country and of the Third World. The patent need underpinned the case for Rowton Simpson to research for, write and complete his *magnum opus* on 'Land Law and Registration'. Concern for that project was my third burden. Rowton, now returned from the Sudan, was attached to the Department of Land Economy and his research studies funded by the Ministry of Overseas Development.

‹§ §›

Writing books on land economy is all of a piece with the discipline of devotion. Finding publishers for them, however, is a lottery. Any substantial work of mine had hitherto been offered to George Allen & Unwin and in 1971 they were wrapping up *Land Use; An Introduction to Proprietary Land Use Analysis*. Within six months of its writing, the text of the history of the Shah's land reforms was looking for a publisher. Diplomacy cautioned me to let Unwins digest what they had. Slim hardback occasional papers were appearing at fairly regular intervals from the University Library printers under the imprint of the Department of Land Economy. They contributed to the promotion side of my obligations. A publisher, none the less, was needed who would contract to service a series of more voluminous texts. The lottery of writing requires patience and the trust that a publisher will be found or will materialise through the cosmic workings of the universe!

Unknown to me those workings were operative in August 1972. Simultaneously, a family holiday was in progress under the clouds and mists of Ben Hope by the Kyle of Tongue. The road south beneath dark far northern skies meant a night at Torridon under the lee of Skye, followed next day by a run to Cringletie House in Peebleshire. At Torridon, the family had packed the car as I marked a Ph.D script – a sample of the kind of 'devotion' pursued on holiday. At Cringletie after dinner drama took over. My briefcase, holding passport and air tickets for Canada, had been left in Torridon deep in the remote Highlands two hundred miles north. In thirty-six hours, Colin Kolbert who was attending the International Geographical Congress in Montreal would be expecting me there. Rescue plans were made, unmade, made again and rejected throughout the small hours. At the crack of dawn,

Jonathan my son raced me, still in holiday gear, to Waverley Station in Edinburgh for the Glasgow train and Abbots Inch Airport. The plan was to fly to Inverness, hire a car, tear across the Highlands to Torridon and double back to Aberdeen. Torridon Hotel, meanwhile, God bless 'em, had sent a message to Cringletie to say my briefcase would be taken by car to the Station Hotel, Inverness. The Torridon messenger, however, had, in haste, left the briefcase in the hotel and had not bothered to find me. Much time was lost before I found it and set off in a hired car for Aberdeen to catch the next flight to Glasgow. My calculations were out. The journey would have meant doing the hundred odd miles in an hour and a half. That forlorn hope was abandoned. A rescheduled journey flew me from Inverness to Heathrow where a suit of clothes was awaiting in the arms of my secretary. From Manchester next day, Air Canada took me across the Atlantic.

That Atlantic flight brought me closer to the unseen, unlooked for and unexpected publisher. Unknown to me, she was 35,000 feet below the jumbo jet that had lifted me off from Montreal, after a night's stop-off, for Winnipeg. Colin Kolbert, now sitting beside me, had found Audrey Clark, the solution of our difficulties, immersed in difficulties of her own; her hotel booking arrangements for the Geographical Congress had gone awry. Mindful of the Denman-Kolbert Rescue Service, he had done what he could and then begged her, with his customary charm, to await our return from Vancouver. Winnipeg meant a day's stopover. My Canadian cousin Mary and her daughter Carol had lived there through three generations. We only had time to snatch a night. It was long enough to experience the odd encounter between myself and two total strangers who had known of each other for a lifetime. According to the winking computers at Winnipeg airport, I was still at Montreal. Computers can't lie! It was foolish of me to say I was in Winnipeg. I wasn't there, so couldn't be put on the onward flight to Vancouver until I had turned up! At the last minute, the stark evidence of flesh and blood prevailed. Despite all computers, we flew over snow-capped Rockies to Vancouver arriving there dead beat and exhausted. The purpose of the visit was as genuinely devotional as the statutory duties of a Cambridge Professor could require. Professor Philip White of the University of British Columbia was cooperating with us in studying the range and depth of the provisions of Canadian Provincial land law as they restricted the acquisition of land by 'aliens'. Vancouver for us was half holiday, half duty and most enjoyable. Richard my second son was there on a Canadian tour with his friend John Street. Business done and contacts made, we returned to Montreal without computer hang-ups. Montreal was unwelcoming. The city introduced itself by a brawl with the receptionists at the Laurentian Hotel. No rooms had been booked, so they said. The Denman-Kolbert Rescue Service was trying to look after itself and was making a poor job of it.

That evening in another hotel we met up with Audrey Clark from Berkhamsted in the UK, the long promised publisher. Audrey was the Chairman and guiding light of Geographical Publications. She deported a trim petite figure alive with electronic alertness and a darting intelligence. Geographical Publications had been founded by the late Sir Dudley Stamp, the eminent geographer. He had left the company in the name of Audrey who had been his soul's delight, helpmate and confident. Geographical Publications had diversified from an almost exclusive preoccupation with the works of Sir Dudley. Here was the 'metaphysical' publisher, ready to consider the Shah's book and other works of the new series. Visits to the Geographical Congress, lunches, dinners and flight delays gave plenty of time to sort out the terms of a publication contract. Before parting, we spent two hours waiting for the Air Canada flight to Shannon. A complex character called Dan Taylor joined us. He was a self-appointed generous host who had established a wellworn rabbit run between the bar and the lounge where we were sitting. His once substantial business profits were rapidly being exchanged for Bourbon and Scotch and the overt signs that in Glaswegian brogue he was 'fu'. As the glasses were emptied, Dan got more and more aggressive towards all and sundry. One by one, he would stand over us and shout, 'Don't point a finger, point a fist!' We reckoned there was less turmoil in Ireland and were thankful for the announcement of the Shannon flight.

⋖§§⋗

The books of the Geographical Publications series were envoys destined for academic and professional bookshelves. There was no standard livery. A Land Economy imprint common to each would launch a book on its way. The books catered to an esoteric readership and probably spent the best part of most days lolling on library shelves. This series and others not dissimilar from colleagues and myself were the beginnings of a necessary literature. Land Economy in a certain aspect is the academic pursuit that answers to the professional practice of the chartered surveyor whose profession has yet to produce an adequate *sui juris* literature to satisfy the basic essentials of a learned profession.

269   *Devotion and Promotion*

The subject enjoyed indirect advertisement when the names of the Cambridge rowing crews or rugger teams were flashed on the TV screens. Otherwise, public illumination came indirectly from newspapers, boardroom periodicals, the professional press, reported conferences and through other conduits of the mass media. The number of my own pieces and slots mounted over the years. When in spate they were averaging between four and five a month. Public knowledge was never directly injected with the serum of Land Economy except on the occasion when, in September 1975, the President of the Education Section of the British Association for the Advancement of Science had chosen 'Education, Development and the Quality of Life' as the theme for the annual congress. To make up its programme, the BA had asked for a Lecture on Land Economy. The opportunity presented a unique public platform. My effort had nothing fancy about it. A simple straight-flung bill announced *Land Economy: An Education and a Career*. The blurb warned that it would treat of the development of Land Economy as an academic discipline in Britain and elsewhere; its potential as a research field; its relevance to the professions of the land and town and country planning; and its contribution to a fuller understanding of the ownership, use and development of land and natural resources as criteria of the human environment and its control.

Acknowledgements and plaudits came from a number of places here and abroad, notably from Professor Francisco Torio of Madrid. He wrote in Spanish with his T's looking like F's and made his letter more difficult to translate than it need have been. The Lecture was worth the effort. The bare lecture room assigned me was dark and pokey. The cafeteria alongside combined to make a somewhat uninviting ambience in keeping with an altogether socially chilly conference on the campus of the University of Surrey.

There was nothing of cohesion about the BA yearly Conference; it was an aviary of birds of passage wherein each was captured for a moment on a Guildford twig to sing a song and depart. Zena Daysh was among them looking lost. No obvious perch had been found for human ecology. Professor Sir Jo Hutchinson was also there. He forced a grin of recognition to his face and turned abruptly in an opposite direction, a gesture that manifested the general cold shoulder of the place. A short distance across the Thames lay Berkhamsted and Geographical Publications. I fled there to leave the manuscript of my Lecture with the warm-hearted publisher, to partake of a hot curry and a real welcome. Zena languished in the 'aviary' at Guildford.

Years of involvement with those aspects of public relations which occupy the middle ground between truth and an ignorant public, and which a sloppy use of language calls 'the media', had born in me respect and mistrust in equal proportions. Society cannot do without the 'media'. Mistrust is doubly sad when generated by wanton bias as I met it in interlocutors on screen and radio. Truth written and published is laid on a Procrustean bed, stretched and 'twisted by knaves to make a trap for fools'. My experience taught me that producers and interviewers on radio and screen tend to fit a speaker's words to predetermined notions of their own. An expert's words which do not bear out the version to be broadcast would be altered or expunged. Some years back, the late Kenneth Alsopp was interviewing me 'live'. He bade me look into the eye of the camera. Below it swung in bold, large lettering an alleged citation from a book of mine.

'You wrote,' began Alsopp and read out the swinging words.

'I wrote no such thing,' I exploded as he paused. 'You are putting lies between my lips.' The interview stopped. Kenneth remained to do it again with another author while I was probably 'blackballed' by the BBC.

Many years later, the 'blackball' forgotten, the BBC called me at breakfast time to be quizzed on a long distance telephone interview concerning another publication I had just written. The publication argued for a balanced judgment of industry's contribution to the quality of life and the environment. The BBC opened with a forceful, leading question designed to hit at industry.

'Professor Denman,' categorically stated the inquisitor, 'you would agree that industry pollutes the environment?'

'That is a loaded question,' I objected, struggling to keep the anger from my voice. 'If I answer Yes, it would make the point you are after. If I answer No, it would confound me before the listeners. I refuse to answer the question.'

'Then we can't go on,' grizzled the interviewer. I was sorry for the man. He was being dug in the back by some wretched 'producer' with other loaded questions.

'Not in this manner,' I agreed tartly and took the earphones from my head, calling into them, 'I'm off to breakfast. Good morning.'

On another occasion, Chris Brasher and a charming lady assistant were given the job of seeking my help over a documentary. They aroused my suspicions by calling all landowners 'landlords'. Both of them appeared to me to be pursuing a vendetta against landlords for alleged unhealthy social behaviour. After two lunches at the Carlton Club – conducive of the truth I thought – my heart warmed to them. Their ideas were so ill-informed, so insouciant and if not corrected could be positively dangerous. They wanted a meeting in the country as background. The best part of a day, therefore, was spent in the Herefordshire countryside walking the bounds of a riverside estate on the banks of the Wye. Cameras and reporters recorded Chris's conversation with me and shot our posturings. In the early evening members of the local Country Landowners Association assembled to hear an address which had been advertised. The talk was live on camera. Although 'slotted' into a published programme, the filming and recordings were never broadcast. It transpired, so I learnt afterwards, that my words and thoughts had run counter to the BBC's 'truth'.

Misquoting can be far more pernicious and difficult to live with than blatant bias. Once to my chagrin and disgust, a text of mine taken out of context was cited to give an inverted image. A zealot for land nationalisation had found an IEA publication of which I was the author making the case against the State takeover of land. My tactic was to list all the sustainable arguments for land nationalisation and then to demolish them one by one. By reading only half way through the article the critic thought he had discovered a Denman *volte-face*. Overjoyed, he rushed into print. Damage was inflicted and nothing much could be done about it.

Much of my extramural lecturing and writing was helpfully sponsored by political interests and pressure groups of various kinds – notably the British Property Federation, the Country Landowners' Association and the Scottish Landowners' Federation – and by professional bodies led by the RICS. Writings for seats of higher learning, polytechnics and Universities worldwide were of course in the nature of things. Some of these 'clients' wanted more than their money's worth, some had no ears for another's views, some were hopelessly pietistic, many were ignorant, many pushy, others docile. Wherever invitations were accepted there 'Land Economy' had an airing and often found a foreign field, alien and away from Cambridge, in which to germinate. The most warming and exciting experiences were connected with actions promoted by industry. Dark and dirty industry, against which the BBC's invective seemed set, would move away from its wells and smells and lend a hand on the land. Matthews Wrightson, the northern engineering group mounted a most successful Rural Land Conference in the Autumn of 1972. It was most praiseworthy and among many similar ventures illustrative of the genuine helpfulness of industry and the concern of its leading men to find solutions for contemporary problems.

Shell International would have wooed me from the University, if ever I had had a mind to leave. Through Charles Lister, they were helpful over my exploits in Persia. Above all they linked me with their most impressive agricultural promotion at Borgo a Mozzano in Italy. Roberto Volpi was the hero of the venture and is deserving of a place in some agricultural Valhalla. As a young man, Volpi went to live among the toiling local peasantry, brought to them the wealth of his knowledge and patiently, carefully and successfully taught them how to apply it. It was at Borgo also that I first met Emrys Jones. He was scraping trout just hooked live from a pool for lunch. His jovial face looked clouded with worry. Cledwyn Hughes, the Minister of Agriculture back in the UK, had posed a policy question to Emrys who was Director of the farming advisory service – should the National Advisory Service for farmers be amalgamated with the National Land Management Service? Emrys had fled from London to get away from it, but it wouldn't vanish. The visit to Borgo, I am sure, helped him say 'Yes'.

Borgo was always a highly-pressured work centre lightened by play days. The play days were more memorable than the days of toil. On one of them when the Persians were there Haleh Afshar had come from Cambridge to translate for me. Haleh and I slipped away for the day. We said nothing to anyone. There were cars and drivers at Borgo. One of these – Haleh could always get anything she wanted – took us south to the coast for a trip across the azure water to Isola Elba. The island was sleeping in the heat haze of midday. Lunch, protracted, wine-filled and leisurely left us, after the last bottle, with an hour or two to spare before the home ferry left for Piombino and the mainland. Behind the fussy little town of Port-ferraio, a rural and aged bus struggled through half the afternoon to reach the Villa Napoleon. It was a close run gamble to get there and back before the sunset boat departed, to leave us stranded for the night. The risk, however, was worth taking. Napoleon's notorious domestic prison stood in architectural grandeur and moulding decay, a furlong or two the other side of high wrought iron gates. The gates bridged the entrance gap in a massive fifteen-foot wall that ringed the tree filled grounds. Bathos ended our aspirations. The gates were locked and progress barred. Suddenly Haleh went missing. She had scaled the wrought iron gates, swung a leg over the parapet wall and dropped with cry of triumph into the weeds below. My nerves tightened as the 'bus, within earshot, started up for the return journey. To stay on the legal side of Napoleon's abode was preferable to being imprisoned in the local *polizia* hut. Haleh's escapade was not for me. Then it happened. A furious yell rent the air in the Italian vulgar tongue, more vulgar than usual. From behind the yellow columns of the *palazzo* and running for her life came Haleh. She was losing her lead over a pack of grizzly hounds led by a ragamuffin waving a cudgel and bellowing out vituperation. A busload of passengers had gathered at the gates in gabbling curiosity. Haleh, the dogs at her heels, reached the wall and began to climb. From half way up the gates, I made useless attempts to grab her. Beneath us, the crowd came to the rescue, drew off the dogs and competed with the frothing *guardiano* in unsocial language and gestures. My leg held and enabled me to pull the grinning morsel of female humanity over the parapet and swing it to safety on the road beneath. Haleh's would-be captor had forgotten his keys. So the gates stayed locked against him, as they were against Napoleon in days gone by.

Promotion of the University and devotion to the academic perfecting of land economy pointed my conscience in an immediate and special direction. Life had never made me a practising chartered surveyor except for the period of war service. My conception of the practitioner, therefore, derived largely from the schooling of him and from the debates and struggles to relate University education to the prescribed texts of qualifying examinations which the profession set its candidates. The 'chartered surveyor' was for me an image to be shaped by didactical skills. Education and the practitioner were always linked in a wobbly articulation. One of the objects of the Cambridge Tripos in Land Economy is to offer a disciplined education as a foundation to practice.

Unfortunately for its academic development, the profession is slow to help itself. Unlike institutions serving other liberal professions, the RICS has never recognised its teachers as qualified members of itself by virtue of their teaching alone. Promising young men and women who after graduation pursue an academic career seldom have time or opportunity to practice professionally and hence remain beyond the pale of RICS membership. If the RICS is ever to lift the profession to high academic standards, accommodation will have to be found for these gifted teachers within its membership.

Howbeit, I was qualified as a surveyor by dint of youth's mud on my boots. From the start, my proclivities leant towards the gown and not the theodolite. In the words of Malcolm Trustrum Eve, I was 'the Great Anomaly' – truly a half and half affair. Land Economy is self-contained, an education in itself and of itself, yet its origins and affinities are with the surveyors' profession and owe much to encouragement from the RICS. My concept of what an educated chartered surveyor could be differed widely from the narrow imprint which practitioners of the estate agency variety make on the popular mind. In the early years of the conception and creation of the Tripos, practise was far from my mind and experience, despite the advocacy of Sir William Cecil Dampier-Whetham's early ideas for empirical training. Time was fully absorbed in research and teaching. The RICS lay in the background. Fortunately, Jeffrey Switzer, schooled, qualified and in every way acceptable to the Institution, was a perfect liaison and deftly fed our ideas into its Council. So it was that at the time of our most seminal thinking, the RICS itself began to stir and change. Annual Summer Meetings became serious Conferences debating fundamental reassessments of education policy and other matters. In reshaping its education policy, the RICS, as we have seen, took steps towards Cambridge and its General Secretary, Robert Steel, wrote his welcome encomium of the new thinking there.

After the Professorship was established, there was time on my agenda for devotion and promotion to take seriously invitations from the central committees in contrast to my slackness towards the local Branch. Committees and councils, however, were never an avid pastime of mine. Whatever its size, a committee is a collective and I am not by nature a 'collective' enthusiast. To be diligent on a committee, one runs the risk of losing one's soul to the collective; to say little or nothing becomes an ineffective waste of time. Pressure, however, mounted after the giants – Chesterton, Pilcher, Wells, Eve, Biscoe, Balch and Battersby in the glow of the RICS Centenary Year 1968 and from the Presidental Chair – ruled that the number of committees, councils, working parties, panels and so forth of the RICS should be multiplied. By the mid-seventies, I had been coopted, invited by other means or enticed to sit on no less than thirteen. More often than not, my name appeared on the Minutes among the members sending apologies for absence. Had I been altruistic, my resignation should have been offered out of regard for others. My name, nonetheless, remained with the membership lists of Divisional Executives, Standing Committees, *ad hoc* affairs and Presidents' panels with the right of participation could I ever attend. On the map of these various bodies my allotted locations were Education and Membership; Research; and International Affairs. Some seats were deliberately kept warmer than others. The *ad hoc* committees were usually the most exciting. They had a start, a finish and a purpose – as chasing the intractable problem of taxing betterment; or finding surveyors to man the Government's land policies. The latter meant on a number of occasions a particularly paradoxical afternoon's sport as the evening of the same day would be spent in the bowels of the House of Commons with Graham Page and his plotters on finding means to oppose the Government's land policies. Hours on the Education Committee were, for me, the most boring. Being a relatively large committee it would ramble and its members seemed more concerned with training and technology than with education – with How to turn a tap on, rather than with What a tap is and Why it is in a particular place, to say nothing of the tap's function in the economics and environmental hazards of water supplies.

Committees tend sooner or later to spin off mafias or ginger groups, sinister and benign. Following Rab Butler's speech to the RICS's Centenary Congress calling for the provision of more degrees in prestige Universities, a ginger group, known to members of the Education Committee and blessed by most of them, formed up behind their official backs with the intent of raising funds and taking action to promote higher learning for the profession in Universities. My sympathies were with this excellent mafia, although Jardine Brown, the Principal of the College of Estate Management, was hostile. Before long, I was commissioned, as the story has already been related, to be their envoy. The banner of the good cause was placed in my hands to be carried with blessings far and wide. For four years the campaign's fortunes waxed and waned. Meetings of Vice-Chancellors, professors, bursars and other dignitaries were held in a number of Universities and a special lunch given in RICS Headquarters for Sir Kenneth Berrill, Chairman of the University Grants Committee. Draft syllabuses were drawn up and staff prospects surveyed. At Liverpool a most impressive glossy Appeal Brochure appeared. Liverpool was all fuss and flurry. Edinburgh under the dour yet kindly measuring and cautious brows of the late Lord Swann, the Vice Chancellor, was cool. When we met, indeed, Lord Swann showed more interest in our mutual affairs than in anything else. He was a prominent Executor of the estate of Sir Sydney Roberts, his late stepfather, and was in the process of selling me the house in Cambridge. Aberdeen, as already recorded, meant business. My travels and diplomacy came to naught at Liverpool and Edinburgh and would have been as fruitless at Aberdeen but for my jibe at Scottish educational policy which I slipped into the epilogue of my RICS Gold Medal address.

Of all the committees, my presence was most secure on those serving international affairs because they cross-fertilised with the doings of the Commonwealth Association of Surveying and Land Economy and similar groupings. The Fédération Internationale des Géomètres (FIG) in the 1970s was waning from the sphere of my interest while the influence of CASLE was waxing. Like most Conferences, the international jamborees hosted by FIG have left no solid memories. The last I attended was in Washington. Only three memories remain clear. One is the realisation that I could distinguish the provenance of negroes by their disposition; if they were American citizens they would probably be sullen, surly and wanting to make their place with one; if Caribbean, they would be more content but lack spontaneous humour; if West African, they would be hearty and humorous, kind and all-roundly human, as people who knew where they belonged. One evening chasing down Connecticut Avenue in Washington, I tested this hypothesis. The cab was in the hands of a jovial, black driver, relaxed and seemingly happy.

'What part of Ghana do you come from?' I leant forward and gently dug him between the shoulders.

'How come you know I'm Ghanaian?' he chirped. His broad white teeth smiled and put to shame anything John F. Kennedy could have shown.

'Instinct,' I replied with a laugh and added for good measure, 'you are probably from Mampong.' We left it there and raced down town swapping Ghanaian stories.

Irritation fathered my second impression, irritation at the rejection of my travellers' cheques by bank after bank. The banks regarded me as an alien nonenity who had probably filched the cheques from some unsuspecting innocent. Their denomination in US dollars made no difference. Only in hotels were the cheques accepted as genuine money vouchers. Without an American Express credit card, one stood nowhere with the banks. A sad reflection on the city. The Ghanaian taximan, in contrast, gave me 'tick' and waited for his cash until we made the hotel.

The third impression was of the architectual beauty of Georgetown, its Regency houses and engaging avenues lying secluded from downtown which at nightfall was sinister, silent and deserted.

12

NEITHER EAST NOR WEST

But there is neither East nor West, Border, nor Breed,
  nor Birth,
When two strong men stand face to face, though they
  come from the ends of the earth!

The Ballad of East and West, Rudyard Kipling

Within a few weeks of Harold Wilson's new administration taking office in 1974, Isabel Peron siezed the reins of government in the Argentine from the hands of Juan Domingo, her dying husband. There were sufficient socialist ingredients in the politics of the British and Argentinian Governments for the new Government in Britain to wish to extend diplomatic gestures of goodwill to the brave lady President-in-waiting of the Argentine. Knowing nothing of these friend and favour politics in high circles, I was perplexed to receive a request via the Foreign Office to help Isabel's government in Buenos Aires over a land question. The request was as sudden as it was unexpected and was subsumed under a category of service well-known in academic circles but which upsets the sense of justice and fair play among private consultants. Because I was an academic, any taint of filthy lucre in my response would have been regarded as *de trop*. An academic is not a professional – nothing so low. His honour is at stake in the honorary appointment – all travel costs paid, of course! Had the solicitation come a year later when the prospect of earning a living as a consultant, after retirement, was on the near side of the horizon, my response could have been different. As it was, I was full of 'satiable curtiosity' like the Elephant's Child and boarded a plane for Madrid *en route* for the Argentine in late July.

The El Dorado continent of Latin America was *terra incognita*. Social sophistication, covert banditry and the reek of cow dung in downtown Buenos Aires struck me as the hallmarks of the Argentinian capital. My home, *pro tempore*, was the handsome mansion of Frank Maynard, the Minister at the British Embassy, and Anne his young wife as alluring as the hills and glens of her native Ireland. Once the heavy, rust-bitten postern had let me into the garden, Frank greeted me with an open hand and wry smile. He was standing at the portico entrance to which a flight of half a dozen steps had lifted me.

'You will find the house cold,' he said. 'It's just been opened up for you. We have been in hiding for a few weeks up country. I'm top of the local bandits' hit list. One is either kidnapped for money or politics here. Money is not the magnet that draws them after me. Anyway, there's little of that. The kidnappers reckon I'm tradeworthy, to be bartered for a frigate from the British Government. Zealots are piling arms to tackle us over the Malvinas. A frigate more on their side would be also a frigate less on ours. Oh! Do come in. There are two machine-guns trained on the gateway and two on the porch.'

Each morning a secret button would operate a sliding wall panel opening upon two highly polished limousines. Frank rode in the rear one of the two with an uncocked revolver on his lap and sat immediately behind a fully-armed police officer. The forward vehicle carried a platoon of armed men. As the garage doors lifted to the street, this primed domestic arsenal would move forwards and fall in behind two other cars likewise armed and waiting outside. The entourage took a different route to the Embassy every morning.

On the first morning there was little to do. Hours were spent at an extensive Agricultural Show in the middle of the busy city hum. The stalled cattle were enormous, especially a Betty Bunter breed of Herefords standing massive and square, on heat for the colossal bulls to rear upon them. The Show conditioned me for a series of evening lectures to be given to members and friends of a kind of Farmers' Union, the *Rural Social*. Receptions and parties were routine obligations of an evening. One evening half way into my visit, Frank Maynard's imposed poker face informed me we were to dine in secret and below ground. Underground or not, the rendezvous and fare surpassed all expectations, elegance and artistry of cuisine.

'The Embassy files record that you are out here to talk to the Government,' Frank whispered sideways. 'Be that as it may. Your hosts tonight are the members-in-waiting of the Government next but two. Time will allow your words to gather weight. Your hosts *prudens futuri* will dwell upon them. Pass the mustard.' Listening and noting were more prudent pastimes than giving opinions. My words were few.

Next day at lunch with Sir Donald Hopson, the British Ambassador, and Lady Analise his beautiful Danish wife, my tongue prattled. The great Agricultural Show down among the city traffic and skyscrapers had been opened by a diminutive Glaswegian, Norman Findley Buchan, a Minister of State from our Ministry of Agriculture, Fisheries and Food. He sat opposite. His raucous socialism stirred the Colonel Blimp within me. Speech flowed freely and with the splash and thunder of a torrent. After coffee, Sir Donald and Lady Analise were at differences over who should see Norman Buchan away to the airport. Each declared roundly to the world at large that, 'It would be a great pleasure to do so!'

Two days were to pass before the exact and ostensible purpose of my mission was explained. The British Embassy couldn't tell me in so many words. Exposition of a kind, the reigning Government's version, was supplied by the local Ministry of Agriculture. Apparently, throughout Argentina, countless *estancias* with vast, rich, rolling *pampas* plains were failing to realise full potential output. Consequently, and contrary to all economic sense, the Government proposed to ginger up production by imposing a tax on the hypothetical potential of each estate. My presence was required for discussions with officialdom over how this taxation gymnastic might be accomplished. Curiously, there was little opposition to the impost, even among the farmers and landowners, probably because no one seriously thought it to be a practical feasibility.

A morning's drive from the city and one was in a land of skyscapes, of clouds racing over studs and farmsteads. Life was lavish, taken at ease and lived in sumptuously appointed mansions surrounded by carpets of carefully tended garden. My visits were notably to Anglo-Argentinians selected for the purpose, open-hearted, open-housed hosts seemingly oblivious of the ravages a biased, wrongly calculated *renta potential* tax could inflict upon them. Our speech was banter, nothing serious as became the country's superficial life style. Frank Maynard fitted in well. He was a handsome fellow of a debonair wit which bubbled like a shoal of fun and laughter over the deeper fears which threatened life. He took delight in instructing me how to avoid the hidden perils of social life in Buenos Aires. We were fleeting back down a stretch of the Pan-American Highway which runs northwards over a grid of latitudes. At intervals of several miles, were attractive small hotels.

'Those,' said Frank and jerked his head towards them, 'have a purpose and a lesson. Hear ye the story of the English gentleman out here on business last month. I, as stand-in for the Commercial Attaché, had introduced the English merchant to a fair lady, influential and knowledgeable in the business world of the city. I had forgotten to warn him of the dangers of shallow English gallantry. Three days later, the poor fellow telephoned in pitiable distress. After a most successful day's negotiation which had ended laden with promises of future contracts, he had invited his newly-won ally to lunch, an occasion that had been a lively and enjoyable one. When he rang her next morning, his call was handled by a secretary who, detached and cool, passed a message to the effect that *Senora* was out and had left instructions to say that all business was off and there was no point in trying to revive it. 'What,' he lamented, 'what have I done? I must have offended her in some grievous way. His misery was pathetic. So I explained further with a saddened heart and calculated care. He should understand, I told him, that in Buenos Aires if a man asks a woman to lunch it is *ipso facto* an invitation to an hotel on the Pan American Highway for an hour or so afterwards. Should he not do this, he would be guilty of insulting his lady guest. The poor fellow had, of course, failed to extend such a courtesy to his lunchtime partner. No wonder she was upset. He should have concluded his business with her in the first place by doffing his hat, bowing low over her hand and waving *adios* – not asked her out to lunch without the post-prandial ravishing.' By the end of the story I began to wonder what indelicacies I might unwittingly perform in this El Dorado city where the pruderies of St James's Street WI are *infra dignitatem*.

At dinner that evening the raconteur in Frank was in full swing. He was entertaining Dutch and Russian friends to whom he had introduced me. Prominent among his stories was a genuine 'Wartime Churchill', one in which Frank himself had been an actor. Stalin and his Generals would in the thick of battle send lists of urgent requirements in code to the British Government. On one occasion a formidable ten pages ended, after its requests for armaments and medical supplies, with a petition for 50,000 condoms. When these lists arrived they would be chopped into sections for separate dispatch to appropriate factories. The end bit to this particular list found its way to a rubber factory in Lancashire. Down at t'mill the locals had translated the Russian request as '50,000 sleeves'. Diligent to the letter, the Lancashire lads produced 50,000 items, three and a half feet long and six inches wide. They were packed and awaiting dispatch orders from Frank in the Foreign Office. Officialdom was distraught. Should they go or not? Better put it to the Old Man. 'Yes!' said Churchill and added in a stentorian voice, 'See that they are stamped "British Made, Medium Size".'

My Report to Isabel's Government may have let another *Senora* down. It suggested that the President scrap all thoughts of *Renta Potential* tax. Let the free market set the farmers' profit margins and tax landownership, not enterprise. She had little time to do anything, however, before falling victim to a military defection some months later. For me, the Argentina episode was an enrichment of knowledge and experience. Beside the exotic excitements it included a home-from-home morning with my friend the Rt Rev Cyril Tucker, Bishop in the Argentine. The Archbishop tried to see all sides of Argentina's predicament.

Despite its happy cheer, my memory of Buenos Aires is shrouded in sad pathos. Farewell to Frank Maynard was cut short at the breakfast table. Halfway through the toast, the front door bell rang. Frank looked up at his wife. The intelligence of a silent understanding passed between them as the housemaid tripped into the room. 'The police, to see you Sir. 'Tis urgent, they say.' Frank arose from the table and declared, 'I'm ninety percent sure they are terrorists in disguise.' Shaking my hand with a hearty 'Cheerio', he made for a hidden lookout in the roof. Sir Donald Hopson was another sadness. As Britain's Ambassador to Peking he had been incarcerated by the communists and had accepted the posting to Buenos Aires as a sinecure. Within a month of my visit he was dead from accidental snakebite.

The Maynards had a love of India that was infectious. Love and memory of the subcontinent coloured the speech of their every day. With the passion of a connoisseur Frank and his wife had collected ornaments, statues, pictures and whatknots in tangible evidence of India's claim upon them. When I hinted of being in Delhi before the end of the following month the news was hailed with joy. CASLE was actively organising the surveyors' profession throughout the Commonwealth and preparing to educate and train its future generations. An all-India Conference had been programmed and I, as official academic adviser, was expected to be there. Delhi lay a long way east of Buenos Aires and my Cambridge home half way between called strongly. The Maynards, in their zeal, however, pushed me. 'Now that you have been to Buenos Aires,' they urged, 'you must see Delhi. Delhi for us means Mahindrapal Singh, the spiritual personification of India. You will go with a written introduction to the guru. Please be our ambassador of goodwill.' What they didn't know was my travel schedule before Delhi. It included Washington and Vienna; and there was the prospect of an exacting winter tour of the Caribbean in the longer offing.

CASLE and the Maynards between them, however, won me over. Come Michaelmas, I was off again. Delhi Airport at 2am of a morning is not to be recommended. Dripping with tiredness and caught up in the vortex of a whirling mass of bleating children trailing distraught, sweat-sticky mothers in utter confusion, I hugged my baggage and squatted fatefully beside it to be met in the thick of the mêlée. A mid-day breakfast and a grin of expectancy on the faces of the CASLE crew met me next morning. The Conference that followed was overpowered by Indian civil servants, a haughty, humourless lot, proud of their British ICS heritage, generalists all and, in their own judgement, superior to surveyors and all cognate lesser mortals. Little was achieved except to learn the lesson that conference talk was not always the best medium through which to communicate CASLE's educational hopes and intentions. Within a few days, the Conference broke up and in self-propelled fragments the CASLE contingent made for Sri Lanka. That was the only time I ever flew on a false passport. Professor Ben Acquaye, who headed the Ghanaian delegation, was never fully himself without a female companion. To this end he wished to stay in Delhi overnight. 'Would I,' he asked with a touch of diffidence, 'mind taking the afternoon flight to Madras and so set free for himself my reserved seat on the over-booked flight next morning?' To establish his claim to my seat Ben needed my passport. His handsome, negroid features looking out of the passport he handed me in exchange ill-matched the pale age-lined face of the elderly Cambridge Professor who boarded the plane to Madras that afternoon. 'Professor Acquaye' indeed!

The prospect of being of full-time service to the CASLE Conference was put in jeopardy soon after arrival in Delhi when I went in search of the Maynards' guru, Mahindrapal Singh. There was royal blood in his veins but he lived modestly in the backyard environs of the city. Delhi at night is a lamp speckled maze of confluent trackways marked out by pin-points of light tracing dimly lit streets. To walk anywhere at dark in that slough of despond, even if one knew the way, would be foolhardy. For a few rupees, the hotel porter would whistle up a bone-shaking taxi, not for exclusive use but to be shared by a motley bunch of customers whose make-up changed with the mileage. My man had taken a load of instructions and, after an hour, had plainly lost his way. The cab portentously sliding in the mud slithered to a stop in a huge puddle, a veritable minor lake. Its offside front tyre was punctured and flat. The taximan had neither tools nor jack. My sole comfort was the thought that Mahindrapal Singh was reputed to be a man of high spiritual qualities and, though over the 'phone had spoken of other engagements that evening, would doubtless wait for his guest with the patience and dignity which became him. Out of the murky blackness that engulfed the puddle and the car, stray cars would appear, salute and vanish. To go in search of an alternative lift in that no-man's-land would be to court disaster. The driver, all the while, sat motionless, waiting for I knew not what. Nearly an hour later, a wobbly Ford, with a driver cast in a Samaritan mould, stopped and offered us a wheel-jack. The mockery of fate was to discover, once the tyre was mended and the wheel in place, that we had broken down but a short walking distance off the guru's abode.

He received me with great poise, accepted with grace my lateness and exonerating tale of woe, then proffered a chair at a kitchen table that could have come from the Caledonian Road. The ambiance was a twilight and stillness disturbed only by his cadenced speech that kept rhythm with his nodding head. It was, nevertheless, hard for him to avoid showing a suppressed impatience. Berobed, his tall stature topped out by an onion-like turban, he towered over a diffident female who may have been his wife and to whom he gave snappy orders for a meal. The tension eased somewhat when I passed him the Maynards' introduction. Our acquaintance, however, was too short for me to discover what attracted them to him. The evening apparently held for the guru the prospect of a spiritual retreat with some lady friends hence his hardly concealed jumpiness. But he found a moment, willingly and no doubt eagerly, to hail me a local taxi and point it in the direction of the mud, the dull lights and the dingy city.

In Colombo, the Sinhalese were less conscious of their self-importance than the Indian civil servants had been. The occasion was no less formal, more so in a way, as Madame Bandaranaike, the President, opened the Conference under an impressive roof raised in her honour. Satisfactory business was done with University and Government Departments. Promises of future courses in Land Economy floated in the wines of reciprocal parties, papers were read and old friendships rekindled. Beneath the felicitous cooperation, however, lay sadness and disillusion for some people attending the Conference like the Ferandos and others of former standing in an erstwhile prosperous Ceylon. The Ferandos' attractive and intelligent daughter was employed at Harrods and, just then, was leading a mannequin troupe in the Continental Hotel, Colombo. Her perform-ance was not solely for the benefit of the CASLE delegates, although it might have appeared that way. The parents lived in a not incommo-dious dwelling where I enjoyed an evening's hospitality and listened to tales of sourness. My book, the King's Vista, had raised hopes for the Ferandos that the Shah of Persia, through my influence, would be able to free them from the shackles of Mrs Bandaranaike's brand of socialism, a local horror personified for me at a recent drinks party by the pompous figure of D.B.G. Keuneman, Minister of Housing and Construction. The President's grey politics had brought the Ferandos low. Yearning for escape they had entered a dream world of false hopes wherein was the hallucination that I might find a way for them to leave

Sri Lanka and live on the Caspian littoral. The promise to keep in touch made as I left them was a genuine but downcast pledge. Arrangements for a future rendezvous with the daughter in the Georgian Restaurant in Harrods were to fall apart and leave a note of regret to taint an otherwise happy memory of Sri Lanka.

~ఇ ఏ~

Generations ago, tracing boundaries and making maps was the job of Sappers and their early civilian counterparts. A civil professionalism that matured behind these beginnings created the vocation of the land surveyor. On the edges of civilisation there was neither a land market nor a monetary tax base in land; landownership lay in seizure and not in grant; autocracy was conjoined with priesthood and town planning was unheard of. Slowly in the march of progress, the techniques of professionalism caught up with land management, valuation, broker- age and planning. The process happened in Britain long before any thing like took place in the colonies. There, nonetheless, whatever its scope, the profession of the surveyor was invariably in expatriate hands. These adumbrations of a qualified profession were the pattern and shape of things when, after 1945, the Commonwealth Foundation and its off-shoot CASLE set about the tasks assigned to them. The Sappers and the army had queered the pitch in the days of colonial administration. The uniformed civil servants of India were descend- ants. CASLE's purpose was to help the Developing Countries of the Commonwealth educate themselves in modern practice as a feature of the development process and to organise national, indigenous Insti- tutes for the better organisation of the surveying profession. Local practitioners, however, whether in private or public service, in most places were in advance of Governments and Universities and not in need of persuasion and illumination. Getting representatives of Government and the teaching establishments to talk across a table in Conference bore fruit if the ground for discussion had been well- prepared beforehand. The successful handling of Governments and Universities in Nigeria and Ghana in the 1960s to back the local, newly-established degree courses and the first African indigenous surveyors' Institute in Ghana provided an instructive prototype. If that were to be followed, a precursory study of attitudes and demands would be a necessary step to any serious debate of future policies and arrangements. India and Sri Lanka had been Conferences without such preparations. Soon after India with a Caribbean Conference in prospect, groundwork was started well in advance. Money for the job, lack of which had held up proper preparations in the past, was now offered by the Commonwealth Fund for Technical Co-operation. The stage was set. The honour and responsibility for the preparatory studies were offered to me. The Caribbean necklace of island countries was far too large a handful for a single tour, especially if the two buckles, Honduras and Guyana, on the mainland, were included. A before and after Christmas 1974 programme was arranged.

A previous visit to Trinidad had warned me to take care. On that occasion careless porters at Lisbon had sent my shirt with other baggage on to Rio in the small hours of the morning. Lance Hamilton Murray, a prince among the cricketers in Trinidad, lent me his shirt on arrival and was rewarded to see it exhibited on television that evening. The visit was a preparation for a preparation. Philip White from the University of British Columbia had flown across the United States to join me and plan a joint operation. Whatever else we did, both of us were determined before departing for home to set foot on Robinson Crusoe's Island and make footprints in the sand along with those of Man Friday. The scheduled flight to Tobago lifted off the tarmac just as we rushed into Trinidad airport. So, in an experience repeated only once again many years later over the Fricksburg Mountains of South Africa, I sat huddled against the pilot in a two-passenger craft, on an impromptu flight to Tobago. Tossing over the sea with storm clouds chasing up the rear was demoralising enough but horror gave way to the stark terror of the landing. Ahead a foreshortened airstrip could be seen. Its near edge was a precipitous 1500 foot cliff that fell sheer to the shingle beneath and foreshortened the runway. The tiny plane flew straight at the cliff, skimmed the knife-edge with barely four inches to spare and came to rest with two badly shaken professors drooped over their rattling seats.

Blue lagoons and silver sands by day and the incessant beat of steel bands in the sultry air by night are the lure and pulse of Caribbean culture. These seductions satisfy the lust of the tourist whose pockets are emptied to support many of the island economies. Hid from the holidaymaker and passing visitor, however, are the Caribbean's immaturity and precarious lifestyle. The motif varies from place to place. The Bahamas with its scattered archipelago belongs in culture more to America than to the Spanish Main. Trinidad is the oil-rich member of the family. Among the Lesser Antilles are *quasi* colonies under the defence shield of Britain. Guyana like Honduras has a short sea-face and broad landlocked hinterlands. The University for the island countries is a trinity of three separate campuses. Guyana has its own seat of higher learning. Jealousies abound, notably between the islands and the mainlands. Lack of qualified manpower is a common want and varies from the acute to the tolerable. Teaching is patchy and at the higher levels tends to look to the United States for pattern and precept rather than to Britain.

No sooner had I alighted in Jamaica than calamity darkened the skies. With the off-shore playground of Paradise Island and the rest of the Bahamas behind me, my itinerary for the coming two weeks stretched far into the truly Caribbean and was vouched for by a substantial folio of air tickets. On the second morning the folio was nowhere to be found. Gloom descended. George MacFarlane, the doyen among my kindly hosts, put the ramifications of his not inconsiderable network of contacts at my disposal. They would have been of little use had not 'good fortune' smiled at the hotel poolside. Her name was Valerie, a throw-back to the Trojan beauties of the Iliad and the best walking advertisement for British Airways in the Pegasus Hotel. Once fully apprised of the tickets loss, she went on a planned string-pulling assault with the aid of George's minions and her inner knowledge of British Airways. Codes buzzed over the transatlantic wires and by the afternoon Valerie presented me a fistful of ticket replacements.

The tramp through Government offices was giving me leaden legs and the froth of Caribbean life had turned sticky. So I proposed to Valerie that we went to sea. The Director of Surveys knew of a boat. *Malheureusement*, it was water-logged! So he took us on an off-chance hunt among the craft bobbing at anchor in the harbour to find a longboat with a motor that worked. At a pull, it jumped into life. Soon the waters of the cay-studded bay were receding behind its bubbling wake. The tang of the historic Caribbean was in the air as the bow pushed over the waves to Morgan's Harbour. Imagination re-wrote history. Here the infamous Pirate King, who was also a local Government worthy of 17th Century Kingston, had ravaged the Spanish Main and sent his buccaneers ensigned under the Jolly Roger to plunder the defenceless. Morgan's priceless, ill-gotten treasure was somewhere about in the sea-girt caverns. Ian Fleming's *Live and Let Die* could almost have been true that morning. At a strategic distance lay Nelson's Quarter Deck. This, again, was no fancy sea-mark but the very spot from whence, with an eyeglass to his one good eye, the great admiral would scourge pirate and Spaniard alike in the name of the British Crown.

Port of Spain in Trinidad carries in its very name the romance of history unknown to the froth of Paradise Island. So it is with Barbados.

That relatively small banana island has a ring in its name far greater than its size warrants. Maybe history has much to do with it. In the late 15th and early 16th Centuries, the island was uninhabited. There were no Caribs to contend with. Aristocracy from England, the Earls of Pembroke and Carlisle, challenged each other over its lordship. When Carlisle won, the Earl of Pembroke quit his bananas and sailed with estate staff, ships and crew northwestwards to a new mainland that lay south towards the sun from the Virgin Queen's sovereignties in Virginia. This new landfall, the Earl's men called the Northern Territories. This history I would sport to considerable effect in Barbados. Each evening, as dusk fell and the lights came up with the moon among the palm trees to sparkle in the glass and silver on the dining tables of Coconut Creek, a loneliness would creep upon me. To dispel it, I would kindle conversation with the next door tables. My neighbours when asked where they had come from would frequently answer 'The States'. 'You mean the Northern Territories?' I suggested with a grin. If of Scandinavian, German, Dutch or other non-British extraction this quip would occasion a sullen sipping of soup and end what might have been a lively conversation. Decendants of Scottish, English and Irish emigrants would see the joke and unfold their family stories. One morning on Coconut Creek beach while I was drafting my Report, a brawny Canadian kicked sand into my discarded *Financial Times* and stirred the *angst* in me with the observation, 'Say, only another 160 points to go and there's no more Stock Exchange!'

At Coconut Creek, arousal at 4.30 am will give you just time to catch the early morning flight to St Lucia. I performed the feat once. St Lucia is bigger than Barbados, is nothing like so developed and has a land tenure bewitched by unreformed English land law, mixed with Code Napoleon and French juridical precepts. The day's trip would have accomplished more but for the hours spent with Richard Big-wood, the United Nations Development Programme's Director of its Caribbean Physical Development Project. Stuffed in cabinets and plastered over the walls were plans, highly imaginative, colourful, certainly original of what the future might hold for St Lucia and many another Caribbean island. On St Lucia, golf courses were to replace fishermen's huts, traditional wharfs and much else dear to the heart and lifestyle of the islanders. 'Have you asked the fishermen, the wharf-holders whether they want their land used in this way?' I enquired of Richard while scrutinising the coloured wall sheets. 'Oh no!' he replied, 'why should we?' So I suggested he shut up shop, went back to the UN in New York and advised them to stop wasting other people's money making plans to gather the dust of ages in some forgotten cupboard. Bigwood, a planner like all planners, thinking he had all power was too imperial to consult the landowners by whom in the last analysis all real, effective decisions are made. The UN had given him subordinates in other small island bases. These minions thought likewise. They were not to blame. If they were not working for UNDP, they would be employed by the World Bank or some other world agency on similar futile exercises. That property power was primary power was a truth yet to be learnt the whole world round. My travels from Vancouver, to Buenos Aires, to the Caribbean, home to London and on to India, Malaysia, Australia and the South Pacific, affirmed the global ignorance.

Apart from St Lucia, it was Guyana where the paradigms of the problems which beset me were most clearly expressed. Its University had given degrees in Land Survey only, had taught and trained that side of the profession throughout the length and breadth of the land from Nassau to Port of Spain. Land Economy was not necessary, in the view of H.D. Hoyte, Minister of Economic Development. The day before seeing him, Choo-Shee-Nam, a hydrographic surveyor of Chinese extraction, had taken me with his family, his half Dutch, half Scots and truly wonderful hybrid children to survey the Black Bush Polder. Engineers had carved this impressive new land out of the Courantyne Delta at New Amsterdam. It was as big as Barbados and was waiting to be settled *de novo*. New farms, new homesteads, new forests and all rural cognate things were wanted. Hoyte reckoned the entire development to be an engineers' job! Nothing would budge him. Because under the British everything had been left to the engineers, so it could remain. My mind flicked back to the Indian civil servants. In Guyana, also, I was given the most instructive VIP treatment. My hosts in their zeal informed me of the local 'Cambridge University' and urged attendance. The 'dons' and other incumbents would be walking the streets from the Tower Hotel to the Pegasus Hotel to encourage 'undergraduates'. At first, my bonehead thought it was but another training centre to be put on the agenda, until one of the kinder among the excellent hosts hinted that Cambridge University was the local bagnio. There were enough problems at the Cambridge I knew to court others in a 'Cambridge' I didn't.

The loneliness in the Caribbean which beleaguered me from time to time was taken care of on my second visit in 1975 when Alison, my secretary, came out to Coconut Creek. Any earlier arrival would have been out of the question and would have coincided with the Regional Surveyors' Conference in Jamaica that had called me there the previous week. Alison worked hard, ate well, delighted in sifting silver sand through her toes and when in full bikini displayed across her back an archipelago of fascinating moles corresponding to the map of the Caribbean Islands. Looking at the more reliable chart, discovered for us a string of the smaller and more beautiful islands yet to be tackled – Grenada, St Vincent, Dominica, Antigua, Anguilla. The agreed plan was to make first for the spice island of Grenada, then go round to St Vincent, Dominica and take off over the northern Antilles towards Anguilla and home.

Anything approaching a land economy surveyor was hard to come by. Two expert valuers from Jamaica and Guyana carved up the entire region between them. The people who looked after us were, therefore, land surveyors. They were busy men and left us of an evening *sans* guidance and counsel. Ignorance brought us to grief on Dominica. A mountain range divides the airport from Roseau where lay the only concentration of dwellings and the Anchorage, its solitary hotel. The Anchorage serves meals under a high thatched panoply below the stars. That evening, the menu offered 'Mountain Chicken'. Somebody must have made an excursion to the mountains to hunt for the bird after we had given an order for it. Much marching of feet and slamming of hidden doors tested our patience for an hour before the 'chicken' steaming hot arrived on the table. Zoology is not my strong point but I've yet to see a live chicken with six frog-like legs. An excellent savour, however, allayed all fears. Sadly, the enjoyment of the meal was cancelled many times over by a night of incessant sickness. The nausea added perceptibly to my knowledge of Dominican wildlife. The Director of Surveys next morning informed my pale and tottering frame that 'mountain chicken' in the local vernacular stands for a sprawly wild toad, an inhabitant of the nearby mountains. Alison survived the night and even the roller-coaster drive back over the habitation of the mountain chicken. As for me that day, death was preferable to any other option and indeed seemed the most likely, if feelings were anything to go by. We were off, however, to Antigua. The last time I had landed there the aircraft burst a tyre and slithered within a few feet of a ducking in the ocean. Now in my present state I wouldn't mind what happened.

❧ ☙

On 7 April 1976 the passage of time had turned me into a *de jure* old age pensioner, without a pension. To qualify it was necessary to be unemployed. The University allowed the able-bodied and foolish to opt for a further two years in harness after reaching sixty-five. This I had done. The coming two years were likely to be critical for the fortunes of Land Economy. There had accumulated in my favour over the years a backlog of unclaimed Sabbatical years; banking these would have allowed me to retire for two years on full pay there and then. Money, however, had marginal significance and was of little consequence. The full Professorial stipend at the time barely paid my incone tax. There were no golden handshakes or enticing redundancy payments. It was the future of the Department that concerned me. The Department had been my life work. Wrong decisions could do great harm. No one knew what the authorities were thinking. Hindsight strongly suggests the question had not even reached their agenda. Maybe, my choice would have been otherwise had relationships with the Secretary General, Treasurer and other lords of the University *demesne* been happier. 'Establishment men', re-elected time on time to the central bodies, saw to it (so I was told by an officer who had been on the inner scene) that the old hostilities were unabated.

A period of tranquillity and goodwill to mark the end of my service was a reasonable hope. It was not to be – as it was at the beginning, so it was to be at the ending. Not only were the corporate bodies – the Board of Land Economy, the General Board, the Trustees of the Development Fund – at loggerheads with each other but officialdom, so it seemed, was out to disparage me personally. On one occasion, they contrived to let it be known that the question of deciding what was to be done with the Department and the Chair would not be seriously faced until I was safely out of the way! But of more immediate urgency was a letter, in late summer 1976, from an officer in the General Board Office. It was addressed to me personally and missed by a hair's breadth being an outright accusation of misappropriation of funds. It called upon me to provide to the General Board, forthwith, a detailed breakdown of all personal and staff travelling expenses over a number of years. On the face of it, the accounts looked somewhat incongruous. Travelling expenses for Departments of comparable scale were usually in the lower £100s while for Land Economy they were well into the £1000s. So be it. Every year the Assistant Treasurer carefully went through the draft Accounts with me and knew the answers to the questions put and, what is more, knew also that all was well. Had the officials really wanted greater detail, courtesy would have suggested a telephone call to the Assistant Treasurer and a cup o' tea with myself. No such diplomacy. Let's dress him down! My travelling round the world on behalf of the

University was fully-warranted and in any case was funded by patrons, sponsors and other external principals or by the Trustees of the Development Fund and not by the University's Education Fund. That source at best, as characteristic of its parsimony, would have supplied but a few £100s. It was not called upon to do so. It was relieved of all just claims by the generosity of the Trustees of the Development Fund so as to present the Department whiter than white. The episode looks petty from a distance but at the time it was hurtful, alarming and unjustified.

About this time, the Treasurer of the University without warning stalked over my doorstep. He was a tall, thin, austere man who also had a complaint. He carried a white card. This he handed to me in silence. The card was an invitation to a dinner party. The card was familiar to me; for indeed, I had drafted the wording. The Treasurer seemed moved by a kind of inner agitation.

'What is the meaning of this?' he enquired in incredulity, as if I had dirtied my pants.

'That's an invitation to Dr Sturrock's Farewell Party', I replied and added, 'Hope you can come'

'The Financial Board has commissioned me to ask who is paying for it.'

'The Trustees of the Development Fund,' I answered more puzzled than ever.

'We thought so. It won't do. You have no authority to misuse the Development Fund in this way.'

'That I know,' I supported him. 'I personally have no authority. But the Trustees have every authority. If you have any query take it up with the Chairman, Sir Oliver Chesterton.' For good measure I added, 'You should know they are also authorising a portrait of myself. Perhaps you know about that?'

'What! You are not serious I hope.' The Treasurer looked as if his oxygen supply had suddenly been cut off.

'Always with you, Treasurer. These are Trustees' issues. Do get in touch with the Chairman, if you must. Nothing to do with me. Have a cup o' tea?' He left without a further word and I watched his shining head shaking in wrath and unbelief disappear up Trumpington Street.

The farewell Dinner Party was a welcome and fully-deserved success. The Treasurer, sadly, did not attend. Some days later, Sir Oliver Chesterton received a letter from the Treasurer. It was fussy, pompous and reproving and rightly irritated the Chairman. The issue was too trivial for anger. After some weeks and a Counsel's Opinion on the Trustees' standing, the Managers of the Development Fund gathered round the Chairman to watch him frame a reply. He didn't need help.

Words twitched his brow and quivered on his lips as he muttered the lines of Hilaire Belloc:

> 'Remote and ineffectual don
> That dared attack my Chesterton.'

Then he wrote in a vein which acknowledged the 'reproof' and advised the Treasurer to see that it didn't happen again; and ended in straight-flung 'Chesterton' to the effect that the Managers were Executive Trustees and the University a mere Custodian. We heard him declaim, straightening his back and quizzing his handiwork, 'If we want to put our money on the 2.30 at Newmarket, no-one can stop us.'

The passage of arms with the Treasurer, the row over travelling and other misunderstandings which soured any milk of human kindness flowing between Land Economy and the University authorities need never have happened. Cambridge officialdom in my experience had always been austere and bureaucratic. They tithe the mint and cummin of Statutes and Ordinances and forget the weightier matters of personal kindness, understanding and reciprocation. This self-righteousness may have been something put on just to vex me. One was aware of a frigidity which if thawed in the slightest degree by the warm spontaneity of social intercourse would be seen as an unseemly mixing of patricians and plebs. The unhappy business of finding a new Professor was exigent. The people who were most concerned, most likely to be affected by it, who understood the issues profoundly were kept at bay, kept waiting, kept guessing and fed on rumour.

History was repeating itself. Staff, undergraduates and the profession were affected for the worse by the dithering uncertainty, delay and half-truth. Certainty, however, was not altogether missing. All the money for the Professionship was coming from the Managers of the Development Fund. That was one thing certain. The Managers had a fiduciary responsibility: that was another certainty. A third was the adamantine determination of the Managers to veto all attempts that might risk the fulfilment of the terms of the Trust. Consultation between them and the University could hardly be improper. On this cooperative note, the Chairman of the Managers again took up his pen, this time to write to the Vice-Chancellor to ask what the authorities had in mind. The question was put fully three years before my retirement. The Chairman's letter was ignored. Two years later, he faced the embarrassment and indignity of having to write a second time to point out to the new Vice-Chancellor the fact of his predecessor's discourtesy.

Painfully the University awoke, and sought light in its darkness by asking a special Committee what they would advise should be done.

The membership of this côterie and its activities were General Board secrets. The only decision eventually known was to appoint another Committee. At no point were the Managers or the Board of Land Economy consulted. Again the Managers acted, for their own sake and to scotch rampant rumour. The Chairman begged the Vice-Chancellor to keep faith with the Trustees. Nothing was patent until the General Board reported to the University. The concealed wickedness was revealed: the General Board intended to walk away from Land Economy as we understood it towards 'urban economics' – whatever that might be. The gauntlet was thrown down. From the Managers there would be no money for urban economics – a nomenclature too ambiguous and suggestive. Reluctantly, the description of the subject was widened and shaped to the Managers' better liking.

All appeared to be arbitrary and clandestine, including the all-important membership of the Advisory Committee which, as on a previous occasion, had been set up to find names of suitable persons to propose for the Chair. The impression was given that the General Board had no intention of consulting the Board of Land Economy, the Managers or anyone else competent to advise. A purposely devised leak to the Managers, later confirmed by letter, let it be known that one of the excellencies on the Committee would be a Past President of the RICS. When the list was published in due course it was too late to object. The only chartered surveyor was an unknown civil servant from MAFF. No one of eminence, knowledge and standing in the profession had been recruited. The mistake was grave because the foolish General Board officials, in pride and self-conceit, had refused to consult anyone who could have advised them properly. When the smoke of battle cleared and the chaos and confusion were revealed, Oliver Chesterton sent another broadside into the Vice-Chancellor. It had little effect; by then the enemy's ships had departed over the horizon.

Of all the hurtful and demeaning treatment suffered by me at the hands of the General Board over my long years of service to the University, the most shameful and bitter was being completely and deliberately ignored by the University when it sought guidance on the future of the subject. My thoughts had indirectly coloured the precepts sent to the General Board by the Board of Land Economy but that was all.

As an out-going incumbent, I did not expect to sit officially on any of the Committees. It was the total disregard of my presence in the University, long before the question of named successors came into the picture, which justifies condign criticism. The treatment evoked from me a suspicion that the authorities were seeking a person who could and would de-Denmanise Land Economy?

≈§ §≈

Puzzled anger, with puzzlement dominating over anger, best describes my state of mind towards the University but never for long. Neither of those dark moods were to cloud out the sunshine from life in the run-up to final retirement. Days were too busy and excitement too insistent for brooding although that forthcoming 'departure' did on occasion cast a foreboding before it. My diary was committed to obligations running to horizons far beyond normal retirement. The Community Land Act 1975 was still on the Statute Book. Throughout the country, in speech, writing and radio it claimed my denunciation. Opposition advisory Committees under Grahame Page, Keith Joseph, Hugh Rossi and others still met in the House of Commons and expected my presence. The old stamping grounds – Ghana, Nigeria, Iran, were to be visited. In Ghana, a Land Administration Research Council (LARC), linked to the Department at Cambridge had been established and entrusted to a geographer-planner of abysmal ignorance in land tenure. LARC was a worry and a time-consumer of my days in Ghana where also Otumphuo, the Asentehene, had commissioned my help over an Estates Office for the Golden Stool. In Iran, with Valian redeveloping the Holy City of Mashad, fresh audiences were to be arranged with the Shah. He had in mind lavish new educational projects. More specifically, I was helping with a new compendium – *Iran Under the Pahlavis* – edited and supervised by George Lenczowski of Stanford University, California. Visits to Italy kept Borgo under review. And unexpected engagements increased the pressure, among them a lecture assignment at Ann Arbor, Michigan, on what to do with America's polluted lands. New developments of great pleasure were the opportunities which came to establish links in Australia, SE Asia and Saudi Arabia where gaps in a world chain of Cambridge-planted Land Economy activities were waiting to be filled.

At thirty-five degrees South, Buenos Aires to the far, far West lies coordinate with Sydney out on the world's far Eastern cheek. At both these extremities there were problems for Land Economy. An appeal for help from New South Wales, as with the Argentine, had come without prior warning. One side of the world was trying to balance the other in my records of countries whose Land Economy problems were known to me. Half a world between made no difference. Land problems were human problems alike the world over, neither East nor West discriminated between them. Or so it seemed to me, when a weighty package from Australia brought a copy of Else Mitchell's *Land Reform Report* and a request beseeching an analysis and comment. The request was couched in an urgency that might just as well have included paper, pen and a stamped addressed envelope. Fortunately, hard on the heels of this surprise, Tom Whipple, from the University of Sydney and his lovely wife Pat came to Cambridge on the way home from the USA. Tom Whipple straightened out my bewilderment.

Goff Whitlam headed the then Labour Government of Australia. Hoping to leave a lasting testimony of achievement, he had appointed Else Mitchell, a prominent QC, to cast an eye over all the vast territories of the Commonwealth of Australia and advise how best the Federal Government might reform the land laws and initiate a land policy. Else Mitchell was no Daniel come to judgment. When launched, his Report terrorised the ranches, the outback and the urban coastal sprawls of the great cities. The Provinces were distracted. At Tom Whipple's suggestion, the Opposition in New South Wales had sent the Report to me and posted him to follow its arrival. Else Mitchell, the learned QC, had mixed for Federal Australia a concoction of poison. Its ingredients contained all the socialist nostrums on land and property which for the last thirty-five years had spelt disaster in this country, from the abortive attempt to vest development value in the State in 1947 to the community land scheme now crumbling to pieces under our feet. Tom, however, held in his briefcase the shapings of a far more formidable undertaking. Days passed, wine was drunk at Pembroke College High Table and in lesser places; my analysis of the Else Mitchell Report was rejoiced over. We became firm friends and, bit by bit, he divulged the hopes in his briefcase: would I come to Australia equipped to give a series of public lectures on property and land? A number of sponsors would be found, including his Department in the University of Sydney. The lectures and my presence would liven the prospect of a Land Economy Degree at Sydney and maybe could provide even a midwifery service.

Another year was to pass before arrangements were made, sponsors found and the town-criers were out in Sydney. An irresistible temptation to see my friends in Iran, once yielded to, came within a hair's breadth of shattering the programme. At Tehran, my Japanese Airways flight to Bangkok was announced to be running three hours late. Delay in Tehran meant more dining time. Half way through dinner, Emami, my cheerful and ready host, 'phoned the airport to learn that the Japanese flight had unexpectedly landed. Quick! Farewells were taken for granted, kisses blown and bodies and baggage swung into a waiting car. The airport was reached before the servants had cleared the half-consumed kebabs. The Japanese plane had gone! There was not a passenger in sight. The clock stood at '1 am', all customs and check-throughs were empty. Then from an open doorway came the hiss of an impatient engine. The Bangkok Flight was still on the tarmac. Flaunting a first class ticket (doctor's orders) the unmanned barriers were taken by storm. Within fifteen minutes of the dinner break-off, the door to the first class cabin closed with a thud behind me.

Another night and a day, and there was Tom Whipple waiting to greet me at Sydney airport. Tom is Anglo-Saxon by derivation and his wife a beauty of the Race. Like other Aussies boasting four-generations and longer pedigrees, his speech has only the slightest hint of Ocker. Unlike many of his fellow countrymen, Tom opens his mouth, it lets the flies in and he doesn't seem to mind. Careerwise his predicament matched mine of earlier years. Diehards, traditionalists and planners were opposing his far-seeing efforts to establish a Department of Land Economy at Sydney University. Opponents were, as I might have supposed, in his own Department of Town Planning where the puissant institution and power of property were only dimly understood. Elsewhere in the higher clouds of opinion the opposition was one of sheer prejudice. He motored me to downtown Kings Cross and the questionable mercies of the Chevron Hotel.

Australia is expansive and too vast to have a soul of her own. There was, nevertheless, a spell which bound me. The two weeks went by in a cinematographical flicker of lectures, social events, drawing pictures, driving hired motor cars, sailing great harbours, scouting the wilderness and visiting the 'cemetery' city of Canberra. The spell bound me tighter. It was reminiscent of something I'd met before. Whatever was it? The resemblance slowly dawned on me: it was the Caribbean and the spell of islands. Yet here in Australia the islands were not scattered over an expanse of a limitless ocean. They were city islands surrounded by endless wilderness and crouched together for protection and comfort along the long latitudes of the seaboard, afraid, so it seemed, to penetrate too far into the wild devouring emptiness behind them, lest they lose the maternal embrace of the ocean. Locked between land and sea, Australia's cities were expanding skywards. Sydney Harbour, a natural esturine 'gem of wonderous beauty, as I saw it from the air, had once offered to the ocean fair virgin lips now ill-fitted with the crude dentures of high-rise office blocks.

On the morning of my arrival the Royal Yacht bore the 'Pommy Queen' and her Prince up the waters of Port Jackson to a cacophony of sirens. Australia welcomed her with the polite reserve of 'a daughter in her Mother's House but mistress in her own'. A fair daughter but yet in youthful maidenhood.

A close-knit fabric of long history had yet to be woven. Australia lacks the enrichment of retrospects. No one had ever held the land of Sydney by Knight's Service of a feudal King, so property men in Sydney hardly knew the inner character of their merchandise. They talked of owning 'buildings', not the abstractions of tenure. They joined in 'The Building Owners & Managers Association of Australia Ltd' (BOMA) with warm hearts and hard hands. The grace of BOMA was Sheila Evans, handmaiden to the President. She looked after them all with great care and diligence and for my short sojourn was the light of my days. Within the University, I spoke of BOMA with men and women who showed either much or little knowing. The academics wanted only to touch the Association's money, not their persons.

There were exceptions, the 'Tom Whipples', blessed men and women whose far vision and wide-open souls wanted to blend the professional and the academic in mutual bonds of understanding and benefit. Comment approved my lectures and led me to believe the lectures had succoured Tom Whipple's cause which ultimately succeeded in the establishment of the first Professorship and Department of Land Economy in the University of Sydney. For myself, I was content to be spellbound by Sydney and the other fair cities of the coast.

The impression that Australia was an archipelago of island cities whose livelihood lay seaward was ratified by a visit to Canberra, the land-locked Federal Capital. It is as beautiful as a well-kept necropolis is beautiful. The city is an outcome of what one would expect of an opportunity which gave planners their head. No one can 'go downtown' in Canberra – there isn't a 'downtown' to go to. No shops, no traffic congestion, no hoardings, no litter, no people, no life – only Government Offices and bureaucrats. Canberra's decentralisation has no peace of the countryside to fill the centre, nor the ease of a garden. A thanatoid plight seemed to occupy it. Any excitement, such as it is, lies out on the circumference. My hosts, great people who lived elsewhere, drove round Denman Drive and on to Capitol Hill. My links with the Denmans of Canberra are either non-existent or at best very tenuous and uncertain. Sir Richard Denman (before his inheritance of the family Peerage) was Governor General of Australia in 1913. His wife, Lady Denman, had laid the first brick and cut the tape to establish the new Federal City. Stories abound of the event. According to one version, Trudy Denman was given a golden trowel to do the job and a golden visiting-card box with the hitherto secret name of the city in it. Trowel and visiting box were to be employed in unison when the guns boomed. Unabashed, she opened the box to the sound of the gunfire and was tongue-tied to pronounce the name. Before her, flanking the middle distance, were two conical hills, Mount Ainslie and the Black Mountain. Inspired by their shape and the valley between them, she called the city 'Cam-Bra'. The golden trowel and box are in the possession of the present Lord Denman; so that much of the story is certainly true. The rest of the tale is almost certainly apocryphal – in 1913 it would have been a bust-bodice anyway!

My lecture series ended in Adelaide. By then exhaustion of mind and body had overcome me. At Canberra, a fainting turn had warned of the state of my health. Even so, a fellow called Bentick, of the Duke of Portland's ancient line, and now a University Lecturer-cum-farmer, insisted on my coming 'down farm'. He lived by himself, rose at 5am of a morning, boiled three eggs finger-fed for breakfast and then walked the hill pastures until midday. My weariness made walking sheer agony; nerve and sinew had almost gone and a lecture stint was still before me back in Adelaide. Verily I believe and truly, the last lecture would never have been given had not Professor Percy Johnson-Marshall from Edinburgh been in the Adelaide audience. He strengthened my resolve. Not only was he a friend, he was also a planner!

Cape York pointing northwards shows the way to Australia's maturity. That maturity lies northwards across the Timor Sea. When the Australian Commonwealth carries the full burden of civilisation in giving aid to her northern neighbours in South East Asia, to satisfy appeals for men, money, books and teachers, maturity will be hers. In that day the contracts that linked me to Malaysia for many years will be unwanted. The Peninsular called me, I did not call it. Once there, however, a few months after returning from Australia, the lure of the East was unrelenting.

> 'If you've 'eard the East a-callin',
> you won't never 'eed naught else.'

The summons, as so often had happened in the past, was unexpected. This time it came from the Registrar of the Technological University of Malaysia. The formal invitation was followed up quickly by a meeting with Azmi, the University's Head of Department, at the June FIG Congress in Stockholm.

Malaysia is a world apart in many ways. She appoints her King *per rota* every quinquennium. To my amazement she also practices voluntary apartheid. The Malays are for the most part *burmaputra* and exercise dominion of choice. The Chinese are clever, hard-working and rich with commerce at their fingertips. The Tamil Indians are the administrators. Each race keeps itself to itself in living and dying, housing and schooling and much else. The British presence, apart from a new generation of go-getters, is largely represented by London-based owners of rubber estates who by neglect are out of touch with current land and other markets and are prey to the easy touch of unscrupulous locals. Land Economy soon took over from what had been before at the University and was welcomed by the Malaysian professionals. Thus, by the pull of its own fascination and a genuine desire, the Peninsular had stimulated me to set up another lively outpost of influence. It was to bind me to the Peninsular for many years after Cambridge University, by virtue of its Statutes and Ordinances, had requested my retirement.

Another new horizon that opened before me on the eve of formal retirement were the burning sands of Saudi Arabia. The cooling light of early dawn from across the desert of *Al Jafurah* was my daylight introduction to the blue, white and gold of Saudi Arabia and to the apocalyptic realisation of having only one more year in the Chair at Cambridge. The morning was Saturday 1 October 1977 and I was a stranger in a bedroom of the Royal Hotel at Khobar. The paradoxical world of life after retirement was casting its shadows before. Age which disqualifies a man for his accustomed job at the same instant qualifies him as a 'consultant' to dispense the wisdom of years to all who might seek it. The Royal Commission of Jubail and Yaabu were under royal command to carpet the desert to the west of the Port of Jubail with the spontaneous creation of a new city to house and employ a 'permanent community' of 400,000 souls. An American architect was working with Colin Buchanan & Partners under the unforgiving eye of Bechtel Incorporated to plan, propagate and plant this new Eden. Anticipating the demands of post-retirement mammon, I had signed a consultancy sub-contract with Colin Buchanan.

My arrival in the unfamiliar Kingdom of Saudi Arabia had been an unhappy one. Fussy customs men whose logic defied all reason were willing to let me in but insisted on banning my new leather folding suitcase as if it were some forbidden fetish. Beyond the barrier, Aramco, the official contact and a kind of American businessmen's embassy neither expected nor helped me. Hours later a sixfooter, with a bumptious gait, shaking desert dust from his trousers glared down at my forlorn figure and slurred, 'Say, I've been watching yoo. Where yer going? Are yer Bechtel's?' Somehow, he retrieved the confiscated baggage. With a toss of the head, he beckoned me to a grimy truck and the outer dark beyond the airport glare. Discomfited by this Saudi welcome and feeling heavy-hearted about Bechtel's crude insouciance, I felt disinclined to try and keep up with this Yahoo.

Colin Kolbert in whose close friendship and company I always rejoice was a partner with me in the Buchanan contract. He was no Arabist but had a smattering of Islamic law and, through the hearsay evidence of others, a grasp of its fundamentals. Being well-grounded in Justinian and the Classics, he had the knack of working from first principles. We met for a dawn breakfast and to catch the Bechtel bus. Fifty miles north over almost featureless desert tracks, markings in the sand spelt out the site of the new city. On the way there, just beyond Khobar, rose the empty, neglected shell of a hotel that never had been. The planners, economists and architects responsible had never been taught the basic lesson that the primary power of decision-making over land lies with property rights and not with planning. No one had enquired of the proprietary pattern of the site before the hotel was built. Bedouin to whom King Feisal had by decree given the land some years back waited until millions had been poured into the site to complete the building and had then moved in to claim ownership. There it stands, a monument to folly. Later I used the object lesson to ram home to the Buchanan planners how essential it was to have a truly reliable proprietary survey or a sound immunity clause in their contract of commission. Caution was necessary, for despite the glaring evidence of the hotel nonsense, the Royal Commission were reluctant to face ownership questions and tended to dismiss my insistence as embarrassing curiosity.

Our brief was novel and to many paradoxical. The Royal Commission wanted a specification of the means by which the land use patterns of Colin Buchanan's Plan could be preserved in the future without Government control over land use. Whatever we proposed had also to accord with the will of Allah the Merciful and with the writ of the Koran. Back in Khobar that evening, as we hammered the sand from our shoes, not in any Biblical sense as a sign of rejection of the place, our mutual thoughts flickered with images of John Adam's New Town to the north of Princes Street in Edinburgh and of much else in the Scottish land scene. Centuries back, Craig and other learned Scots lawyers had divided land ownership (*dominium*) in Scotland into a two-tier system: the *dominium directum* vested in the feudal lord and the *dominium utile* in his vassal. What ho! Here surely could be the solution to the Saudi conundrum. Whether Islamic law would countenance the division of ownership in this fashion was problematical. If it were permissible, the King could hold in his person the superior *directum* over the lands of the city and all private rights would be derivative yet constituting full ownership and subject to the royal proprietary directives. We liked the idea as a formula and in the fullness of time our recommendations were to fashion it into a code for Jubail.

The chaos of my arrival was as nothing to the high drama of departure. Wined and dined by Bechtel's generous Jane and John Robb, my spirit was high when, just before midnight, the customs at Dhahran were negotiated without let or hindrance. Suddenly all was pitch black. A few pocket torches caught in their puny searchlights the panels of the departure lounge doors as they closed to cut off massed humanity shoving against them. Blue and red lantern lights dimmed to a 20-wattage glimmer appeared after some time to make bewildered shadowy forms grotesque as they fumbled and struggled for the few precious seats available. Speech was *sotto voce* and freighted with rumour. My British Airways plane destined for Heathrow had been diverted to Bahrain. A Japanese airliner had been hijacked and was prowling over Dhahran demanding permission to land. Its pleas were rejected time and again and the hours wound on. Tension eventually eased but not the impatience. It was fully 3am before all was clear. Hope slowly returned as panic abated. The chaos, however, remained. It was standing room only as a polyglot mob scrambled for the first scheduled flight to land at Dhahran and the sun came up in crimson out of Persia across the Gulf.

13

TAM SOLI QUAM AQUAE

With cloudy mien and furrowed brows
You take the riches fate allows.
Your wisdom's worth is crowned and prized,
And sea and shore are harmonised.

Faust, Part Two, Goethe

Retirement meant longer breakfasts, a carelessness in the strict sense of the word towards passing hours and weeks of ordered disorder. There was time to tease my watercolours, to try new tricks and techniques. Time for the tedium of one-finger tap-typing. There was no fair secretary to space out the diary, to see to air flight bookings, to answer the telephone to 'nasties', to enter the caveat of her kindly amending smile to check a boss's folly. Psychologically, retirement meant an image change; a new assessment of myself wrought in traumatic inner spasms, especially at four o'clock in the morning. Altogether it was an unpleasant experience. In compensation, there was the release from the inhibiting rectitude of office. Most gratifying of all was the realisation that retirement is a free city where liberties of speech and pen can match its citizens' thoughts with relative impunity.

On 30 September 1978, I walked over the threshold of the familiar doorway in Silver Street, Cambridge for the last time. There were certain loose ends to tie up. These were attended to at home in Chaucer Road where the semblance of an office had been set up, because in the interests of my successor, I placed myself under a vow never to enter the Department of Land Economy again unless invited to to so. At that stage, the vow was a matter of principle. The shadowy figure of the next Professor of Land Economy was in the offing. Only an esoteric few knew his name. It was never revealed to 'strangers' like myself. There were longstanding commitments in the name of the Department that were personal to myself. Some were to run out with a full stop at my retirement. Others merely noted the change as it were by a coma and I 'took the business with me'. And there were to be new sentences framed with events and obligations in the years ahead.

Ghana was of the full stop variety. Fifteen years of careful nurturing, plus the outstanding abilities of Ben Acquaye its Head, had brought the schooling in the Department of Land Economy at Kumasi to pre-eminence in the Third World, a deservedly rich reward. Ben, the pioneer Professor, was acclaimed from Port of Spain to Suva and elected in a spirit of jubilee to the Presidency of CASLE. In recent years Kumasi has sent top graduates to Cambridge to carry back with them Doctorates, Master Degrees and other ensigns of higher learning. The Department had grown up. The umbilical cord was a vestigial remain. The International University Council understandably was not a little embarrassed at financing what looked like discarded baby linen. Money needed for Kumasi University would have to come from a different aid vote. Thus it was that my departure from the Chair at Cambridge signalled a logical *finale* to many years of successful academic and professional cooperation. The bestowal of a D.Sc. degree by Kumasi University was a generous and fitting valedictory tribute.

Unlike the legal profession, the surveying profession had still to build a credible academic base. Research was essential. With me it had always taken precedence over practice whenever, in my view, the two were incompatible. As the future of formal academic life shortened, however, my conscience in the matter became less sensitive. Unashamedly, I permitted the word to run in professional circles of my readiness to consider consultancy and even to listen to the highest bidders. No formal plans were made. Offers came of their own accord. Some were from afar. Others came through the home post. The Lord of the Manor of Broxhead was waging war against his commoners and ready to pay for the services of someone with expert knowledge of the history of manorial customs. Matthew Ridley of Blagdon in Northumberland wanted help with tax problems. A Herefordshire inheritance had to be sorted out. Operations of this kind ran no risk of my breaking faith with those practitioners who had helped over the years with my academic research.

An amusing instance of a one-off unexpected commission happened on the very eve of retirement. FAO were up to the neck in turmoil and confusion trying to organise a grand worldwide land reform Congress. An entreaty came from the Rome Headquarters. A distraught Indian official called Nehemiah was inundated under a deluge of papers. They fell in a thousand different languages flooding his desk. Female scribes rewrote them in the three accepted languages of the Congress. Colin Kolbert, myself and later Derek Nicholls from the Department of Land Economy at Cambridge were each given a desk in FAO and asked to paraphrase the English versions. We were literary snoopers commissioned to seek out and remove all tautologies, solecisms, misspellings, ambiguities and, especially, to watch for any chance indecorous double entendre. There was nothing inherently jocose about FAO English. But one morning Colin Kolbert suddenly collapsed in a paroxysm of mirth. Clutching his chair in one hand, he waved a paper towards me in the other. 'I can't believe it,' he almost wept. The author of an American text under the Heading 'Participation' was desirous of persuading the world audience that participation was critical to land reform. Ideally everyone should 'participate', from the boss to the office boy, women equally with men. The vapid Paper was deadpan serious and wholly insensitive to the double entendre of its humourless author. 'Participation' it declared in stark indecorum, 'is a matter of sex. You can have top-down participation; or bottom-up participation. And sometimes top-down participation leads to bottom-up participation.' A hundred copies of this *faute* were distributed through the FAO Palazzo as a model example of a risqué inadvertence.

Going half professional meant a reversion to school. Lessons never learnt or forgotten had to be picked up. It was too late to master them, apart from suffering my innate reluctance to do so. Thirty years of life, thought and action in an academic ethos had shaped habits, familiarised images and ingrained postulates. Truth, as one saw it there, was outspoken and expectant of contradiction, confrontation, rebuttal, denunciation and criticism. Words were not trimmed nor ideas double-thought. Straight flung speech was never considered impolite. The professional world, on the contrary, appeared to confuse politeness with deference. The shopkeeper's code, the customer is always right, was the aphorism to work by. Should a principal or client wish to think that black is white, don't disillusion him – you might lose a fee! What the French call *prévenance* held precedence over a hammered-out truth. This shibboleth of professional practice ran up against my academic temper. On the other side of the coin, my customary forthrightness grated at times on the sensitivities of colleagues. It was difficult for me to accept with a kind grace and without a sense of shame that parties could negotiate with each other in clouds of unknowing and bandy nonsense, providing each tongue spoke in equal ignorance with the other. My temptation in such circumstances was to play the pedagogue and that was unforgiveable, especially if the parties were speaking to a client. Of all the lessons faced up to, the most disturbing was to realise how narrow a gap could divide the professional man from the businessman. 'Practice' is a euphemistic substitute for 'firm'. Whatever expert knowledge a professional may 'profess', the first claim upon the use of it was to make money, especially if operating in the world of property. If by so doing the clouds of unknowing part for a season and reveal a clear blue sky of truth, its dissemination is no more than an incidental bonus.

These lessons were painful to me. To disregard them did little harm where there was no continuing privity between myself and a principal. Offence would find its way to the dustbin of forgotten memories. Where wounds had been inflicted, however, time could open deeper the broken flesh, if the healing grace of give and take did not intervene. Many leading practices, even in the surveyors' conservative profession, were quickened by the enchantment of post-War change and were eager to shake off a parochial image. Some would blunder into the new international mêlée unaware of the fact that the argot of Wimbledon High Street was an incongruity among the sheiks of Araby, the *obas* of Nigeria and the turbans and bandanna of India. Their credulity seemed to accept the schooling of the College of Estate Management to be the natural philosophy of Creation to which all men everywhere sub- scribed. To a few of these aspiring practices, my worldwide wanderings spelt valuable contacts. The senior partners sought my services. Careful choice was necessary. Practices whose domains were unlikely to overlap with others only could be entertained.

Edward Fleming-Smith, the wartime Land Commissioner for the north of England with whom I had cemented a lasting friendship in those far-off days, came back into my life. When the Ministry of Agriculture, Fisheries and Food defaulted in giving him his just desserts and passed the Directorship of the Land Service to a less deserving man, Edward rightly stood them up. He went private and became senior partner of Smiths Gore, a practice of renown whose traditions were nearly as deep-rooted as those of the Crown, the Church and the landed gentry it served. In an unguarded moment, Edward enquired whether the notion of being an overseas consultant with Smiths Gore would hold any attraction for me. In many ways my temperament ill-fitted the prospect. For generations, Smiths Gore had walked with the picked delicacy of an Agag towards clients, notably the Crown Commissioners. My academic abrasiveness might not fit. On the plus side, their professional domain was far removed from the interests of Cyril Sweett and Partners, the pre-eminent quantity surveyors of the City of London, who also had asked for my cooperation overseas. Edward Fleming-Smith's persuasion prevailed and led me to throw in my lot with him and his colleagues for a few years. The specialities of Smiths Gore and Cyril Sweett were so precisely circumscribed that room was even left for me, without fear of cynically serving three masters, to help extend the international urban development and brokerage prospects of Hillier, Parker, May and Rowden.

The winds that carry the vicissitudes of life gust through the boardrooms and offices of businessmen and professionals with a fiercer intensity than they blow through academic corridors. Cherished hopes are swept away by merciless tornadoes of misfortune, just as deals are being clinched. Such gusts blew away fair prospects in Malaysia for Smiths Gore and in Iran for Cyril Sweett & Partners.

In Malaysia, the British-owned ancient rubber estates were in danger of being rooked of their assets by local sharp practice. A reliable managerial service was greatly needed to serve the absentee owners and curry business between Mark Lane in the City of London and the plantations in the Peninsular. Sir Claude Fenner, one of the handsomest men it has been my privilege to meet, godlike in limb and psychic force and former Chief Commissioner of Police in Malaysia, was now Chairman of a corporation representing the rubber estate owners. He was avid to get something going. Azmi, the professional in charge of land economy at the University, had with a little push from me linked up with Smiths Gore. This London-linked firm, hand in hand with Sir Claude, was perfectly poised to organise the much needed managerial service. Both parties were eager. As a meeting of partners from both countries was convening in London to sign the contract, news came over the international wires that Sir Claude had died of a heart attack a few hours previously. No one could replace him. The deal collapsed also.

The Cyril Sweett Partnership had their eyes on Iran. They shone with the lustre of hopeful anticipation at the mention of the new city being built round the Holy Shrine at Mashad for which Valian was responsible. An impressive glossy prospectus written in Farsi for the firm was slipped among the papers in my briefcase on yet another trip to Iran. When the vodka was flowing at a banquet Valian had arranged in my honour in the Holy City, the moment seemed propitious to hand His Excellency the prospectus and pray that he take to bed and read it. Next morning, Valian tossed it back with a supercilious grin and the comment that he wasn't interested in 'Quantity map-makers'. The joker in the pack, the ambiguous 'surveyor', which in the Farsi had been translated 'map-maker' had played us false. The deal, though shaken, was not off but never came to fruition. Fate had greater furies to command.

Visits to Mashad were only partially on serious business. Valian and I had become friends and he liked to see me. As soon as Ayatollah Khomeni reigned in Tehran, Valian was put behind bars. My heart was heavy with sorrow. The *Daily Telegraph* one February morning carried the news that the grizzly-hearted ruler of Iran had shot six generals and two leading civilians – one of whom was A.A. Valian. Six months later, in my study in Cambridge, a ringing telephone announced a call from California. Now my ears were none too reliable as a recent mastoidectomy had removed the inner ear on the right side, so I doubted my hearing when a voice said,

'Hullo! 'eese that Proessor Déman? Valian speaking. Valian!'

With exemplorary foresight and cunning, he had flogged carpets from the Holy Shrine and with the gathered millions had bribed the jailer for a faked passport and a key to the cell door. As he made his way to the Turkish border, Valian became more closely identified with the peasants than ever he had been as Minister of Land Reform and Rural Cooperation.

⋖৪ ৪⋗

Through Taveuni Island in the Pacific, East and West kiss each other; 180° East and 180° West lie on one and the same meridian. At such a spot, on the other cheek of the Globe, as far from Cambridge as geography could devise, destiny had placed me within days of crossing the retirement meridian. Calculations for this extreme removal had started two years earlier when the grand Council of the Commonwealth Association of Surveying and Land Economy were scratching their heads at a seminar in Fiji's warm November sunshine. That their academic adviser was leaving his Chair at Cambridge to another incumbent was hardly a matter for the minutes. The changeover at Cambridge in their view was merely a hiatus in the progress of my affairs and would probably render me a freer agent to help. While opportunity offered, they struck. CASLE wanted a manpower study of the South Pacific and an assessment of its findings; what was the evidence for or against setting up university degree courses at Suva in the best interests of the surveyors' profession and the economies of the South Pacific?

The Caribbean Report was to be the model. Consequently, after giving a positive answer to the question, 'Could you do something similar for the South Pacific?', the opening week of October 1979 saw me embark on an Odyssey of the Southern Seas. Prologues to the expedition proper included a day or two in Kuala Lumpur to tie the new knots tighter; and a prolonged stop-off in Sydney to squeeze in a rushed lecture or two, to see Sheila Evans and get up to date with Tom Whipple.

The Southern Seas are a collage of island beauty and expansive ocean running into a long, long sky, a vaster Caribbean without the mark of Spain upon it. Limpid blue lagoons held in cups of coral rimmed by palm-tufted silver sands dream the days away under a sultry sun. At night the planets move above among unfamiliar stars. Fiji at centre stage offered on arrival no promise of the famed sapphire seas and skies. Back in the days of Queen Victoria, Fiji had petitioned the Queen Empress to take the country beneath the ample folds of her imperial skirts. The Widow of Windsor, so it struck me, had left an imprint on the place which passing generations had not entirely erased. Skies heavy with drizzle were in dolorous tune with the morose filigree ironwork of the Victorian architecture of the Grand Pacific Hotel and the pealing stucco façades of old Suva. That gloom, however, passed with the night. I awoke expectant, yet oblivious of the nine thousand miles of island-hopping to come.

If all the Fijians on the mainland of Australia and New Zealand were to return home, as the Jews flock to Israel, the island economy and society would enjoy a more relaxed, wider vision of the outside world and of themselves. As it is, the indigenous Fijians are fiercely insular with a pride that gives a peculiar twist to their land problems. Dr James Maraj, a Trinidadian scholar whose intellect chews up nonsense as a mechanical dust-cart devours rubbish, and later to be Fiji's Ambassador to India, was then Vice-Chancellor of the University. He poured me wine at a welcome party and warned me of the peculiar Fijian social problem. Fiji resembles Malaysia in its practice of voluntary apartheid. Local Fijians are a variety of *burmaputra*. All land belongs to them; a rich endowment unmatched by brains to administer it. Land, its use, ownership and production would be in a more parlous state than they were in fact but for the fatherly eye of the Native Lands Trust Board. Intellectually, the Indians who had come to the country as servants ages ago, have the mental advantage but not the land advantage. Land Economy courses, therefore, if not carefully laundered, could clothe the clever Indians in academic dress, bestow upon them expert knowledge and leave the Fijians possessed of the land but devoid of cap and gown. Maraj's caveat was carefully noted. The reasons for it, however, much intrigued me. At lunch soon after the Maraj welcoming party, an opportunity arose to ask the British High Commissioner recently posted from Pretoria what he made of the difference between the apartheid he had left and the variety he had come to. He answered in one word, 'legislation'. 'The fools in South Africa put it in writing on their Statute Book. Here, in Fiji, the practice persists as a voluntary way of life.'

If a globally conscious world could possess tucked-away corners, the

South Pacific would be one of them. Life is lived by the decorum of 'The Pacific Way'. Booklets, rather like the Red Book of Chairman Mao, introduce the uninitiated to the intangible mystique of the Way. Neither code nor creed nor criterion determine it. Waking, sleeping, shopping or waiting through long hours in island airports, it is there. Among its priorities, garlands and sea shells come before GNP. Time has little relevance to life. Certainly the Way defies economic analysis. In association with a sloppy Polynesian 'socialism' it accounts for the engaging backwardness of a region whose peoples superficially pay lip-service to development.

The Solomon Islands displayed this quaint paradox in its forestry and land problems. Sea-girt wastelands now denuded of their once tall virgin forest are abandoned to the mortmain of nature. Re-planting is possible and practicable. Land tenure, however, stands in the way of silvicultural regeneration. After independence, the Government acquired title to and possession of the erstwhile Crown freeholds of Honiara. Some of these freeholds could be traded in exchange for re-planting clauses in felling contracts with the international timber-felling companies. No one acts. An embryonic socialism and the Pacific Way suppose the peoples' paradise will come with patience. Meantime present wasteland is preferred to future conservation and sustained home-generated welfare.

Similar propensity to consume the present and forget the future loomed stark and ugly in the tiny island of Nauru. Plum on the Equator, this once solid outcrop of phosphate is a macabre monument to man's greed. Years of exploitation have turned it into an open catacomb. Grotesque, ill-shaped stalagmites of white rock stripped of all natural phosphate tower house high from the barren island shelf. An attempt was being made to find the natives a new homeland in New Zealand against the day when the natural wealth of the island is exhausted. Laurie Stott, the Director of Lands and Surveys, whose wife runs the one and only hostel on the island, was sceptical of the operation and of the votes of the islanders in support of it. 'Madame Nauru, her sisters, mother, aunts, other female relatives and friends,' he observed with a cynical smile, 'prefer to spend their regular phosphate royalties cruising to the clothiers of Auckland and returning to suffer the pinch of pride sweltering in new mink coats in the broiling equatorial sun of Nauru.'

A small precarious economy, a surveyors' profession known only by its land surveyors, leagues of sea between neighbour and neighbour and the introxicating liberties of recent independence were common features of the scattered island States. They gave a certain homogeneity to the Polynesian approaches of the mighty Pacific. Yet each, after its own custom, presented a vignette of the Pacific Way. The mosaic of differences had its general fascination but each island visited staged a particular experience that marked the place. New Hebrides, now Vanuatu, from above the pocket-handkerchief airport of Vila was as beautiful as Nauru was ugly. Among its loveliness was enacted a diplomatic farce run on Gilbert and Sullivan lines between the British and French Residencies. New Hebrides was still a condominium, half French, half British with the French having things their own way whenever the not infrequent clashes were 'settled' by *parlez*. Speech was a patois of English nouns and French verbs. Chartered surveyors were unknown. Local *geometrics* had cobbled together a work-a-day cadastre which determined the fortunes of the land market with the help of a dice and a planning officer. Altogether a not unhappy land, daily torn, indeed, by petty 'Agincourts', but one I left all too soon for Western Samoa.

Samoa is a lotus-land of quick forgetfulness. It has a regal air breathed by beautiful princesses. No wonder Robert Louis Stevenson chose to be buried there. There are four royal lines. A capricious sovereignty moves from one family to another and waits upon fickle fortune's favours. The University of the South Pacific and my ostensible business were banished to the half light of the auditorium when, on my first evening, Ma'aafa Fetani's dinner party in honour of five princesses upstaged me above the footlights in Samoa's regal society. Compared with the comedy of the New Hebrides, Samoan society radiated true opera. Our hostess was not unknown to me. Some months previous, we had met at Sir Hugh and Lady Springer's house where she had sported a yellow carnation in her hair, a *punctum* to her constellation of wit, grace and beauty. She now welcomed me to her homeland. Western Samoa is a Garden of Eden where Eve is in command. Besides Ma'aafa Fetani, a royal in her own right, and other princesses three generations born, there were Aunty Mary and Aggie Grey. These two remarkable sisters dominated the catering trade of Apia. No one has been to Apia, the capital, who has not partaken of Aggie Grey's hospitality and been to Aunty Mary's *Apian Way*. Their two restaurants are run with a panache which measures up to a regality of its own. Samoa was unique in a number of other ways, not least in its *long-houses*, the high-thatched, oblong open colonnades under which the well-to-do lived unembarrassed though watched incessantly by the prying eyes of neighbours and passers-by. Robert Louis Stevenson's house, a titular tourists' attraction, was an incongruity and detraction out of step with the Pacific Way.

Tonga, by contrast, was low key. Secretaries stood in for their royal masters. The Dateline Hotel was a tug-boat affair compared with the liner Queen Elizabeth II's elegance and bustle of Aggie Grey's. The higher the office, notably that of the Deputy Prime Minister, the thicker was the official head of he who occupied it. Besides these peculiarities, in Tonga it rained and rained and rained. As we drove to the Agricultural Research Station among the banana trees, the roads became chains of elongated lagoons each with its brood of lesser puddles, deep in mud and barely passable. The ceaseless rain took possession of the evening and no one appeared to mind acting-out the old refrain *I'm Singing in the Rain*. A jolly company, graced by Adrea McIllroy of the USP Centre, took me *á la* Pacific Way to a full dress dinner and dance. Two yards from tables sheltered only by flimsy canopies, bucketing rain flooded the dance floor and drenched the whirling couples who steamed across it, hot, damp and happy. Tonga's wet was an integral part of life. For me Tonga raindrops took their place along with the other logos of the South Pacific – the denuded wastelands of the Solomon Islands, the phosphate mines of Nauru and princesses of Samoa.

The thousands of miles flown among the islands had shown me at first hand the widely scattered Receiving Centres of the communications network operated by the University of the South Pacific. To these Centres, scheduled lectures from Suva were daily relayed by satellite. The process was a monologue. The microphone, like a church congregation, doesn't answer back. Educationally, the lack of dialogue was a radical weakness. Misunderstandings once launched over radio waves can float in the mind for years, for a lifetime even. Outweighing the system's limitations by far were the benefits of learning brought to thousands of students throughout Polynesia, students who never in a lifetime could have travelled the sea leagues to Suva. Many able undergraduates, satisfying all qualifying expectations for University entry, would have to face a return sea voyage or expensive air journey of thousands of miles to enjoy academic residence in the University. There was a sustainable case for Land Economy courses to be injected into the system, as the self-contained management of the island states generated an overall demand for more trained minds among the alert, able islanders.

Amid farewells to Fiji and rushed last minute interviews, an event occurred whose significance, unrecognised at the time, was to affect the pattern of my life over the coming ten years. Tim Davey, the mastermind behind many of the doings of the Native Lands Trust Board who had arranged with great care and efficiency every detail and most of the interviews of my own extensive programme, had invited Colin Floyd, a UN planner, and me to a bite of lunch. Floyd brought with him a map, prepared in idle moments, to satisfy an itching curiosity. Its message struck me with the force of a hurricane. Round the tiny islands that peppered the ocean and their bigger neighbours, Floyd had indicated the boundaries of the sovereignties which each sea-bound State would assume were it to follow the fashion of declaring a Territorial Sea two hundred miles from a coastal baseline. The result was a revelation. The Gilbert and Ellice Islands, with the Phoenix cluster between and far-flung Christmas Island, all of the same family, would have a submarine domain and spread of super-jacent waters to match in width the United States of America. Surely, the curves and circles on that map marked the parameters of future seabed economies creating wealth untold in Davy Jones's Locker? One thing was certain: my estimates of demand for hydrographic surveyors would have to be revised. A second certainty was the existence of similar patterns in other seas, other oceans, the world over. I saw a new world, a submarine world, awaiting the energies and wisdom of Land Economy to help reap its harvests.

In addition to Colin's map, Tim Davey's programmes had provided other unexpected dividends of satisfaction. Just before leaving Fiji, there was a social chat with Ian Thompson, Chairman of Fiji Sugar, and his son. It put right a wholly erroneous history which they had absorbed of the origins of the Development Courses under Henry West and others in the Land Economy Department at Cambridge. The younger Thompson and a friend had just returned from attending the courses. They were under the misapprehension that the progenitor of the courses was a Paul Howell, a sociologist who had been sent to Cambridge as a liason with the Overseas Development Administration. After straightening out this little misunderstanding, there was just time for last minute shopping and the purchase of a commodious, red leather, floppy portmanteau. Deprived, as I had been, of the company of a secretary my memory had fortified itself by accumulating a vast local literature. This was stuffed into the new purchase, until the ribbing squeeked and the locks were strained to bursting. Chats, interviews, recordings, admonitions from Maraj and other social engagements were cut short by the summons of a hasty departure flight to Auckland.

At Auckland, I had to hump my overstretched luggage to the end of an interminable grey corridor to be confronted by passport control and an irritating health official who wanted to know if I had arrived from Birmingham. That great city at the time was suffering an epidemic of scarlet fever and health officials were jumpy all round the globe. With a supercilious grin I offered the assurance that, 'You always know a Brummie by the shamrock in his turban'. There was no response. The beauties of New Zealand must have anaesthetised the humour of its inhabitants. At Wellington next morning, as I grabbed my bits from the merry-go-round at 'Baggage Reclaim' the large red over-strained portmanteau burst asunder and disgorged paperbacks, leaflets, memoranda, loose files, confidential letters and much else in a river of disordered literature. My cup of woe was full. Wellington was in the lull of its Sabbath rest. The shops were shut. No one would redress the calamity. The disintegrated bag was carried bottom-up under aching arms. Thank God for Pat Whipple who coped with it two days later in Sydney; and for the kindness of Tom Whipple and Sheila Evans who would not allow me to depart on the flight to Heathrow until they had unwound my South Pacific neurosis!

❧ ☙

One evening in Malta, an amateur but gifted lady chiromancer read the character lines written in my hands. Some of what she said was certainly recognisable, especially her comment about the gap that can open between the expression of my thoughts and people's understanding of them. It is the thoughts themselves that are the puzzles not the way they are put. She was right. Some ten years or so after my retirement, a leading economist who at the time was Master of a traditional Cambridge College, inclined his head towards me at a public dinner and muttered, 'We are only now understanding the economic significance of property rights which you tried in vain to teach us years ago?' It was going to be the same with the perception now forming in my mind of the economic significance of the seabed, its relevance to Land Economy and the critical part the surveying profession should be playing in seabed management. As the old jurists used to put the matter when discoursing on the king's rights over the seas – *tam soli quam aquae*, as with the land so with the waters.

Opportunity to ventilate my whirling seabed thoughts came in January 1979, very shortly after my return from the South Pacific. Bob Steel, the vigorous Secretary General of the RICS, had reminded me of an obligation I had to chair and find a speaker for the Geometers. The Geometers is a surveyors' dining club constituted on the lines of the old rotten boroughs. It was born out of the turbulent womb of FIG when the British surveyors were responsible for running the triennial Congress. Members of the Geometers take it in turn to Chair a dinner, frame a topic and invite a guest who is not a chartered surveyor to talk. Here was the tailor-made occasion to introduce my seabed vision. In a reply to Bob Steel, I put the question to him: 'What about using our imagination and trying to mount a discussion on the future responsibility for the management of the resources of the ocean bed and the part that the chartered surveyors should be preparing themselves to play?' My heart sank at his trying-to-be-enthusiastic reply which was to the effect that, 'The theme of hydrographic surveying would, I am sure, be an interesting topic . . .'. He had missed the point. Hydrographic surveying would be but an incidental preliminary to the management of seabed resources. If surveyors can manage the land, they can manage the seabed – *tam soli quam aquae*. Again my thoughts were not understood. The Maltese palmist was so right.

Good fortune, however, attended my way. Within days, other obligations had me sharing a lecture programme at the School of Oriental and African Studies with Paul Ashcroft. To my amazement, he was legal adviser to the Department of Energy and was speaking on the legal intricacies of seabed wayleaves, licences and landfalls. Here was my guest for the Geometers. Having accepted my invitation he had, on the evening, to face the discomfort of his host not turning up on time. Great Aunt Bessie had died in Malta. Her death had taken me there. With a day in hand before the Geometers' evening, I was stamping the departure lounge at Luqa Airport when British Airways announced the cancellation of all scheduled flights until the next day. On the flight back in the morning our pilot was caught in a circus of aircraft over Paris and ran low of petrol. He was forced to land. So my day in hand was lost. At the moment when Paul Ashcroft and the Geometers were reading the menu, the Malta flight with myself aboard landed at Gatwick. Panting, undressed for dinner and distraught I arrived for the sweet course. Paul spoke well and opened the eyes of my surveyor chums. Jimmy James, President-elect of the RICS, promised to take the seabed on to his agenda in his coming Presidental year. Much had been accomplished by my inadvertent absence.

For the surveyor, as for myself, the seabed was a new frontier. In general, this was universally true as very few people thought of it as a domain of real estate. Nevertheless, ever since President Truman declared a Continental Shelf off the American seaboard in the 1940s with a view to laying sovereign hands on the newly suspected wealth in polymetallic nodules on the seabed, international activity had grown apace. The UN took up the theme and set up a UN forum to debate the terms of a draft Law of the Sea Convention. Scientists of various ilks, inventors and the manufacturers of novel exploration devices were already on the scene. The surveyor had a long way to go. Accountants and other professions were there before him. This very fact was a spur and a challenge.

The Elsevier Scientific Publishing Co. of Amsterdam backed by the 'trade' had organised a grand international seabed conference a year or so before and were now in full spate for '0180' as the 1980 conference of the series was to be called. The Metropole Hotel, Brighton where years ago my innocent speech had upset the advocates of a Land Commission had been turned into a spacious seabed exhibition hall. Elsevier were welcoming speakers, especially to the mines and minerals side of the show. With a mind as innocent of mines and minerals as the Brighton seascape at the time was free of oil rigs, I offered a Paper on seabed management. My offer was not without a perceptible trepidation lest my contribution should border on chicanery. The object of the contribution, apart from its inherent content, was to make a public declaration of my professional interest linked to the promotion of the idea that the seabed should be managed as land – *tam soli quam aquae*.

The Paper was accepted widely. A stake had been driven into the seabed with my name upon it. Round this outpost and led by Smiths Gore there gathered over the following months a consortium – *Marine Resource Development* – to offer a joint service to all takers. This gallant band were a coterie of quantity surveyors, civil engineers, town planners, mining surveyors, hydrographic surveyors, fishery develolpment companies and international lawyers. Leading names included Sir William Halcrow & Partners, Clyde Surveys and Cyril Sweett & Partners. The arrangement was too loosely knit. Besides which it was premature as specialists in seabed management had yet to master coordination. Eventually, some amalgamated with others to mutual benefit, notably the civil engineers absorbed the hydrographic surveyors. When it came to action each constituent acted on its own. In this respect, benefits were not dependent upon cooperation but were peculiar to each individual firm and so justified the original venture expenditure.

Jurisdiction over the seabed was riddled with ambiguity. Britain was no exception. The Government in 1964 had passed the Continental Shelf Act which, *inter alia*, gave the Crown 'sovereign rights' over the Continental Shelf. Parliament had been decidedly cautious and studiously avoided the word 'sovereignty'. To have given the Crown 'sovereignty' would have raised the two hundred miles Continental Shelf zone to the status of a Territorial Sea and interfered with the sacred 'freedom of the seas' round our coast. The new law aggravated the ambiguity of the authority of the Crown Estate Commissioners over the seabed.

Numerous fundamental questions were left unanswered. Were property rights subsumed under the 'sovereign rights' of the new Act? Did the Crown Estate Commissioners have any management right over the seabed beyond the low water mark, let alone over the Continental Shelf? Had the Crown any valid dominion even over the seabed of the Territorial Sea, as Geoffrey Marston has questioned in his masterly book *The Marginal Seabed*. Giving the benefit of all possible doubt to the Crown Commissioners, my mind was yet uneasy on their managerial philosophy. Leading firms of chartered surveyors, including Smiths Gore, had, to the extent of almost creating an established usage, been appointed by the Crown Commissioners as Receivers, a species of sub-agent, to manage some of the Crown inland estates. Surely under the new look and in the interest of efficiency, areas of the foreshore and seabed should be treated on similar lines?

Certain friends of mine who were also Crown Estate Commissioners were ready to wash their hands of the sea. They were not indifferent but confessed to lack of expert knowledge of the management of Davy Jones's Locker. 'I have a leaning towards a five-barred gate, enjoy the sniff of new-mown hay, like oysters but know nothing about bloody oyster beds', is a quote that epitomises their attitude. They gave me the impression, sometimes by implication and occasionally by expression, of a willingness to welcome any reliable help available. At the same time most of them were happy to leave such side issues as the seabed in the hands of the Second Crown Estate Commissioner. Fortified by these responses, I proposed that my colleagues at Smiths Gore approach this august personage directly and enquire how they, if at all, might help specifically in seabed management. As a friendly gesture we invited him to lunch for a chat. With the self-conceit and arrogance of a pompous civil servant, the Commissioner dismissed the management idea. On the contrary and trying to make amends for his rejection of it, he helpfully suggested that Smiths Gore offer their seabed management services to the private holders of Baronies round the Scottish coast and other coastal landowners. Everyone voted the lunch a success. When next Smiths Gore produced their yearly glossy prospectus, it carried a carefully written specification of the firm's general seabed management functions. For some inexplicable reason, the First Crown Estate Commissioner took exception to this advertisement of the firm's perfectly proper liberty. Blame for the trouble was assigned to me! The whole episode was an amusing clash between the logic of theory and the expediency of practice. Academic thinking and its conclusions were manifestly *de trop*.

Memories were still tender on this score when a Fisheries Conference, an annual event in Oban, invited me to speak on the ownership and management of the seabed. Amid a liberal and welcoming display of spontaneous laughter, I referred to the Crown's alleged ownership of the seabed as one of the few condoned major monopolies remaining under Margaret Thatcher. The Conference accepted this given fact of life and I put the question whether the national seabed management polity in the hands of ten or a dozen well-groomed delightful fellows sitting round a table in Carlton House Terrace in the Sassenach capital was the best available. There was an explosion somewhere in the direction of Carlton House Terrace! When the dust settled, the outcome was most unexpected. Smiths Gore quietly cancelled my consultancy for being a dangerous maverick who dared to question the Establishment; a role I had practised most of my life. But the Crown Estate Commissioners, in exchange for Denman's head on a charger, appointed Smiths Gore as the very first firm of chartered surveyors to be given charge, under them, of managing the foreshore and adjoining seabed from Inverness to the Firth of Forth. Whatever was delivered upon the charger, I had 'kept my head when all about me were losing theirs.'

By the time the Crown Estate Commissioners had come round to letting their Receivers help with the seabed, the University of Wales Institute of Science and Technology were offering degrees in maritime geography. These courses were manipulated and manicured to make them acceptable to the Education Committee of the RICS. The new courses would become a route to a specialist qualification for entry to the profession. The degree was an end product of a promise made by JJN James long ago at the Geometers dinner in January 1980. Once in the President's Chair, Jimmy James coaxed the Royal Insitution to set up a Working Party to follow up his undertaking. Its brief was one of unknown images, opportunities new and missed, new laws, treasure trove, international conventions and other novelties. As with Pandora's Box of the legend, there flew out from under its raised lid new hope. They regarded me as the *fons et origo*. It was up to me to justify my expectations. Bob Steel, the Secretary of the Royal Institution, was converted by then. Like all converts, he eventually allowed the chariot of enthusiasm to carry him away. Soon, Bob was leading a crusade on seabed management to the Third World. He wished me well, anywhere but as a member of a new Marine Resources Committee of the RICS which was to take on from the Working Party.

The pioneer thinkers in the RICS saw clearly that educationally the institution was lacking seabed wisdom. The courses at the University of Wales were ulitmately chosen from a number of other options. Groping after knowledge brought me the welcome and profitable friendship of Professor Donald Cameron Watt of the LSE. He had a seat in the Cabinet Office dealing with shipping, the Greenwich Forum under his belt and a mind stuffed with a potpourri of dried wit and fragrant knowledge. His mind, indeed, was a counterpart of the miscellany of books, papers, files, discarded ties, pens, pencils and inks piled to the ceiling in one of the most delightfully untidy offices it had ever been my joy to enter. Time and energy had rewarded Professor Watt with a promising School of Maritime Studies at LSE. Growth would be stunted without money, research staff, teachers and a full Professorship. That was the message to be publicly declared at the LSE's Annual Lecture. The Professor offered me the honour of giving it and making the case for a Chair in Seabed Management. Next day *The Times'* Second Leader took up the plea. The Editor pulled no punches but as so often happens with me, failed to follow the precise gist of what had been said. The Leader rightly informed its readers of the argument expanding the epithet, *tam soli quam aquae*, but then went on to infer that its focus was not on the seabed but on the waters. The Malta palmist was right; was it something in me or in the dull wits of *The Times* Leader writer that missed a mutual understanding?

Going under water to find new worlds can have for small land-locked countries the enticement of the slippery soap on the nursery floor for the infant in the bathtub trying to reach it.

The post-bag informed me of my growing reputation of someone who had knowledge of the new seabed perspective but was uninvolved in international politics or befogged by too great a learning in international law. Hungary, it would be reasonable to suppose, had little cause to debate the UN Law of the Sea Convention. Not a bit of it. Friends at Pécs University wanted an exegesis of the LOS Draft treating of land-locked countries. Was there hope in Hungary of maritime access through the Port of Trieste? Could the University therefore hear my expert views? Trieste or no Trieste, there was the ever-flowing Donau dividing Buda from Pest. There is plenty of water in that mighty river. Had anyone seriously scoured the riverbed? I was no expert on land-locked countries or on riverbed economics. Nevertheless, through much practice, I had perfected the academic art of lecturing on subjects in almost total ignorance of them – the art of asking questions, rather than giving answers! The lecture and myself were well received in Pécs. The visit also provided another opportunity to visit the incongruous cathedral of Pécs, crowned with the Cross and the Crescent standing together to mark where in centuries past the Ottoman and Holy Roman Empires met.

Switzerland another small country far away from ports and seabed yet rich in *lacs* and *zees* was deep in underwater exploration of an exacting and entirely novel kind. A generation back, the world had hailed the brilliant Swiss scientist Professor Auguste Piccard when he took a balloon ten miles high into the stratosphere. Now his son, Dr Jacques Piccard, encapsulated in an amazing submarine bathyscope, was exploring the deepest undersea gorges in the world. A pioneering *Marine Resource Development* brochure had cleverly found its way to Piccard's *Foundation pour l'Etude et la Protection de la Mer et des Lacs* in Cully. Piccard's friend and *homme d'affaires*, Ivan Princeps, a graduate of Pembroke College, Cambridge picked it up. Over the international telephone he asked for more. Meetings were arranged, competitive in kudos and ill-defined hopes. The best we could do was a Feast at Ivan's old College which he greatly enjoyed. Ivan, however, had the trump card and offered an excursion over the bed of Lake Geneva. It amounted to *un plongée dans FA-Forel (PX-28)*, Piccard's *le sous-marin de recherche*. The Indian Ambassador to Switzerland had accepted a similar invitation. The costs of preparation were not negligible and Princeps offered a free ride if I would join a party with the Ambassador at a stated day and time.

Research since retiring had taken me across Europe, Africa and America accompanied by Iris Elkington who in her best mood could summon the energy of *Hermes* and the gifts of *Artemis* to augument a natural grace of humour, intellect and beauty.

For twenty-five years, Iris had aided the genius and buttressed the spirit of Sydney, my twin. On his retirement, this *Artemis* had swopped twins. What was a dash to Geneva compared with the other journeys now behind us! To roll over the bottom of a lake in a bubble would be exciting when left to the imagination. In reality, a gripping apprehension riveted my body to the tiny seat within the glass orb as I squeezed, knee to knee, with Piccard and his assistant.

A transparent lid was fastened over our heads *pour le plongée*. The bathyscope bubble floated six inches above the lake bed; two intricate forceps worked from an inside control panel waved like massive antennae out in front. At last before my eyes spread what I had been talking about for four years — the bed of a sea, albeit an inland one. Unfortunately, neither my colleagues nor I had the monies to hire, let alone to buy, the amazingly fascinating bathyscope.

Half way through 1982, the Institute of Economic Affairs listened attentively to assurances that a book on the seabed would be timely. Elsevier Publishing were eager. Fair enough, but old friends came first. Thus a new Hobart Paperback was published to challenge the seabed monopolies of the world. Financially, the

Boarding Jacques Piccard's one passenger *sous-marin de recherche* on Lake Geneva

research depended on my own money until handsome support came from Pembroke College. Again the old jurist's tag, *tam soli quam aquae*, was guide, advocate and conscience consoler. The Grosvenor Fund, established in the College for land studies, could surely, free of qualms, be used to fund research into seabed usage. Much travel, much sweat, hours and hours of revision and finicky editing plus agonising arguments over the title went into the writing and production process. Some two and a half years later, *Markets under the Sea*, was published.

Placarded by the IEA, the Paperback took the accustomed course of making new friends and new enemies. Harbour folk in Britain took heart. Harbour Boards don't own the seabed of harbours or of their approaches but get blamed if navigation channels are not dredged and cleared. Permission for these operations means time-wasting negotiations with the Crown Estate Commissioners which a seabed vested in harbour boards would obviate. The book won me the friendship of Nicholas Finney, the hero of the abolition of the perfidious Dock Labour Scheme. Many a matey lunch followed at the Associated British Ports annual functions. After the tradition of the IEA, the book was not written as a political polemic. Such animus as it deployed was reserved for the policies embedded in the Laws of the Sea Draft Convention whose biased text deals with the bed of the Abyss, the area of deep sea bed beyond national shelves and waters. Economists were puzzled and unfriendly. Even Jack Wiseman, then Professor of Economics at York University, who wrote a lengthy Prologue to the book couldn't get his brainwashed economist's outlook to accept the *a posteriori* consequence of the existing fact. He wanted the seabed placed on the supply side of a market governed by 'Public Choice'. He was indifferent to the practical fact that to supply any market, destined to break seabed monopolies, would require, *a priori*, an arbitrary division of the seabed into marketable prioprietary units. To put the undivided seabed on the market, even if public choice responded to it, would perpetuate the monopoly. Later on, when he had learnt to breathe the air of Land Economy, he followed the pragmatism of my text. All ended on a joyful note. Jack Wiseman and I made a joint video from the seabed story. The IEA commissioned it as second in line of a successful series.

The Law of the Sea Draft Convention debates under the United Nations were regarded by America, ourselves and sundry other industrial nations as either worthy discourse or harmless claptrap. Concern arose when the Convention showed signs of conceiving, bringing forth and breeding a 20th Century international leviathan of immense and ugly proportions under the deceivingly innocuous title of the International Seabed Authority. That grim collective would have been dominated, under the terms of the then current Draft, by the Group of 77, made up of the Third World and the USSR. It could have held the West to ransom while expecting the West to carry the cost. The womb carrying the embryo of this monster was Part XI of the LOS Draft Convention. More than half the cost of midwifery and the cost of nurturing the creature would be met by the industrial nations and outstandingly so by the US. With one vote per nation, the US and other major financing nations would have had no greater say in controlling the International Seabed Authority than would Gambia, Senegal and other minnows in the pond. The USA wanted Britain's shoulder to be level with its own, in a determined stand against such lopsided, politically and commercially noxious commitments.

Working with the Pentagon and the White House was the splendid Admiral Bill Mott, former legal adviser to the US Navy. He had just retired from his Navy post. Over here, Bill's affinities were close to and harmonious with people of a like stomach, notably Michael Ivens of Aims of Industry. The University of Southern California and the Lincoln Land Institute of Cambridge, Massachusetts had not counted my retirement a real event and outstanding lecture arrangements with both bodies were to send me to the USA in New Year 1982. Michael Ivens saw an opportunity to press Bill Mott into my schedule.

British Airways lifted me over the pearl white snows of the Arctic into the golden warmth of San Francisco and Los Angeles. From thence, a number of US air companies squabbled for my custom to Boston. A week later the winner flew me out of the golden warmth of California into the snows of winter-bound Massachusetts. Meantime, the Hoover Institution of Stanford University wanted to have a chat about my 'papers' for its archives. So, *en route* to Los Angeles, some accumulated jet lag was slept off at near-by Walnut Creek where cousin Mini lived with Maurice, her chrysanthemum-growing husband. The Los Angeles Conference was a routine affair. It left no deeper imprint than the memory of George Lefcoe's bathroom and a walk with Donald Sharp. George Lefcoe, the Professor in charge of events, had the speakers to lunch in his Hollywood-fashioned mansion overlooking distant Pasadena. Ostensibly, we came to lunch. Covertly, we came to admire the high-domed bathroom. That exotic cavern, lit by shafts of green and blue light playing upon the moving waters of a permanently filled *étang* could have come straight out of Rider Haggard. Donald Sharp, a lecturer in residence, had wandered across the campus with me to ask about an economist called Kirwan, a visitor of some weeks back from the Department of Land Economy in Cambridge. Donald was obviously concerned as Kirwan in his eyes was probably a sick man. 'Why do you think so?' I asked. His comments had perturbed me. Kirwan was unknown to me but being associated closely with Land Economy, Donald's worry rubbed off on me. His explanation was as amusing as it was unexpected. 'Kirwan talked to us about British Socialism, OK. But the poor guy really believed in it! He must be very sick.'

Cambridge, Massachusetts, late on the Sunday evening when another US airways competitor got me there, was snowbound, cold, uninviting and lacking such basic amentities as running hot water. The place was something out of Dicken's *Christmas Carol* but lacking the Christmas Spirit. Despite American advanced technology, when I came to fly from Boston to Washington on the following Tuesday, take off was delayed for hours on a snow-carpeted runway.

Stalactites of ice weighing the wings of the aircraft to the ground had to be hammered off. Boston airport has meant hazard for me ever since. The ice perils of January were followed in the autumn by a near calamity when a British Airways jumbo, with Iris Elkington and myself aboard, slithered on landing within inches of Boston Harbour which edged the main runway – the hydraulic brakes had failed! In Washington on the first visit the politicking began. The Army and Navy Club has a likeness to the Carlton without its lightsome *gravitas*. Nevertheless, it was a joy and privilege to spend a few nights in the comfort of the Club's homeliness after the bleak experience of gaunt Massa-chusetts. It was even a greater honour to share a table with Reagan's men. Like King Arthur's table of Wessex mythology, the lunch table was round and allowed admirals and diplomats to share an easy informality. Ambassador Malone, Reagan's Spokesman at the Law of the Sea Convention, had brought with him Admiral Bruce Harlow from the Pentagon. Admiral Mott was the pivot of proceedings. We three met a number of times. Ultimately a missive was prepared setting out the President's case against the USA signing the Law of the Sea Convention. It was pressed into my hand with many a 'say', 'sure' and 'OK' to strengthen their pleading to get it to 10 Downing Street. Once back in real Cambridge, the cogent points of Malone's private entreaty on Reagan's behalf to Margaret Thatcher were carefully annotated. My rubrics linked them with the actual text of the Draft Convention. Michael Ivens had the resulting text printed and distributed to the Foreign Office, to some Ministers and to the Department of Industry. A special copy went to 10 Downing Street with my personal letter to the Prime Minister. She fully backed the American missive. Pointing out that we had been there before them, the Prime Minister informed me that the text had gone to Douglas Hurd at the Foreign Office who, at the time, was number two in that enigmatic Ministry. Michael Ivens called a Press Conference and Malone's copy with our notes were sent humming among the Press. Some days later, Michael and I were summoned to the Foreign Office. We met on the doorstep. Both of us, to our mutual amazement, had turned up garbed alike in identical dowdy 'flasher' macks. Inside waiting our arrival were a dapper Douglas Hurd, a well-groomed Malcolm Rifkind and a similar John MacGregor, then a Minister of State in the Department of Industry. MacGregor with a wry smile on his face and a glance of humour in his eyes came forward and greeted our drab identical macks with, 'Ah! Army issue, I see.'

The macks had been good for a laugh in a light-hearted meeting. The FO responded to our beachcomber rigouts with a display of diplomatic levity and gave enough encouragement for me to dispatch letters of hope and assurance to Ambassador Malone and the Admirals. Malone came in person to Chatham House in St James Square to peddle his wares in the Spring. Bruce Harlow followed to reinforce the pressure. The British Government had no intention of signing the Law of the Sea Convention but kept the Americans and others guessing up to the eleventh hour. My great reward in all this was friendship with Bill Mott. His weight was thrown behind the seabed research and justified an *ad hoc* round of high level talks in Washington in November.

It was on that occasion that British Airways nearly landed Iris Elkington and me on the seabed of Boston Harbour. Debts and credits were evenly balanced between Bill Mott and myself over dinner one evening at the Maison Blanche Restaurant in Washington. Bill's new appointment was to keep an eye on special aliens. He would wake at nights, so he confessed to me, wondering where the latest Russian dissident granted permission to reside in the US had got to. One of his fattest files was labelled 'Svetlana Alliluyeva'. Svetlana was Joseph Stalin's daughter. Sixteen years previously she had defected to the USA, married an American architect and presented the former Dictator of the USSR with an American grand-daughter. This month Svetlana had slipped out of the USA to take her daughter to Britain. Bill Mott had lost track of them. 'No idea where they are.' He commiserated with himself, nodded his lionesque head and munched the Maison Blanche chicken kebab. There was a clutter of dropped cutlery when I coolly remarked with a certain calculated aplomb.

'I do. They are living with me'.

In the summer through a chain of ecclesiastical links which involved Canterbury Cathedral and other Church of England diplomats, including Terry Waite, my wife and I had been asked if we would offer a penthouse flat in the roof of our house to 'an American lady of Russian extraction'. No name would be given until the enquirer had our assurance.

'Does the name "*Svetlana*", mean anything to you?' was the only hint dropped.

'Yes,' I replied. 'Malcolm Muggeridge. She was interviewed recently on television by him. Say no more. The answer is "Yes".' The story was recounted to Bill Mott who, understandably wondering whether or not I was an inveterate liar, gave me the benefit of the doubt and went home to a good night's sleep.

14

ST CHAD'S EVE

> On St. Valentine's Day
> Cast beans in clay;
> But on St. Chad
> Sowe good or bad.
>
> *Goode Husbandrie*, Thomas Tusser

When as a student of farming history my reading included Thomas Tusser, the medieval penman, whose truly organic husbandry went by proverbs and Saints' Days, it was noted that one of the fundamentals his disciples had to learn was the date for the last sowing of beans:

> 'On St Valentine's Day cast beans in clay
> But on St. Chad sowe good or bad.'

On St Valentine's Day there was time to fuss about weather, timing, tillage and the prospects of harvest. By St Chad, 21 March, time had gone. Don't 'Hum' and 'Hah', says Tusser, 'throw the beans in the clay!' The St Valentine Days of my career stopped in September 1978. From then, St Chad was in the offing. My beans would no longer be sown with a harvest prospect in view. They were to be cast away with honorary gestures. Moreover, I had time to muse upon the St Valentine Days of yore whose harvests had been both good and bad.

Adjustment to the new sowing pattern did not come easily. Colleagues, especially overseas, were loath to divest me of a kind of imperatorial directive on which they had come to rely. Formerly, it had been possible, staff willing, to arrange secondments round the world. Such facility was no longer available. Directive had to give way to suggestion. The limitation diminished somewhat my effectiveness. Albeit, CASLE discounting what 'emeritus' can do to a man and looking to the Caribbean and South Seas precedents, asked me to take on yet another manpower investigation. It looked daunting, was more extensive than ever before and was expected to cover the 'Shield' of southern Africa. The brief ran from Malawi, south through Zimbabwe to Swaziland, Lesotho and back through Botswana, Zambia and

Tanzania. Barclays Bank International put up the money. They drew the line, however, at providing for a companion Secretary. Remembering my South Seas struggles, I was insistent that such help was essential and should be provided. My tape recorder had been tossed into Gehenna. The deadlock was broken when Dr Ben Epega, whose postgraduate studies I had supervised and who was now both a close friend and senior partner of one of the leading firms of chartered surveyors in Nigeria, came forward with funds to finance the cost of Iris Elkington coming with me. It was a magnificent gesture, long to be remembered in the annals of the profession, and made all the difference.

The countries of the southern Shield of Africa were unknown to me. Time was in ample supply. I could afford both the time and the money to make a preparatory flip to Zambia. Derek Nicholls and I were joint examiners at Enugu that summer – our dual harness was a feature of the adjustment process that followed my retirement from the Department. He travelled in comfort with three vacant Economy Class seats to stretch out on, while my fate in the First Class Cabin was a sleepless night cramped between self-indulgent executives puffing and squirming in pitch darkness – the lighting system had blown its fuses.

The easiest and often the quickest way to travel from one side of Africa to the other is to fly back to London, Paris or Frankfurt and return. Ignorant of this apparent illogicality and thinking more economically than had been my wont in the past, an attempt was made to fly across Africa from Enugu in Nigeria via the Congo to Zambia. The flight was an appointment with fear and acute discomfort. There were no jets, only ancient flying craft whose propellers gave the impression of spinning on wound elastic bands stretched to the tail. Once inside, breathing was difficult and peace only the gift of death. Regardless of their own and everyone else's safety, disorganised passengers fought and scrambled aboard. Somebody was working the cockpit controls but who was not known. Of stewards and stewardesses, of staff to control the rabble there was no sign. The sheer weight of protesting humanity made the prospect of lifting above the paw-paw trees a hit-and-miss affair as the plane, blown about like chaff in the wind, left Enugu. At Calabar and Douala no one disembarked. On the contrary, more bodies squeezed among the reeking mass of growling unfortunates on board. To the din of babbling, wide-bosomed mammies, with crying piccaninnies strapped to their backs was added the yelling of other children scrabbling on the floor between the women's skirts, the squawks of caged and free-ranging fowls and the sniffing grunts of piglets. They were all aboard the clipper! Six hours of this indescribable misery, noise and hazard were to pass before the threat of near death was removed on arrival at Nairobi.

The wobbly journey across the Congo had brought me to a different and unknown Africa. The streets of Lusaka, capital of Zambia, had all the fine finish and cleanliness of the clinical areas of Harrogate or some other Hydro in England. Compared with Kumasi or Enugu, there was a touch of palid morbidity about them, a still-life as if they had been left on an architect's drawing board. And the East African negro compared with his Western cousin had a deep sadness at the back of his eyes. The East laughed but not with the spontaneity of the West. The dictatorship of Kenneth Kaunda over Zambia was palpable but kindly. There was none of the ruthlessness of Nkrumah. The kindliness was, perhaps, the finger of a mad monarch, of a man who had not fully understood what his capricious authoritarianism ultimately led to. Even on this quick *tour d'horizon* examples of it cropped up. With the flick of a pen, Kenneth Kaunda had pronounced all land to be vested in the State as from a named date. His dictum coincided with my brief which required me to assess Zambia's need for qualified Land Economy officials. The new policy threw the national Land Commission into immediate disarray. Overnight the demand for qualified land commissioners rose from a handful to hundreds. Indigenous qualified men were not there to fill the bill, nor likely to be for a generation. Competent professionals would have to be expatriates, but the numbers required would be far beyond the resources of the Zambian economy to afford. Absolute title to land was now held by the State. All occupiers of land, farms, houses, shops, factories and so on became leaseholds overnight. But there was no one to draw the leases, let alone supervise the outworking of the resulting landlord and tenant relationship. Kaunda's was another example of airy-fairy fancy socialism blundering along its unenlightened way.

For me the camaraderie of West Africa was a want East Africa never met. In Zambia, I was a stranger often left for hours to kick my heels. One morning the waiting went on from breakfast through midday to the afternoon. Professor Kelly, Pro Vice Chancellor of Zambia University, was waiting for me in the lounge of the Padowski Hotel. I was upstairs waiting for him! We never met because Reception was inert. Some hours later a call was put through to his home number. A voice affirmed that Professor Kelly was in the garden but was praying and was not to be disturbed! Eventually he turned up at the hotel.

Kelly was humdinger Irish, a man of bubbling friendliness, kindly courtesy and rapid wit which his brogue sharpened to a sheer delight. 'Tell me now' he enquired after a long hot sightseeing survey of Lusaka town, 'would you like to come home now fer er cup o' tay?'

Home, to my astonishment, turned out to be a Roman Catholic

seminary in the charge of Father Kelly. We found two chairs in the sparsely furnished retreat and the mystery of the praying Professor was cleared up. Kelly dropped three lumps of sugar into a brimming cup and as if pursuing our business of half an hour before suddenly said,

'Wheel now. Did yer know that the Risen Lord has appeared to Pope John Paul?'

'I don't read the secular press. So I wouldn't know.' My reply was a nonsense to parry something coming next which beggared my wits to guess.

'Yes, to be sure, 'twas on Easter Morn as ever was,' he continued, with a far from holy smile. 'The Pope took advantage.'

'"May I ask you a question, Holy Lord?" John Paul, you see, is ever a man to take a vantage point.

"What is it, John Paul?"

"Will there ever be married clergy in Holy Church, Risen Lord?"

"Not in your lifetime, John Paul".

"Risen Lord, please bear with me. One more question. Will there ever be ordained women in Holy Church?"

"Not in your lifetime, John Paul".

"Oh Lord. One more question. Will there ever be another Polish Pope?"

"Not in My Lifetime, John Paul."'

The handshake that parted us back at the hotel I vowed would not be the last. Alas, nevertheless, it was to be so. We never met again. By the time of my return some months later, Professor Kelly had left his post at the University.

The Big Trek in the Autumn was arranged in two parties. First, I was alone and followed a trail through Malawi where Dr Banda, the President, was still flicking flies off his nose with the notorious switch he carried as an emblem of authority. Iris Elkington, replete with a new-fashioned short hair-do, *très chic*, joined forces in Zimbabwe. The files in our bags were as various an assortment as the wares of travelling hawkers. There were pedigrees of Red Pole cattle from Petworth to be sold to Zambia; proposals for new farming colleges; the hopes of surveyors and land agents who were probing for soft spots overseas and much else to be borne in mind and above all sales talk for Cambridge, all additional to the manpower study proper. Unknown to us, among the papers had been implanted a 'time-bomb', set to go off in Zambia.

Oblivious of this lethal undiscovered menace, the journey began and continued in high spirits. The streets of Harare were clean, beautiful and had a similar touch of the clinical cleanliness noticed in Lusaka.

The soul of erstwhile Rhodesia was glimpsed now and again. At dinner one evening a doctor's family of some years domicile expressed confusion, chagrin and surprise at the criticism of local attitudes they had received from 'Home' (they had all derived from the UK). 'We always treat the indigenous black people with every kindness. What else is to be expected, pray!' It was a grudging, somewhat sullen defence. 'So you do the cat', was the best my weary wits could muster to clear the fog from the minds of these kind hosts who were so very far out of range of the modern African. Before leaving on the next leg of the expedition, my job took me to the Secretary of the local RICS and both of us to an interview with Dr Chidzero, the Minister of Economic Planning and Development. The courtesy and patent competence of His Excellency drained all prejudice from my fair-minded colleague. 'My!' he exclaimed, as we descended the marble stairway from the Ministry. 'If they are all like him, what's the worry.' The revelation that an African could be made of courtesy, consideration, wit, grit and commanding intellect provided my companion with a light which would never shine from the arm's distance approach he and other four generation white Rhodesians had been adopting.

The campfire singsongs of my boyhood taught me and other scouting ragamuffins to chant tuneless canticles into the flames. One of these, possessed of a great swing and resonance, was top of the recent campfire pops in the 1920s and rang out in those years in a gibberish no one understood: 'Ngwenyama, gwenyama, enverboo. Yaboo, Yaboo, enverboo'. Over in Swaziland and trying to find my way between Swazi Nation Land, public land and plain man's land, an official who was helping me alerted my unbelieving ears with the observation, 'Oh, that's within the jurisdiction of Ngwenyama.' My mind switched back to the campfires. So, Ngwenyama is the King of the Swazis! The campfire song was an anthem in his praise. Swaziland is a unique sovereignty and has an ethos of kingly regard innate in the breasts of Ngwenyama's subjects. Nothing much could happen there without the nod of Ngwenyama, his family and his court. Tradition and custom ruled palace and people but not without a female warrant interwoven in the procedures and processes. Ngwenyama as a reigning monarch had, by custom, to take a new wife every year. The man on the throne, Sobhuza II, was in his eighties. Eighty or more wifely princesses held court for him. At the death of Ngwenyama, there is no 'The king is dead, long live the king' to ensure an immediate accession. One of the females of the line, by a formula of inheritance, becomes the Great She Elephant and gathers to her bosom all the powers of the

A breakfast 'sitting' with sketch-book in Lesotho

throne until by nods, nudges and, perhaps, a touch of necromancy one of the royal princes is proclaimed Ngwenyama.

The General Accident, Fire and Life Assurance Co of Perth would put my life premiums up fourfold had they witnessed our flight from Swaziland to the enclave of Lesotho. My bottom was getting used to the nobbly seats of tiny airplanes. The last experience was in a four-seater skimming the treetops between Lusaka and Kitwe in Zambia. Now, with a look of dejection in its drooping wings, here was a six-seater waiting for us and pointing southwards over Swaziland to the barren Vrede and snow-capped Maseru. The pilot who changed airlines much as a python changes its skin appeared nervous – he knew what lay ahead. His main anxiety were the dark clouds of a thunderstorm brewing up from the north. There were only three people in the departure lounge to board the aircraft. Halfway over the mountains, the pilot near whom we were crouched started to mutter warnings about having to land. His eyes looked in fear backwards over the tail to the blackness coming up behind like the wrath of God, then forwards to the snow on the high peaks in front. As the wind rose, the frail object to which we had entrusted our all began to behave like a kite, swirling, dipping and tossing over trackless waste and roaring rivers. My bowels responded to the movement. Sickness came on with the mounting darkness. Iris Elkington seemed to be practising a form of yoga. Flying over the Congo and up to the cliff face in Tobago was first class travel compared with this devilry. If only the pilot would land, my poor nausea-wracked body could crawl into a ditch. The high mountains loomed from the south and we appeared to fly at a perpendicular angle to their menacing rock face. The ground below was death, as treacherous as the air above. The pilot abandoned the plane to the elements and God's mercy. Eventually, the hideous nightmare of the mountains ended with a thud and thanksgiving, as our wheels hit the Mersuru airfield. My carcase lay in a crumpled heap by the doorway of the battered aircraft. I could get no further and was past caring. Luggage, customs, tickets and life itself were of no consequence.

Lesotho is a pocket-handkerchief of a country with an inflated pride. On ill-judged assumptions about its resources and scholars, the University that very summer had gone independent and cut itself off from the ties that had bound it to the University Colleges of Swaziland and Botswana. At weekends, the little kingdom abandons itself to dissipation. Hosts of workers, well-paid in South African currency, surge across the Border to squander it with prodigal disregard of the outcome in Lesotho's casinos and gambling halls. The small country can never shake free of its giant neighbour and was behaving like a fractious infant. Of one thing we were certain: we would never again fly in a moth kite over the Drakenburg. We would leave by some other route. A kindly surveyor, Chris Aitken, offered a lift to Bloemfontein in Afrikaner country as a way out by road. So we crossed the border to a land where the sadness behind the eyes of the negroes was deeper than ever. The President Hotel in Bloemfontein inhaled the sadness. Its spacious architecture and luxurious apartments were no antidotes for the prevailing unhappiness. Fortunately, we had no business to transact other than to engineer a quick getaway.

At daybreak next morning, the intoxicating air lifted our spirits as we took the road westwards towards Kimberley. It was as cool as Brighton beach in mid-June and the sun peeping above retreating Bloemfontein shone brightly through the mirror above the windscreen. Ours was the veldt for the next twelve hours as we followed the tracks of Kipling's Elephant's Child, by Karma Country on towards 'the Great Grey Green Greasy Limpopo River all set about with fever trees.' Avis, the car hire folk, had produced a smart red Ford Escort which got us to the diamond pit at Kimberley for breakfast. Beyond the low scrub and disordered townships of Bophuthatswana, the sun came up in scorching heat. At Mafeking, the so-called civilisation of South

Africa ran out and with it the doleful help of negroes manning petrol pumps. Trouble set in at the Ramatlabama Border Post. The hut was abandoned but a sound of revelry came from behind. Some local fire water was burning in the veins of the two swaying border guards who shuffled into the hut to stamp the passports. At the Lesotho departure point the previous day officials had taken some care to work the accepted border trickery designed to camouflage the fact that the permitted traveller through South Africa had not been there. A detachable white slip of paper affixed to the passport bore the stamp of a temporary visiting visa. Woe and betide the passport and its bearer should an official stamp miss the detachable slip and mark the passport with the indelible proof that the bearer had passed through South Africa. The drunks at Ramatlabama did just that. Neither they nor we cared at the time. Only months later when Iris had hopes of helping me in Nigeria did fateful doom overtake her. Not only was the passport branded but she herself was listed among the suspicious, should she ever afterwards attempt to enter Africa, however clean the passport in her hand might be. From the Border onwards the road was lost. Open trackways in the dust crossed and recrossed each other in mounting confusion like the railway lines at Clapham Junction. The afternoon heavens showed no guiding stars. So keeping Pheobus to the left of the windscreen, we pushed on more in despair than hope. The first settlements at Gaborone came to us out of a gathering dusk. There were neither flags nor bunting. No one met us, no one greeted or welcomed what in our eyes were two deserving heroes. Raffle, who had been detailed to shout salutations, had been rushed to Jo'Burg for an emergency operation.

Gaborone is little more than a small modern shopping centre, cushioned against the vast Kalahari Desert and its famous game reserves by a slender periphery of comfortable residences and the usual appendages of native housing areas where, in this case, somewhat unusually, all the do-it-yourself structures had the loos standing by the front doors. There was much general business to be done, the Ford Escort to be returned to a collecting base and a visit to a teaching establishment which had the pretence of being a University College. New friends who were arranging the programme left us to our fate one afternoon. We used the time trying to find and gain access to the Great Grey Green Greasy Limpopo River, forgetting that it was 'all set about with fever trees'. Sorely tempted by the spreading waters over the mudflats of the wide oozing river, I stripped to the buff and went for a plunge. Help! The mud, a treacherous quicksand, seized me. Feet, legs and ears sank in the slime. The only hope of extraction was to lie out full length and push to land like an inverted turtle. Iris Elkington was paddling in the fetid waters down stream. On return to Gaborone, we were met with horrified, unbelieving cries of 'Bilharzia! You are bound to have got it. Anyone who paddles in the river, let alone be so barmy as to swim in it, can't escape. Cheer up! There's nothing to be done. Symptoms don't appear for six months. Then, your eyes turn to darkness, your urine to blood and your liver suffers *cirrhosis*. Have a drink.'

Full of forboding, we left next day for Zambia. There the 'time-bomb' takes over the tale. A friend of mine, Christopher Mitchell-Heggs, who is an international lawyer of some repute and practices in Paris, had an idea. He had been born in Rhodesia and knew it well. Jointly with a French consortium, he wanted to promote in Africa the chance of setting up indigenous Agricultural Colleges geared to practical farming and eventual farming careers for the graduates. My trip was an opportunity of the first rank and not to be missed. A briefing trip to Paris furnished a clutch of letters addressed to all manner of dignitaries in Zambia; above all, those who would have influence with Kenneth Kaunda. Not wanting to disappoint a friend but having little share in Christopher's sanguine outlook, the letters were stuffed in my bags and a promise given to seek out the contacts and exercise my salesmen's patter. As the financial sponsors of my manpower project were Barclays International, one of the first engagements in Zambia was a morning with the Bank Manager and Chairman of Barclays Bank in Lusaka. Business done, I sought the Manager's help in finding the Michell-Heggs' *di maiores*. Ministers and other lesser fry were as nothing compared with the ace of the pack whose whereabouts no one seemed to know. His name, among the others in one of the letters, was pointed out to the Manager. The man jumped from his chair and threw the missive on the floor.

'Not him!' He exploded in a cry of unbelief and dismay. 'Only this morning that guy fled the country for the Congo, leaving behind in his house a cache of arms and blatant evidence of an armed coup to overthrow Kenneth Kaunda. That fellow's name is death to himself and to anyone known to be or appearing to be associated with him. In the hands of an official or Minister this letter would certainly indict you and Iris. You both could be in one of Zambia's infamous jails facing sentences of summary execution.'

Bilharzia and the dire glooms of Botswana were the joys of heaven compared with this awful revelation. The offensive letter was torn to shreds, burnt and flushed down the loo. The 'time-bomb' had been defused but only just in time. That was the last that Zambia or any other place in Africa heard of the new dawn of farming through the Mitchell-Heggs Agricultural Colleges. Christopher was warned by telephone to keep low. We moved on, thankful but badly shaken, to recuperate on the golden beaches of Dar es Salaam where only mosquitoes threatened disease and destruction.

❦

My endeavours in southern Africa were a late sowing well into my St Chad's Eve. Even so, there was a detectable harvest. The reaper was Isaac Ofori whom we have previously met as Ghana's Commissioner of Rural Development. Ofori established the courses and teaching in Land Economy at the University of Lusaka's outreach at Kitwe on the Copper Belt. The outcome matched in style and substance the rich harvests sown in the St Valentine years. The other beans which I now tossed at random in the clay were of a different order. They were not meant to seed career harvests. The beans were sown in trenches of another man's digging not for my reaping or fortune but as gifts in the cause of another's aspirations. Many were simple friendships but with little of my heart in the cause.

Sitting on the five-barred gate of retirement, swinging my legs in contentment and looking over the harvests of past St Valentine Days, there was one harvest that yielded unexpected merriment and not a little gain. The crop was reaped in the 1980s, well into St Chad's Eve. It was to do with my house in Chaucer Road and my encounter long ago with Fred Willey, Harold Wilson's Minister of Land and Natural Resources. My criticism that Fred Willey's first Land Reform Bill would have enriched some of the most well-to-do had prompted him to limit the benefits of leasehold reform to houses below a specified rateable value. My leasehold house in Chaucer Road had a rateable value well above the Willey's limit. My landlords, the University, laughed. Fred Willey's stopper was the University's protection against any pretence from me. Now, from infancy I had been able to read, an early sign of the precocious. This simple accomplishment the University hardly believed! I, therefore, looked at the Leasehold Reform Act 1967. Tenants who had improved their houses should not be penalised by allowing the improvements which they had made to the house to be reflected in its rateable value. Sir Sydney Roberts, from whom I had bought the house, had improved the place by creating three independent flats. Friends who had managed the place for the University in the past were called to my aid. Three pages of 'tenant's improvements' were prepared from what they remembered of Sir Sydney's changes and other improvements. This impressive list, however, was impotent to aid my case unless the University agreed to it. Apart from the long list, it was noted that the Act defined a tenant's improvement as anything the tenant had constructed or built on the messuage. The original lease of 1906 unequivocally stated that 'in consideration of the tenant having built the house, the landlord hereby grants him a 99-year lease'. Surely, then, the whole house was itself an improvement! Let's have two schedules: one, the list of three pages; and the other, a simple statment, 'Tenant's Improvement, one house'. This ingenuity was solemnly activated by my lawyer. Some months went by. Nothing happened. We lost patience and entered prosecution against the University in the Crown Court – 'Denman v. the Chancellor and Scholars of the University of Cambridge. Vernon McElroy, who had taken on the mantle from Mills of Head of the Estate Management Service, hailed me in Trumpington Street some days later. 'The Financial Board' he said 'have fluttering hearts and pale faces. Will you parley?' If the 'one house = one improvement' formula were upheld at law, there would be but very few houses remaining outside the benefits of the Act. Echoes of McElroy came through the University's solicitors to the effect that they would accept the three pages of improvements if we would withdraw the action. The Financial Board were scared of the legal outcome. We compromised by agreeing to postpone the action while the District Valuer looked at the three pages of listed improvements. All parties took time to do things, especially the solicitors for the University. Waiting for a pot to boil always draws out time towards eternity. So it was in this instance. Other demands were pressing. The dear folk at Borgo a Mozzano in Italy wanted a text in English of their outstanding achievements, not a mere translation but a full history in English. Colin Kolbert and I had undertaken the job. It was in keeping with my St Chad's Eve programmes. Thus it came about that the Borgo a Mozzano telephone exchange took the call from Cambridge with the news that the District Valuer's figures had put 12 Chaucer Road firmly within the beneficial arena of the Leasehold Reform Act. For us, therefore, to continue the court action would have been a mere academic and costly exercise. It was called off and the matter left to enrich the research files of the Department of Land Economy. Fred Willey's Act had proved to be as fair as an unfair Act could be. My work at Borgo a Mozzano flowed along happily and smoothly to produce *The Fountain Principle: A guide to new positive rural development planning, 1982.* I returned home to redeem my leasehold.

The grinding screech of heavy machinery, soulless concrete lead-roofed barns, arrow-like roads straight and hedgeless, these are symptoms of heart disease in a countryside where health lies with the lichen rust of ancient tiles, the reek of wood smoke and meandering homeward lanes. Artists and poets think this way. Deep in the irrational recesses of my soul there's agreement with them. No wonder then, when Henry Moore the sculptor proposed me alongside himself as a patron of the Small Farmers Association, it was a joy and an honour to fall in beside him. The two of us were quickly joined by His Grace the Duke of Buccleuch. Henry Moore's proposal reached me by letter from David Hunter-Smith, the Chairman and founder of the small farmers' group.

No one seemed to know what a small farmer was when the originators were staking out the Association's claim to recognition in the farming community. After some domestic debate, the SFA declared themselves champions of the family farm. It was a bold, vision-led and timely innovation rewarded by the enthusiasm of an effective band of supporters from the true heart of the countryside. The farm, home and livelihood of a working family, the epitome of a true husbandry, was under threat of extinction from competition with massive agro-business and from national and international policies devised at Westminster and Brussels to benefit the big boys whose voices were vocal in the corridors of power. Someone had to speak up for the family farmer. The Common Agricultural Policy (CAP) from Brussels with its questionable price supports gave the big farmer new purchasing power to buy up land and squeeze out the valiant 'little Hampdens', as Grey, in his *Elegy in a Churchyard*, would have called them. The family-owned farm was dear to the heart of Jefferson when he framed the Constitution of the USA. To him it was the bulwark of democracy.

God speed the Small Farmers Association. I certainly would do what in me lay. As a Patron no more was expected of me than to show a presence at Conferences and AGMs. For me, at first, more time and money were put into my Patronage than an orthodox performance would have required. The invitation caused me to face with serious regard the meaning and status of Patron and the demands of an honorary position. The former meant giving public recognition and support to the object of my patronage. The latter, so I found by experience, could be a form of giving and acting from motives other than a wholehearted affinity with the sense and object of the cause. With the Small Farmers Association there was no honorary post to hold, little more than my name on the letterhead. So I settled for a quiet Patronage along with the Duke and the sculptor.

That the Duke of Buccleuch, the largest landowner in Britain, should be a Patron of the Small Farmers Association seemed in every way a fitting relationship, wholly in keeping with the true heartbeat of the countryside. It rekindled and deepened a friendship begun in former years when the Duke's father and others in the Upper House, looking over their shoulders to catch my watchful eye in the Visitors' Enclosure, were doing battle against the socialist Government on the Land Commission Bill. Of Henry Moore nothing was seen nor heard, a miss and a sadness. In the Spring of 1980, the Small Farmers Association held its first Conference. For reasons never fully understood, perhaps because I was not the owner of 300,000 acres of farmland and villages, the Executive Committee invited me to open the proceedings with a lead Paper. The assembly was a classroom affair rather than a woolly Conference. For me it was a delightful début into the responsibilities expected of a properly-conducted St Chad's Eve.

To cynics there is no such thing as a free lunch. They are probably right. It is all a matter of values. When, therefore, a little after my retirement, the International Conference for the Unity of the Sciences (ICUS) handed out free travel, free accommodation and free meals to help what they called the pursuit of Absolute Values, the letter of invitation looked like an irrefutable condemnation of the cynics. The context, moreover, was heavy with academic kudos; a previous Conference had been chaired by Lord Adrian, supported by an assortment of Nobel Prize Winners. When turned over, the letter had no suspicious small print on the back vitiating its apparent genuineness. A similar offer had reached Colin Kolbert. Both were followed up on our part by lunching Brian Wijeratne, the signatory to the letters. His ingenuous charm captivated us and silenced any misgivings. Eventually the extraordinary promise was made good. In November 1979, the luxurious accommodation of the Central Plaza Hotel in Los Angeles opened its doors to me. Two days of attending lectures and social gatherings passed without qualms. All seemed above board. Were there truly no strings attached to the deal? On the morning of the third day while at ease listening to someone describe the world as God sees it, a gentle female hand tapped me on the shoulder and stuffed a note into my top pocket. The note read, 'You are a most important person. Can we meet? I would like to talk to you.'

Smiling behind me as I turned to acknowledge the note was a most delectable maiden in early youth. An hour later, over a gin for one and coffee for two, the price-tag on the 'free' offer was displayed. The girl was proselytizing on behalf of the Unification Church, the Moonies! She came from a village in the Middle West. 'So you won't know much about America, my dear, will you?' My comment interrupted the flow of her carefully rehearsed patter which stopped in mid-flow as I added, 'One has to see a country from the outside to understand fully what is going on inside.' The barb was a shame. She was too innocent, too insouciant and unsophisticated for it to catch hold. Wilting somewhat under my continued avuncular advice which touched upon her not wasting her youth misleading herself and others into thinking Absolute Truth was to be had in the Unification Church, she gave me a wan smile and we parted company.

Another item on the price tag was my lecture. Someone had thought up a title – *'The Economics of World Hunger'*. Its delivery stirred the audience to much disputation. My opening line made the point that the subject was *non est* because there wasn't any world hunger! A starving country here, another there, a well-fed country in another place and an over-fed continent did not add up to world hunger. Search for Absolute Values was on the same plain of falsehood as the mistaken notion of world hunger. In the finite mind the search could only lead to distortion like all 'world' conceptions. The holistic approach amounts to trying to see the world as God sees it. When John Buchan (Lord Tweedsmuir) bade farewell to his friend Lawrence of Arabia for the last time, he commented, so his autobiography, *Memory Hold the Door*, tells us, on how bronzed and well Lawrence looked. Lady Tweedsmuir disagreed. 'He tries to see the world as God sees it' she objected. 'No man can do that and live.' Within a fortnight Lawrence was dead. The ICUS was trying to see the world as God sees it. I despaired. Moreover, the whole proceedings greatly angered me. Ultimately all pretence fell away. The finale was a great banquet with the Rev Sun Myung Moon and his family apostolate playing hosts to the entire Conference. My name had been implicated with the Moonies and without my acquiescence. The price of the 'free' lunch had been high indeed. The only consolation was the experience. The duplicity that fooled me can fool others. To have knowledge of it is to be forwarned of it. Look thrice, oh academic colleagues, at all invitations to participate in an International Conference of the Unity of the Sciences, the Professors' World Peace Academy, the World Media Association and the Global Economic Action Institute.

Seeing the world as God sees it is an intellectual bent that has become widespread among 20th Century theorists. Maynard Keynes' revolutionary *General Theory of Employment, Interest and Money* could not have been written without postulating in some measure the mind of 'mass' man.

Concepts which beggar the competence of an individual finite mind to apprehend the truth, 'see' the world, the nation, the region, the 'British People'. Whatever is "beheld" can only be depicted in the abstract, by symbols, maps, charts, statistics and so forth – no one can see the reality. A land use map is within the competence of a good draftsman, if he is given the appropriate data. He can draw the symbol – the map. Where the thinker goes astray is in supposing he alone or with others of like mind can do anything directly to alter what is 'seen'. The conglomerate symbolised in the map is the outcome of myriads upon myriads of minor decisions by those who have effective command over themselves, other people and resources. Today's Grand Illusion personifies the collective, 'the community', 'the countryside', 'the British People', 'the nation', as if the collective had a persona like unto the human personality, capable of knowing good and evil. Hitler personalised Der Herrenfolk by projecting his own wicked desires, wishes, judgments and actions into the collective concept. To Hitler, the Holocaust, which annihilated millions of Jews, was for 'the good' of the mass – the Herrenfolk. Any man who wants to be absolute thinks and acts like this. Someone wants something done and in imagination projects his personal wants on to a collective, thereby giving it the faculty of having a human face, a human want; and what is, therefore, **good** for the wisher become the **good** of the community. The community is given a persona. No man can in fact see into another's soul, let alone into the souls of millions. To think he can do so, is to delude himself into thinking he can comprehend the Whole as God does. 'No man can do that and live' – Lady Tweedsmuir was surely right.

Modern macro-planning is all of a piece with trying to think as God thinks. That is why I have always been sceptical of it. And why in 1979 when a group of well-meaning folk drew me to their company which became the Land Decade Council with the aim of improving the land use of Britain, I was at best a happy sceptic among them. First among priorities for the Council was a standing Land Use Map of Britain kept up to date by regular periodic amendment. We would meet on the premises of the *Architectural Journal* wherein was housed a domestic pub, The Bride of Denmark. How Professor Dudley Stamp, the great progenitor of the Land Use Map of Britain, would have loved the pub and the plotting!

Prospects were ambitious: conferences, counter-conferences, education drives, surveys, Government accord, international cooperation, shifting targets, a programme full enough to afford a failure here and there.

When my colleagues, anticipating possession of the national data, supposed they would be able to do something with it to alter the pattern, my enthusiasm began to wane – they were about to take the 'God thought' tack. With the passage of time, governance passed to the hands of a small bench of *Septem Viri*. These valiants tried sally after sally, scheme after scheme, before yielding to the inevitable conclusion of the futility of hoping to conceive an optimum land use pattern of Great Britain. Who were we, unmandated busybodies, to tell ourselves and others what was best for the nation? The Land Council was in danger of joining Lawrence of Arabia in the shades along with others who tried to think as God thinks. The years in the company of Graham Moss, Leslie Fairweather and the Earl of Selbourne were, nonetheless, a period of enduring friendship. These Land Council's elders were more God-like than they knew.

Thinking as I do, it was a discerning fate that moved Sir Ian Mactaggart to offer me a seat on the Board of his Grove End Housing Association. Here was something truly micro. Admittedly it was a kind of collective, but one small enough to correct itself. The housing association ideal carries within itself a destructive contradiction. In principle, members of a Housing Association should never be called upon to pay more for their home in rent, management fees or purchase price than the bare cost of these essentials. Profit pursuit is banned. However, there is no legal ceiling to expenditure. Tenants are expected to face rents which will 'wash the face' of the Association and fund its debts. Management facing heavy loans at high interest rates could on this principle set legal rents at levels above what would be claimable under the Rent Restriction Acts.

So it was with Grove End Housing Association. Two generations back, the reigning baronet in Glasgow had acquired the property, a sizeable block of flats in north west London. Rent restriction legislation threw shadows before it of eventual bankruptcy and slumdom. It happened that Westminster City Council were on the lookout for houses or flats to help curtail the growing queues of "hopefuls". The Council put up money to finance an *ad hoc* Housing Association as a bargain for a proportion of the flats as they fell vacant. Grove End Gardens were sold by the Mactaggart family to the Association for something approaching a price which, discounted to annual revenue, produced an income from rents nearer free market values while the tenants were assured of well-managed and repaired dwellings.

Grove End Housing Association paid its way handsomely and discharged all dues by adjusting rents within legal limits. Only the directors were expected to give their services *gratis*, even to the extent of meeting all personal expenditures. Such one-sided policy was hard to defend in an institution where the customer met all running costs. With me it was a case of honorary service given and little heart in the cause. My loyalty was to Sir Ian, the Chairman. Many a time we stood together on one side of divided opinion, against an opposition of solicitors, bankers, accountants and representatives from Westminster City Council. Because my schedule of duties called me overseas frequently and at times when the Housing Association Board was meeting, the opposition moved me off the Board on the grounds of my laxity of attendance. Years later after much jockeying behind the scenes my seat was restored. Success from early good management and consolidation of reserves induced in the Grove End Housing Association managers a reckless lurch towards prodigal investment. A large country mansion at Middleton Stoney was acquired to ape what was happening at Grove End Gardens.

More money was spent than garnered as demand was slack. Businessmen running a non-profit Housing Association are in my experience as hardhearted and ruthless, probably more so, than a board of directors supervising a profit-geared company and are prone to investment ventures for their own sake. When Sir Ian died my resignation was gleefully accepted.

My years with the Land Decade Council and the Grove End Housing Association wrought in me the realisation that loyalty to friends or to some special person would often keep me hammering away at their side, voluntarily and honorarily, while for various and widely differing reasons my heart and mind were not wholly in tune with each other. Heart and head pulling in the same direction harmonise truly with honorary status. There was no fear of wearing a Janus face, of looking two ways at once, when Alfred Sherman of the Centre for Policy Studies asked me to take the Chairmanship of the Centre's Land Group. The only difficulty was timing. The group with little time to spare met in a cramped cabin off Wilfred Street. It was New Year 1979 with a General Election looming and every prospect of the Government changing before the clocks changed with the coming of spring.

In opposition, the Tory Party had been too engaged tearing down the Government's land policies to give much time to thinking up positive alternatives, although Sir Hugh Rossi's working committee, of which I had the honour to be a member, met regularly and to a purpose. With private property and free market aspirations to the fore, grounds for State controlled land policies were hardly necessary.

Nevertheless, the Tories are not votaries of dogmatic *laissez-faire*. If public controls can be shown to be necessary to safeguard our "green and pleasant land", can be policed without too much disturbance of quiet possession and costs match benefits, they will give thought to finding the best forms. Alfred Sherman with a stentorian rasp in his voice would declare his version of the way, the truth and the life and then walk away from the table. The cogitators would be left to wonder and mumble over what to do with conclusions once we had come to them. Waiting upon our work were the two Olympians – Margaret Thatcher and Keith Joseph. They met with us only at social sherry parties arranged to shake hands, pat backs and denounce consensus politics. The Land Group were hand-picked partly by myself and partly by the Centre.

Eventually we realised that ideas and proposals should be committed to a published report. The timing of publication was as inopportune as its editing was delightful. Draft texts were put up as deliberate 'Aunt Sallies' to be shied at, amended and await the editing skills of Patricia Kerwen – that was the delightful part. The timing upset was caused by the General Election. The Report, *Land in a Free Society*, was written with the Election in prospect. By the time of publication, a Conservative Government was in power. Albeit, the Report prescribed basic ideals and principles to guide future policy and an explicit schedule of actions. These proved to be useful up to a point. Some took time to mature. Of the eight action items three were pursued immediately by Michael Heseltine, the new Secretary for the Environment who,

(a) repealed the Community Land Act 1975 with its power to bring all development land into public ownership;

(b) required all government departments and public authorities to keep public records of land owned by them and of the circumstances of its compulsory acquisition;

(c) introduced legislation to require all government departments and public authorities who had acquired land compulsorily and which subsequently had proved surplus to make provision for passing it back to the market either by way of public auction or private treaty.

In the same category as the Centre for Policy Studies where heart and mind were in tune with honorary service has been my directorship of the Westminster Industrial Brief company.

Commitment to Toryism and the Party found me with a directorship of this small company operating with an awful (in the true sense of the word) virility out of all proportion to its size. Lord Chelmer, when Treasurer of the Conservative Party, had proposed the incorporation of a company to buy research material from Central Office and retail it in the form of a journal, *Westminster Industrial Brief*, to the City and companies generally.

Costs would be kept within the purchasing margins of customers.

There would be two foundation directors, Richard Davenport and myself. As with the Housing Association, profit-making was barred and shares held at nominal value. The gusto and successful accomplishments of WIB had little to do with my contribution. Volunteer well-wishers, notably Alistair McAlpine (now Lord McAlpine) cavorted into boardrooms and drove the new venture on its way.

Being a kind of para-political initiative, WIB's fortunes followed, at the outset, those of the Party. Years later, with a growing demand to propel it forward, the company distanced itself somewhat from its parentage. The passage of years have not condemned but blessed its name. Links with Central Office became less immediate. The *Brief* also fed on material gleaned from elsewhere. WIB from the early 1980s organised discreet *fora*, discussion occasions where Ministers of the Crown in person spoke and exchanged views with subscribers. These 'hair-down' meetings have become popular and greatly applauded features of WIB life. Richard Davenport died. A succession of generous hearted peers and others have held the reigns of WIB. Out of unfeigned love of the girls who have done all the work, I have continued as a grandfatherly figure, hoary with age and unremittingly amazed that so much could be accomplished by so little effort on my part.

Honorary service knows no retirement, only resignation. Since the far off days of CHEC's Malta Conference my association with Zena Daysh and her Council has died down and rekindled like a guttering candle. At one stage, one of CHEC's ephemeral sprouts, a Finance Committee, was looking to me as its Chairman. Sir Hugh Springer, in whose luminous prestige the Council basked and without whose introduction its standing with the Commonwealth Foundation would never have had credibility, was still CHEC's Chairman. Once or twice a year my conscience was moved and the Executive Committee's Minutes would record my presence in the cross-talk of its ding-dong meetings. Zena badgered me to write 'place' Papers, especially for the biennial meetings of the Commonwealth Heads of Government. Above all others, participation in CHEC's activities was an honorary commitment out of friendship with Zena and not from wholehearted acceptance of human ecology.

Tension between heart and head tautened when on the elevation of Sir Hugh Springer to be Governor General of Barbados the Governing Board placed me in the Chair. I started taking the wrappers off its constitution, its finances, its membership, its very *raison d'être*. The process was like peeling an onion, skin after skin fell to the floor and no core was to be found; CHEC was *de facto* not a Council but a person – Zena Daysh. If Zena were to resign or pass away, CHEC would go with her. She was both soma and psyche. For lesser mortals, like myself, trying to make sense of what CHEC was getting at, its purpose looked shadowy, contradictory and foggy. Zena professed to know what human ecology was when many about her held different and divergent views. The term is so amorphous, so imprecise, suggestive of a concoction of vagaries that, like democracy and other generalisations, definitions are indispensable to rational action. Clear-cut definitions of human ecology itself, of its aspirations and of its work are just what CHEC never had. Through all the years of my association there ran a series of solemn meetings each called to explore, anatomise and define its viscera – all to no lasting effect. Human Ecology in academic circles was not unknown but was circumscribed in meaning and was given a berth in some respected haven of prescribed knowledge – medicine, for example, at Cambridge University.

Halfway through my term as Chairman (at the time the halfway stage was not apparent) I made a silent vow to take steps, for Zena's sake, to help secure CHEC's future. Clearly at her age and facing the threat of *àpres moi le deluge*, the future would have to be based on firm finance. No one would take on CHEC as Zena had done without a stipend to exist from hand to mouth on a shoestring of uncertain length. CHEC would have to acquire a capital fund sufficient to secure a constant, reliable income flow. Reliable income was essential not to finance projects, such as they were, but to buy the bread and butter of daily administration. CHEC's existence so far had depended upon the goodwill of the Commonwealth Foundation and that goodwill upon the noble Hugh Springer. Paradoxically, the Council had survived so long that many had come to the false supposition that there was hidden gold where none existed. Donations to fund administration costs are anathema to the run-of-the-mill charity. Objects seeking funding must be precisely defined and of unwavering purpose. The Governing Board was an amorphous mass of moving amoeba, as indeterminate as CHEC itself. Time would be wasted in seeking to attract support from the business world and the Funds for such elusive imprecision.

Slowly the idea dawned of treating support for the Council as a project in itself for which in principle the Funds could with reason be approached. Conditioning the thinking was my concern about the "green movement" to which human ecology was close cousin. The movement could become a political menace unless private enterprise and voluntary response were entrusted with its command. My thoughts turned to industry. Responsibility for the environment had hardly at that time burdened the industrial conscience.

The day was some way off when the politics of the "green

movement" were to be dubbed "tomato politics" – they start green and turn red! Had that day been closer, my approach to industry would have been easier. The political outlook and these two actualities provoked me to do something about it, to help industry into the limelight of accepted responsibility and ask for money as a reward. I founded the Human Ecology Foundation (HEF). Its object would be to raise money from industry and elsewhere to finance capital for environmentally-orientated charities like CHEC on the understanding that the Council itself should be the primary beneficiary. The hill before me was long and steep. Much encouragement was given by Lord Home of the Hirsel. He helped with finding Patrons; and once the preliminary steps had been taken gave a Reception in the House of Lords to launch the venture. The response was slow at first but over the long-term the outcome by any reckoning has been a success. The leading Funds, Wates, Rank, Laing and so on supported us. The principle and purpose of the HEF was unlike CHEC in that it was international and not confined to the Commonwealth. To put the right stamp upon it, a Third World figure of renown was wanted for Chairman. Heaven sent General Gowon, the former Head of State of Nigeria, to occupy the seat.

CHEC and HEF worked cheek by jowl. When CHEC was invited to send representatives to the Commonwealth Parliamentarians Association Conference in St Stephen's Hall in the Houses of Parliament, General Gowon and myself joined Zena Daysh. The grandees of the Commonwealth nations sat side by side under a stage which on normal days is the vestibule to the debating Chambers of Lords and Commons but on that day was a crescent of Household Buglers encircling the seated majesty of the Queen and Ministers of State.

After the assembly in St Stephen's Hall, there was just time in the run up to lunch to pop over to the Queen Elizabeth II Hall and check up on the CHEC bookstall. To our surprise, the building was teeming with security guards and police. Without asking who we were, the police herded us straight into a lift. We were trapped. The lift bumped open at the top floor and precipitated us into the arms of the Prime Minister who was waiting to receive her guests. 'Keep mum,' I whispered to Zena and the General. 'Go forward'. With wine glasses filled to overflowing we became 'Parliamentarians' along with the rest of them. All the Cabinet were there, the front row of the Opposition and MPs two a penny. We clicked glasses with Healey, Kinnock, Callaghan and numerous other stars in the firmament of Commonwealth politics. Suddenly there was a stirring by a doorway where drinkers had clustered at the far end of the room. Out of the mêlée, as if from a

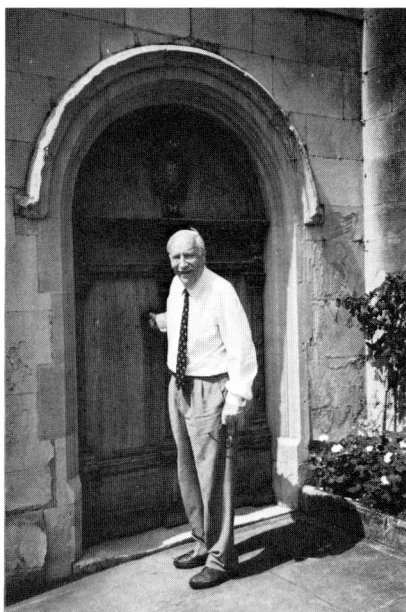

Lengthening shadows under a Pembroke College archway

rugger scrum, the Queen emerged. Walter Hamilton, the Queen's Equerry, knobbled General Gowon and led him over to her. 'Ah,' she exclaimed with great surprise, 'I haven't seen you General for a long time.' We felt we were no longer gatecrashers. Even so, life was a little precarious. Zena and I made for the safety of the lift and left Jack Gowon in the radiance of the royal smiles. The lift bumped open on the ground floor and we were once again in the presence of Margaret Thatcher. She was heading a guard of honour to bid farewell to the Queen, unaware that Her Majesty was chatting with Gowon upstairs.

❧ ❦

The story of the University Registrar who on being asked by the Chairman of a Charity Dinner if he could suggest the name of a wit to propose the toast of the evening, responded to the effect that the University had no wit to offer but could come up with two half-wits, besides belonging to my anthology of stale jokes, strikes a chord of a somewhat different significance. It raises questions whether two halves always make a whole; can a half by itself be a whole? Life cast me in a half role as one who looked exactly like a selfsame 'other'. Waiting

347   *St Chad's Eve*

delivery into this world, hand palm to hand palm, eye to eye, his legs aligned with mine, I doubled in mould identical with Sydney my twin. Until a short while ago, my half life was shared with his unifying spirit which animated us to think alike as well as look alike. Nevertheless, my half life has in its own consciousness been a whole life, golden and windy, rich and turbulent, lived and enjoyed with God and man. My friend Norris McWhirter, who shared his life with Ross his identical twin until that noble man was brutally murdered in Enfield, has spoken of the choice that befell him after that tragedy. The wholeness of two in unity was no longer his. Either he had to live the life of a half man or the lives of two men in one. My twinhood may account for an inward strange awareness which seems to cut off the fullness of life in one direction to make room for a half life approaching it from another. More than one pair of halves have made the legend of life recounted in the foregoing chronicle. As a professional chartered surveyor, my way took the path of the academic world and in the eyes of my professional colleagues made me so 'heavenly' minded as to be no earthly good. As an academic, my thoughts were too near the outlook of the practical professional to be wholly absorbed in academic abstraction and made me at times impatient of intellectual niggling. Although a voting Tory and rejoicing in the respect and confidence of the Party, the other half of me is a fundamental Whig, ready any day to sacrifice conformity and conservatism to popular liberty. As a Christian, a daily walk with Christ was conducted by a casuistry that lead my steps too close to the world's desires and values. A Professor of Law who had read some of my writings once judged me to be a lawyer; while an academic economist of high repute acknowledged my particular handling of the institution of property as an innovatory contribution to economic thought. The truth is, I am neither a lawyer nor an economist. Had I been wholly one or wholly the other, Land Economy would probably never have been conceived and developed.

This two-way life, this half and half affair, has made of me neither an optimist nor a pessimist. These two existences are attitudes to life which in the last analysis do not belong to my experience of reality. The optimist looks forward to the future good things of life, the pessimist to the future sad and sorrowful. Both "look forward" to the future and so away from where I am. One can in existential reality only be in and enjoy the present moment. Anything sadder than the present can only belong to the imagination and not to reality. So it is for happiness; no one can be happier than they are at the present moment. I am despondent because I have lost my wallet. My despondency derives not from what I experience at the present moment, a certain lightness in the back pocket, but from what I imagine I shall not be able to do tomorrow. Mine is an imagined destitution. When the wallet is found, I am "happier"; not because of the experience of the present moment, the heavier back pocket, but because of the imagined shopping and spending spree to come.

The process of seeing, hearing and thinking takes time, however infinitesimal its measurements. Time carries experience into the past. All thoughts of the future are creatures of the imagination. Imagination itself is a process of time and is carried back into the past. The paradox, indeed, puts the future into the past. Only "thou" can fill the present moment, not as an object – as one beloved or one abhorred – but as 'thou', unconditional. To be conditioned is to be adjectival and adjectives carry thought into the past. The present moment is filled only in so far as I, myself, go to meet 'thou' in love's self-giving. Hate cannot fill the present moment, for hate cannot meet 'thou'. Love's impact of 'I and thou' can pass away as and if I lose 'thou'. Only I and Thou Absolute can wholly apprehend the present moment and defeat all hazards of time. Faith alone knows that Encounter.

Thus reason, for me, points and confirms Faith. Nothing is present save I and Thou. Through God alone can God be known. He is the Present Moment to fulfil it and banish all loneliness, that consequence of being divorced from Thou. The ultimate sinner is the soul that goes out turning away from the Redeemer who alone fulfils the Present and overcomes its loneliness.

The richness of life for me has been my God and my neighbour and love's dialectic – *agape*, friendship, affection, eros – that ever fills the loneliness of the otherwise empty present. These thoughts are no metaphysics, they belong to the physical, to the real; for me so real that love's 'thous' can be named: Sydney my *alter ego* who shared life and love with me from the beginning; Winifred my first love in Eve's fair form; Colin the steadfast; Jeffrey the stalwart; Oliver whose lead I'd follow to the Well at the World's End; Irene of Esk Water; those wonders my secretaries – Peggy, Jennifer, Carol, Kenna, Pat, Alison One and Alison Two; Iris who twinned with Syd and me in joy and devotion, Jessie Hope whose soul is knit with mine in one; my mother; Jonathan and Richard, my sons, and their children who have come after them. And paramount, Thou Absolute. These and many, many another I love. These have conquered the loneliness in me. These have shown me the ways and wonders of life, have judged me, not in open court but in Chambers where *coeur à coeur loquitur*, 'and the poets poured us wine'.

God put me in an Eden alongside Adam my neighbour, there to sow my beans in the clay. Come St Chad's end, I shall need currency and a passport to take me through the Gateway of the Flaming Sword to the Other Side of Eden. Passport and currency have been provided me by love's "thous". With the epigram 'You brought nothing into this world, it is certain you will take nothing out', I thoroughly disagree. I shall take out what I never brought in – a love of God and my neighbour. These are passport and these are currency – the only currency valid on the other side of Eden.

INDEX

Aberdeen University, 203
Abu, J.Y., 232
Academic and professional
    qualifications, 80
Academic aspirations and progress,
    80–1
Acquaye, Professor Ben, 169, 248,
    283–4, 304
Afshar, Amir Khosvow, 229
Afshar, Dr Haleh, 208, 273
Afshar, Professor Hassan, 208
Agrarwirtschaft, 133
Agricultural Extension Centre, Shell
    International's at Borgo a Mozzano,
    220–1
*Agricultural Law and the Ownership
    of Land in England,* 149
Agriculture and Fisheries, Min. of, 73
Agriculture Fisheries and Food, Min.
    of, 191
Aims of Industry, 161
Air Ministry, 56–65
        Lands Officers, 58
        Land Service, 58
        Regional H.Q, Cambridge, 61–5
        and St John's College, 61
Alliluyeva, Svetlana, 325
Allsop, B.G.K., 126, 128, 129, 132
ALPO, *see* Land and Property Owners
    Assn. of
Annan, Noel (Lord), 114, 156
Asante, Yaw, 169
Asentehene, Prempah II, 2
Asentehenes, 249, 295
Ashby, Sir Eric (Lord), 168, 187
Ashcroft, Paul, 315
Aspatria Hall, Carlisle, 77–9
Auckland, 313
Azikiwe, Nnamdi, 167
Azmi, 299

Balfour, R.D., 169
Banda, Dr, 329

Barnes, Letty (Mother), 12; Family,
    20–1
Battersby, Edward, 234, 235
Bauer, Peter (Lord), 119
Beament Cttee., 190, 192
BBC radio, 103, 104, 271
Bective, Earl of 93
Berrill, Sir Kenneth, 184, 276
Betterment Levy, 160, 163, 165, 166
*Bibliography of Rural Land Economy
    and Landownership, 1900–1957,*
    132
Black Shirts, 53–4
Bodkin Adams, Dr John 51
Bolla, Professor Giangestone, 146,
    150
Bolney Wood Development, 38–9
BOMA, *see* Building Owners and
    Managers Assn. of Australia
Borgo a Mozano, 220–1, 273, 295,
    336
Boyd-Carpenter, John (Lord), 160,
    164
Boy Scouts, 23–8
Bradfield, John, 166
Bradshaw, Carol, 224–5, 227
Brighton Dome, Lecture in, 159
British Association for the
    Advancement of Science, 270
British Columbia, University of
    (UBC), 204–5
British Property Federation, 272
Brooke of Cumnor, Lord, 162, 165
Buccleugh, Duke of, 337, 338
Buchan, Norman Findley, 279, 280
Buchanan and Partners, Colin, 300
Buckingham University, 263–5
Building Owners and Managers Assn.
    of Australia (BOMA), 298
Burger, Dr Anna, 201
Butler, Sir Monty, 61–2
Butler, Rab (Lord), 58, 160, 166,
    203

Cadastres, 266
Caldecote, Lord, 117
Cambridge, Mass., 323
Cambridge Colleges, wartime
  requisition of, 61–2
Cambridge Inter-Collegiate Christian
  Union (CICCU), 100
Cambridge, University of, 81
    Agricultural Studies, Board of,
  107, 108, 109
    Agriculture, Faculty of, 85
    Agriculture, School of, 106 et seq
    Agriculture and Forestry Board
  of, 108
    Directorate of Lands and
  Buildings, 151–2
    disillusion with, 82, 83
    first appointment, 83
    Forestry, Readership in, 108
    Forestry, School of, 106
    General Board of the Faculties,
  117–9, 184–5, 190, 191, 258–9,
  294
    Horticulture, University
  Lectureship in, 108
    invitation to, 81
    Old Schools, 184, 185
    Overseas Studies Cttee, 193
    Professional duties, 258
    Regent House, 108, 154, 156
    Rural Economy, School of,
  proposed, 106–7
    See also Land Economy,
  Cambridge
CAP, see Common Agricultural Policy
CASLE, see Commonwealth
  Association of Surveying and Land
  Economy
Centre for Policy Studies, 342
    Land Group, 343
Chadwick, John, 255
CHEC, see Commonwealth Human
  Ecology Council
Chelmer, Lord, 343
Chesterton, Sir Oliver, 292–4
Chestertons, 106
Chidzero, Dr, 330
Christ's College, Finchley 24–32
Church Assembly, 101–3
CICCU, see Cambridge Inter-
  Collegiate Christian Union
CLA, see Country Landowners Assn.
Clark, Audrey, 269

Cleary, Fred, 258, 260
Cohen, Sir Andrew, 168
Coming of age celebrations, 42–3
Common Agricultural Policy (CAP),
  337
Common Land, Royal Commission
  on, 150
Commons and Village Greens south of
  the Border, research into, 150
Commonwealth Association of
  Surveying and Land Economy
  (CASLE), 197, 199, 202, 256, 277,
  285
    Adviser on Education, 202
    Caribbean islands, 286–9
    Colombo, 284–5
    Delhi, 282
    East Africa, 328
    Southern Africa, 327
    South Pacific, 308–13
Commonwealth Foundation, 197,
  256, 285
Commonwealth Human Ecology
  Council (CHEC), 254 et seq, 344
    and Human Ecology Foundation,
  346
Commonwealth Fund for Technical
  Cooperation, 286
Commonwealth Universities Assn.,
  256
Communist régime and
    Czechoslovakia, 138, 139
    Germany, East (DDR), 143–7
    Hungary, 199–202
    Poland, 140–1, 175–81
    Romania, 201
    Russia, 133
Community Land Act, 295
Conservative Party, 159 et seq
    links with, 186
Consultancies after retirement, 304
Continental Shelf Act, 316
Corfield, Sir Frederick, 162
Costain, Albert, 162
Country Landowners Assn. (CLA),
  272
County War Agricultural Executive
  Committee, 73
    and classification of farms, 74
Crowden, James, 93
Crown Estate Commissioners, 306,
  317, 318
Cumberland, life in, 69 et seq

Cumberland War Agricultural
  Executive Committee, 65, 73
    Deputy Executive Officer, 65
    duties with, 73, 76
CWAEC, *see* County War
  Agricultural Executive Commitee,
  73

Dalton, Hugh, 121
Dampier-Whetham, Sir William Cecil,
  105–9, *passim*, 154
Dartmonth, Raine Lady, 257–8
Davenport, Richard, 343
Davey, Tim, 312, 313
Daysh, Zena, 255, 270, 344
Dean, Noel, 84, 109, 112, 118, 120,
  125, 129, 132, 151, 152
Denman and Denman, 38
Denman, Bessie (aunt), 12
Denman, Elizabeth Jane
  (grandmother) 13, 16
    Clarence Gardens, 15
Denman, Ellison (uncle), 12, 15, 16,
  33
Denman, Jonathan (eldest son), 88, 97
Denman, Richard (younger son), 98,
  203
Denman, Sir Richard, 298
Denman, Robert (father), 11–16
  passim
Denman, Roland, (brother), 12, 38,
  58,
    marriage, 65
Denman, Ruth (aunt), 11,16
Denman, Sydney (twin brother), 11 *et
  seq*
    and Elaine Graham-Brown, 65
    identification confusion, 75,
  122–3
Dinwoodie, Laird of, 148–9
Diocesan Lay Reader, 98, 101
Doctoral Thesis, London University,
  loss of, 76
Dorman, Sir Maurice, 255
Down to Earth, 260
Drudy, Patrick, 261
Dudley Ward, P., 89

East, Barry, 260
Eastbourne
    Berrys Hotel, 41
    encounter with women, 41, 44–8
    estate agency in, 36–9

housing development in, 38–9
    motorbike débacle, 41
    religious enlightenment, 50
Ecology, 254
    human, 254–5
Economic Affairs, Inst. of, 161, 263,
  320
The Economics of World Hunger, 339
Edgson, Stanley, 124, 125
Edinburgh University, 203
Edward VIII, King, abdication of, 55
Elkington, Iris, 320, 325, 327, 329,
  332, 334
Elliott, Neil, 126, 148, 175, 177
Elmshurst, Leonard, 132
Emami, Mansour, 209, 210, 212
Enugu, University Campus, 168, 237
Epega, Dr Ben, 327
Estate Capital, 121, 147, 149
Estate capital, 136, 147
Estate Economy, 136–7
    property rights and, 136
Estate Management, College of
  (London), 35, 276
Estate Management, Cambridge,
  University of, 81, 82, 83–5, 107 *et
  seq*
    Board of, 83, 84, 111, 112, 121,
  132, 151, 184
    Dept. of, 116 *et seq*, 150–1; post-
  1952 Financial backing, 116–7,
  129, 132
      Development Fund, 132
      Lectureship in, 117
      Ordinary Degree in, 110
      Readership in, 107, 108, 119
Evans, Sheila, 298

Fabry, Prof. (Karlovy University,
  Prague), 202
FA – Forel (PX-28), 320
FAO, *see* Food and Agriculture
  Organization
Farming Today, BBC, 104
Federation Internationale des
  Géomètres (FIG), 197, 198, 199,
  200, 202, 276
Fenner, Sir Claude, 307
FIG *see* Fédération Internationale des
  Géomètres
Fiji, Native Lands Trust Board, 309
Filton, extension of, 59–60

Finchley, 11–13, 24–32, 42
    housing development, 13
Finchley High School, 24–5
Finland, international conference in,
    132–5
Finney, Nicholas, 321
First Footprints, 103
Five to Ten, 104
Fleming-Smith, Edward, 306
Florence, 147
Floyd, Colin, 313
Flysheets, 105, 156
Földes, Iván, 199–200, 201
Food and Agriculture Organization
    (FAO), 199, 304
*The Fountain Principle: A Guide to
    new positive rural development
    planning*, 1982, 336
Freemasonry, 52–3

Gardiner, Gerald (Lord Chancellor),
    166
Garth-Moore, Q.C. Evelyn, 102
Geographical Publications Ltd, 269
Geometers Dining Club, 314
Ghana, 167, 169, 172–4, 231–5, 248
    academic advice to, 169
    Land Secretariat, 169
    Surveyors, Institution of, 233;
    Hon. Fellowships, 233, 248
Gilbey Lectures, 110
Gowon, Gen. Dr Yakubu, 231, 346
Graduation, London University, 39
Gray, Norman, 51–2
Grove End Housing Assn., 341
Grosvenor Estate, 125
Grosvenor Fund, 321

Hamson, Prof Jack, 208
Harlow, Admiral Bruce, 324
Harriman, Leslie, 229, 241, 243
Harriman, Chief Hope 229, 249
Harris, Ralph (Lord), 263
Harrison, Gabriel, 261
Hart, Dame Judith, 229
Haslemere Estates, 259
Hayek, Prof. 79
Heath, Sir Edward, 159, 163, 225–6,
    262
Hill, Alison, 224–5
Hillier, Parker, May and Rowden, 306
Hodge, Sir William, 181
Hogg, Quintin (Lord Hailsham), 100

Home of the Hirsel, Lord (Dundas,
    Lord), 58, 346
Honeymoon, 66–9
Hopson, Sir Donald, 279–80
Hoyte, H.D., 289
Hudson, Lord, 74–5
Human Ecology Foundation, 346
Human Environment, UN Conference
    on, Stockholm, 257
Hunter-Smith, David, 337
Huntings Surveys, 246, 248
Hurd, Douglas, 324
Hutchinson, Prof. Sir Joseph, 185,
    187, 190

Ibo people, 231, 237, 238, 240
    culture, 171–2
    Ozo ceremony, 238–40, 242
    marriage, 250–1
    widow's rights, 250
ICUS, *see* International Conference for
    Unity of the Sciences
Ikpa Mission, 171
Independent University, 263–4
Infancy, 14–16
International and Comparative
    Agrarian Law, Inst. of, 146
International Conference for Unity of
    the Sciences (ICUS), 338
International Geographical Congress,
    Montreal, 267
International Seabed Authority, 322
Inter-University Council, 168
Iran, 209–30, 295
    Homayoun, Distinguished Order
    of, 230
    land reform in, 209, 216–7
    Land Reform and Rural
    cooperation, Min. of, 214
    Rural Affairs, Min. of, 214
    Rural Research Centre, 216, 217,
    218
    Shahanshah, 219, 221–3
    tour in, 209
Iranian Embassy, 229
Iranian National Land Reform
    Institute, 214
Iran Under the Pahlavis, 295
Ivens, Michael, 187, 322, 324

James, Jimmy (RICS), 315
Jennings, Sir Ivor, 153, 183
Johnson-Marshall, Prof Percy, 299

Jones, Arthur, 159
Joseph, Sir Keith, 343
Jubail and Yaabu, Royal Commission
    of, 300–2
Ju Ju masks, 171

Kahn, Robert (Lord), 153
Karl Marx Universität, Leipzig, 142
Kaunda, Kenneth, 328
Kelly, Prof., 328
Kennett of the Dean, Lord, 165, 254
Kerwen, Patricia, 343
Khomeni, Ayatollah 307
Kings Vista, The: a Land Reform
    which has changed the Face of
    Persia, 220, 224
Knight, Dame Jill, 164
Kolbert, Colin, 175, 201, 236, 248,
    259, 301
Kwame Nkumah University of Science
    and Technology (KNUST) (Kumasi
    University), 169, 248–9, 304
        Hon. Doctorate, 249
Kwapong, Alex, 89

Lambton, Prof. Ann, 220
Land Administration Research
    Council (LARC), 295
Land Commission, 186
Land Commission Bill, 161, 162, 167
        opposition to, in Lords, 165, 166
        White Paper, 161
Land Decade Council, 340
Land Economy, Cambridge, Univ. of
        after 1 October 1961, 152
        Board of, 191
        Dept. of, 181, 189; opposition to,
    189–91
        Development Fund, 186, 259,
    261, 262, 292, 294
        Farm Economics Branch,
    (Agricultural Economics Unit), 191,
    258
        Fellowships in, 259–61; at
    St Catharine's, 260; at Magdalene,
    259; at University (Wolfson)
    College, 261
        Funding, 258–61
        Professor of, 181, 185, 186–8,
    189, 191
        promotion of, 272, 292, 295,
    296; see also Lecture tours etc.
        Readership in, 185

staff, recruitment of additional,
    258
        Tripos, 137, 152–3, 184
        antagonism to, 154–7
        establishment, 5 May 1962, 156
Land Economy and development
    policies in Third World, 192
Land Economy: An Education and a
    Career, 270
Land in a Free Society, 343
Land Law and Registration, 194, 267
Land management, 254
Land in the Market, 160
Land and Natural Resources, 162
Landownership
        early research in, 123, 124
        economy, basis of, 131
        academic discipline, basis of, 146,
    147
        resource distribution, basis of, 254
Land Ownership and Resources, 132,
    147
Land planning, post-Second World
    War, 254
Land Policy Group, 160–1 et seq
Land and Property Owners, Assn. of,
    (ALPO), 260
Land Securities, 129
Land Use: An Introduction to
    Proprietary Land Use Analysis,
    220
LARC, see Land Administration
    Research Council
Laslett, Peter, 154–6
Law of the Sea Draft Convention,
    321–2, 324
Lay Readership, see Diocesan Lay
    Reader
Leasehold Reform Act 1967, 163,
    335–6
Lectures, extra-mural, 272: see also
    Lecture tours etc.
Lecture tours, conferences and
    research travel overseas
        America North 295, 323
        Argentina, 278–82
        Asia, Southeast, 295, 299
        Australia, 295–9
        Caribbean, 286
        Czechoslovakia, 138
        Fiji, 308, 310
        Finland, 132–5
        Florence, 147

Lecture tours – *continued*
    Germany, East, (DDR), 142,
  143–6
      Ghana, 169, 172–4, 231–5, 295
      Hungary, 199, 319
      India, 282
      Iran, 208, *et seq*
      Italy, 198, 199
      Lesotho, 331
      Malawi, 329
      Malaysia, 299–300, 307
      Malta, 255
      New Zealand, 313
      Nigeria, 168–72, 231, 235, 237,
  244–5, 295
      Norway, 198
      Poland, 140–1, 175
      Romania, 201
      Saudi Arabia, 295, 300
      South Pacific, 308–13
      Sri Lanka, 284–5
      Sudan, 194–7
      Swaziland, 330
      Switzerland, 320
      Tchad, 246–8
      Zambia, 327–8, 334
      Zimbabwe, 329
Lecturing, 94
Lefcoe, Prof. George, 323
Leslie Abott and Co., 36
Limpopo, River, 333–4
Lincoln Land Inst., Cambridge, Mass.
  322
Lister, Charles, 213
Lichfield, Prof. Nathaniel, 188
Liverpool University, 203
Llewellyn, Dr David, 225, 227
Lloyd, David, 194
Local Govt. Conference, 17th, Church
  House, 164
Lofthouse, Reggie, 75
London, University of
    Convocation Library, 80
    MSc., 81
    PhD., 81
    Estate Management external
  degree in, establishment of, 106
Loveday Cttee. On Higher
  Agricultural Education, 130–1
Loveday Report, 131
Loveless, Rev. Bill, 250
Lusaka, Univ. of, at Kitwe, 335

Ma'aafa Fetani, 311
McAlindon, Peggy, 84, 87
McAlpine, Alistair (Lord), 344
MacGregor, John, 161, 324
MacRobert Trust, 203
Mactaggart, Sir Ian, 341
McWhirter, Norris, 348
Magdalene College, 259, 260
Mahindrapal Singh, 282, 283
Maitland, F.W., 115
Malone, Ambassador, 324
Malta, 255
Mammy Water, 241
Maraj, Dr James, 309
Marine Resource Development
  Consortium, 316
Maritime Studies, School of, 319
Markets under the Sea, 321
Marston, Geoffrey, 317
Martin, Irene, 203
Maudling, Rt. Hon. Reginald, 160
Maxwell, Robert, 265
Maynard, Frank, 278–82
Media, encounters with, 271–2
Middle East Oil Consortium, 213
Miles, Professor C.W.N., 85–6
Mills, E.F. (John), 182
Mitchell, Else, 296
Mitchell-Heggs, Christopher, 334
Moon, Rev. Sun Myung, 339
Moonies, the, 338
Moore, Henry, 337
Morrison, John, 115
Mosley, Oswald, 53–4
Mott, Admiral Bill, 322, 324
Municipal Corporations, Assn. of,
  158

Nagy, Fercue, 200–1
Nauru, 310
Ndi Nze ceremonies, 238, 240
Newcastle University, 189
New Zealand, 313
Ngozi, The Blessing 239, 241–5.
  marriage to Okwudilli, David, 243
Ngwengama, 330
Nicholls, Derek, 304, 327
Nigeria, 167, 168–72, 231, 235, 237,
  248
    academic advice to, 168–9
    missionaries in, 171
    University of, 167, 168

Nigerian College of Arts, Science and
    Technology, 168
    at Enugu, 168
    at Zaria, 168
Nkrumah, Kwame (Ossafago), 167,
    172
Nsukka, University of, 237
Nuffield College, Oxford, Fellowship,
    151
Nuffield Foundation, 150
Nugent, Dick (Lord), 162
Nwunye Ozo (Ibo title wife), 238–45
    passim

Obo of Benim, 89
Ofo, 240
Ofori, Isaac, 232, 335
Ogbuefi, 248
Ojukwu, 232
Old age pensioner, 290
Orde-Powlett, Richard, 92
Oriental and African Studies, School
    of, 315
Origins of Ownership, 120
Oswiecim, 176
Otaniemi conference, 132
Overseas Development, Ministry of,
    168, 193, 229
Owen and Co., 37
Ozo titles, 239, 240

Pacific Way, the, 310
Page, Graham, 162
Parents, 11; *see also* Denman, Robert
    (father) and Barnes, Letty (mother)
Parker, Lord Chief Justice, 147
Peron, Isabel, 278
Pettit, Geoffrey, 93
Pembroke College, Cambridge, 62,
    261, 321
    Fellowship, 181–2
Pepys Library, 260
Persepolis, 224
The Persian Land Reform 1962–66,
    220
Piccard, Dr Jacques, 320, 321
Pogjucki, Land Policy Adviser, Ghana,
    172–3
Pooley, Fred, 264–5
Popkiewicz, Dr Josef, 138, 140, 144,
    175
Portrait, by Richard Stone, 249
Powell, Enoch, 162

Prior, Jessie Hope (Twinks) (Wife),
    66–9, 96–9
    and bronchiectasis, 96
    engagement to, 65
    and Jonathan, 96–7
    mother of, 97, 98
Prior, Margaret (mother of Jessie
    Hope) 88, 97, 98
Prior, Peggy (sister of Jessie Hope) 65
Prior, J.M.I. (Baron Prior of
    Brampton), 93, 221
Prior, Richard Henry (brother-in-law),
    170
*'Private Property and Community
    Needs'*, 159
Proceeds Tax, 161
Proprietary land unit, 150
    Proprietary structures, calculus,
    82
Published works
    articles and papers, 104, 149, 158
    books and pamplets, 82, 103–4,
    120, 121, 132, 161, 220, 321
    for children, 163–4; and BBC
    radio, 103–4

Reading University, 203
Religious experience, 50
    faith, 205–7
Residences
    Bedfordwell Road, Eastbourne,
    55
    The Bells, Hemingford Grey, 88
    Bradwells Court, Cambridge, 86
    12 Chaucer Road, Cambridge,
    90–1, 276, 335, 336
    Church Street, Salcombe, 18
    7 Regent Terrace, Cambridge, 86
    Rose Cottage, Cumberland, 70
    Strathleven, Finchley, 18
    Teigngrace, Church End,
    Finchley, 18–22
    Woodside Park Gardens,
    Finchley, 12
    Retirement, 290 *et seq*
RICS, *see* Royal Institution of
    Chartered Surveyors
Ridley, George, 125
Rifkind, Malcolm, 324
Rippon, Geoffrey, 162, 165
Rivet, Pat, 198
Roberts, Charles, 73, 79
Roberts, Hervina, 126

Roberts, R.P.F., 85
Roberts, Sir Sydney, 90, 276
Rossi, Sir Hugh, 162, 266, 342
Royal Institution of Chartered
  Surveyors (RICS), 120, 137, 197,
  198, 203, 272
    International Cttee., 192
    and maritime geography, 318
    Marine Resources Cttee., 318
    promotion, 274–6

Samoa, 311
Salcombe, 17, 18, 20
Samii, Mme. Zahna, 209, 214
Samuel Harold (Lord), 129, 130
Samuel benefaction, 129, 261
Sandford, Margaret
  (Mrs Maurice Wood), 97
Sawyerr, Harry, 234
School days, 17–32
    Salcombe, 17, 20
    St Mary's Convent, 19
    Finchley, 24–32
Scottish Landowners Federation, 272
Seabed, economic importance of, 314,
  315; management of, 318
Seldon, Arthur, 263
Sey, Prof. Sam, 234
Shell International, 272
Sherman, Sir Alfred, 342
Simpson, Dr Rowton S., 193 and
  Sudan, 194–7
    Land, Law and Registration, 267
Skeffington, Arthur, 163
Small Farmers Assn., 337
Smith, Col. Frank Braybrooke, 107
Smiths Gore, 306, 316, 317, 318
Southern California, Univ. of, 322
Spencer Barnard, Charles, 209
Springer, Sir Hugh, 255, 311, 344–5
Stallard, Robin, 127
Stanford University, Calif., 295, 323
Steel, Robert (RICS), 314
Stinking Lilly, 79
Sturrock, Ford, 134
Stys, Prof., 111, 177
Swann, Lord, 276
Sweett and Partners, Cyril, 306
Switzer, J.F.Q. (Jeffrey), 110, 13,
  137–8, 151, 175, 187, 275
Sydney, University of, 296
Synod, 101

Tchad, Lake, 247
  irrigation project, 246
Tenant-Right Valuation in History
  and Modern Practice, 82
Territorial Sea, 313
Thatcher, Mrs Margaret (Baroness),
  143, 164, 202, 229, 324, 343
Thomas, Trevor, 132, 185
Thompson, C.H. (Timber), 85
The Times, 1960, leaders in, 158
Tonga, 312
Torio, Prof. Francisco, 270
Trustham-Eve, Sir Malcolm, 102
Tucker, Rt. Rev. Cyril, 282
Turner, C.W., 118
Twum-Barima, 169, 173

UGC, see Universities Grants
  Committee
Umeh, John, 168, 231, 237–8, 245
United Nations, 197
University Grants Committee (UGC),
  117, 118, 124, 190, 262

Valian Dr A.A., 209, 214, 217, 221–3
  et seq, 307
Vanuatu, 311
Vernay, Sir Ralph, 265
Volpi, Prof. Roberto, 273
Volta Dam, 174

Waite, Terry, 325
Wales, University of, Inst. of Science
  and Technology, 318
Walker, Peter (Lord), 229
Warburton Lecture, Manchester, 266
Watercolours, 303
Watt, Prof Donald Cameron, 319
Wellbourne, Edward, 114
Weller, E.P. (Sam), 84–5, 109, 110,
  112, 132, 152
Wells, Sir Henry, 136
Wells, Sir Sydney, Russell, 106
Wells, Sir William, 106, 109
Wells Report, 136–7
West, Dr Henry, 194, 313
West African Pilot, 167
Westminster, Duke of, 261
Westminster Industrial Brief, 343
Wetham, Edith, 110
Whipple, Prof. Tom, 297, 298
White, Sir Graham, 59
White, Prof. Philip, 204, 268, 286

WHO, *see* World Health
    Organization
Wijeratne, Brian, 338
Willey, Rt. Hon. Frederick, 163
Williams, Tom, 121
William Temple Society, 100
Wilson, Dr, Master of Clare College,
    89–90
Wilson, Harold (Lord), 164
Wiseman, Air Cmdre, 58
Wiseman, Prof. Jack, 322
Wolfenden, Sir John, 190
Women, early encounters, with, 41,
    44–8
Women's Land Army, 77, 78
Woodcock, Stanley, 33

Woodcock and Son, 33–6
World Health Organization
    (WHO), 199
World War, First, 15–18 *passim*
World War, Second, 56
    classification of farms in, 74
    farm dispossession in, 74
    farming in, 73 *et seq.*
Wrocław, Technical University of, 138

Yoruba, 231

Zambia, University of, 328
Ziman, John, 154, 156
Zuckerman, Sir Solly, 147